Research and Theory in Advancing
Spatial Data Infrastructure Concepts

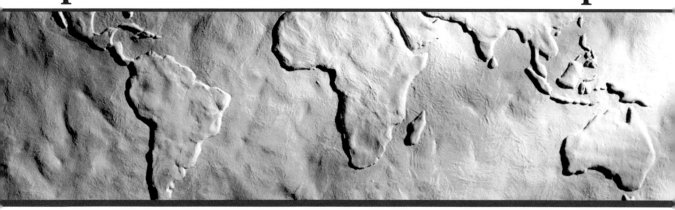

Harlan Onsrud, Editor

ESRI Press
REDLANDS, CALIFORNIA

ESRI Press, 380 New York Street, Redlands, California 92373-8100

Copyright © 2007 ESRI

All rights reserved. First edition 2007
10 09 08 07   1 2 3 4 5 6 7 8 9 10

Printed in the United States of America

Library of Congress Cataloging-in-Publication Data
Research and theory in advancing spatial data infastructure concepts / Harland Onsrud, editor.—1st ed.
    p. cm.
   Includes bibliographical references and index.
   ISBN 978-1-58948-162-6 (pbk. : alk. paper)
   1. Geodatabases. 2. Global Spatial Data Infrastructure (Organization) I. Onsrud, Harlan Joseph.
G70.212.R473 2007
025.06'91—dc22   2007015117

Ask for ESRI Press titles at your local bookstore or order by calling 1-800-447-9778. You can also shop online at www.esri.com/esripress. Outside the United States, contact your local ESRI distributor.

ESRI Press titles are distributed to the trade by the following:

In North America:
Ingram Publisher Services
Toll-free telephone: (800) 648-3104
Toll-free fax: (800) 838-1149
E-mail: customerservice@ingrampublisherservices.com

In the United Kingdom, Europe, and the Middle East:
Transatlantic Publishers Group Ltd.
Telephone: 44 20 7373 2515
Fax: 44 20 7244 1018
E-mail: richard@tpgltd.co.uk

Cover and interior design by Savitri Brant

Contents

Foreword

This book is a project of the Global Spatial Data Infrastructure (GSDI) Association for which contributions were solicited in conjunction with the GSDI-9 Conference held in Santiago, Chile, in November 2006. The GSDI Association promotes international cooperation and collaboration in support of local, national, and international spatial data infrastructure developments that will allow nations to better address social, economic, and environmental issues of pressing importance. The Association (1) serves as a point of contact for those in the global community involved in developing, implementing, and advancing spatial data infrastructure concepts; (2) fosters spatial data infrastructures that support sustainable social, economic, and environmental systems integrated on scales ranging local to global, and (3) promotes the informed and responsible use of geographic information and spatial technologies for the benefit of society.

As part of its mission of advancing new knowledge, the GSDI Association invited scholars from around the world to submit research articles that could be presented at the GSDI-9 Conference as well as considered for inclusion in this volume. Authors were requested to address fundamental theory or research in any area of spatial data infrastructure advancement. We invited reports of research studies that might

- test or analyze innovative approaches in addressing technical, legal, economic, or institutional challenges;
- critically assess current spatial data infrastructure initiatives;
- document and analyze successes and challenges in standardization and data harmonization efforts and case histories;
- describe conceptual models that incorporate emerging or future technological, institutional, economic, legal, or combined solutions in overcoming spatial data sharing impediments;
- compare or analyze existing alternative approaches or models for planning, financing, and implementing SDI or related initiatives in different countries or regions of the world and assess the effects of policy and technical choices in addressing cultural, social, and economic issues;
- assess whether SDI projects are achieving programmatic or broader goals such as supporting national economic competitiveness; increasing the efficiency and effectiveness of government; and advancing health, safety, and social well-being;
- critically examine best practices in terms of their policy, technological, institutional, and financial approaches and their ultimate effects on improving efficiency, effectiveness, and equity; and
- identify and assess the implications of various practices and approaches for local, state, provincial, national, transnational, and global stakeholders with a particular emphasis on developing nations.

We believe this volume has been successful in addressing these topics. We thank all researchers who submitted articles to the peer review process and all members of the Peer Review Board.

We also thank ESRI Press, and we are grateful to Jack Dangermond, Jeanne Foust, and Carmelle Terborgh at ESRI for their enthusiastic and persistent support for publishing this volume as an open-access book. We believe this is a landmark volume in that it is the first open-access scholarly book on geographic information science produced by an established publisher. Because of the desire to make this volume available to our peers in very poor nations as well as wealthy nations, the book uses open-access licensing, which allows each author to retain copyright in their work but requires each to grant an open-access license to the rest of the world to use and reproduce their work under specific liberal conditions. Rewriting all of the normal legal forms to allow this project to proceed and gaining permissions from other publishers to allow their graphics to be incorporated into this open-access volume was no small task. We greatly appreciate the financial contribution and willingness of ESRI to explore this new publishing frontier by making scholarly materials that advance geographic information science concepts more accessible globally.

Harlan J. Onsrud, President
GSDI Association (2005–2006)

Luis A. Alegría Matta, President
Local Organizing Committee for GSDI-9
Director, Instituto Geografico Militar, Chile

GSDI-9 Peer Review Board for Research and Theory in Advancing Spatial Data Infrastructures

Professor Harlan Onsrud, University of Maine, Editor

Introduction

HARLAN ONSRUD

UNIVERSITY OF MAINE, ORONO, MAINE, UNITED STATES

At a high level, the term "spatial data infrastructure" (SDI) is largely self-explanatory. Yet when applied in practice, the concept is complex and has attracted varying definitions in different institutional and social contexts. The Global Spatial Data Infrastructure (GSDI) Association has stated in its April 2006 GSDI newsletter that:

Spatial data infrastructures provide a basis for spatial data discovery, evaluation, and application, and include the following elements:

Geographic data: the actual digital geographic data and information.

Metadata: the data describing the data (content, quality, condition, and other characteristics). It permits structured searches and comparison of data in different clearinghouses and gives the user adequate information to find data and use it in an appropriate context.

Framework: includes base layers, which will probably differ from location to location. It also includes mechanisms for identifying, describing, and sharing the data using features, attributes, and attribute values, as well as mechanisms for updating the data without complete re-collection.

Services: to help discover and interact with data.

Clearinghouse: to actually obtain the data. Clearinghouses support uniform, distributed search through a single user interface; they allow the user to obtain data directly, or they direct the user to another source.

Standards: created and accepted at local, national, and global levels.

Partnerships: the glue that holds it together. Partnerships reduce duplication and the cost of collection and leverage local/national/global technology and skills.

Education and Communication: allowing individual citizens, scientists, administrators, private companies, government agencies, nongovernment organizations, and academic institutions with local to global interests to communicate with and learn from each other.

Not all definitions of SDI include all of the above elements, and some definitions may contain other elements. Furthermore, the list above says little about who will or should take responsibility for supplying and maintaining the elements. In many nations the assumption is that an SDI is the geographic information technology component of electronic governance. In these nations there may be a strong focus on government itself actively supplying or at least facilitating the creation and maintenance of each of the above elements in order to achieve government objectives. In other nations, government may consciously take a much more passive role in regard to provisioning elements such as geographic data, framework data, metadata, services, and clearinghouses. Rather, governments in these nations may focus on promoting standards, capacity building, and interoperable frameworks and technologies in order to encourage actors other than government to create and maintain SDI elements that will meet broader societal and economic goals in addition to the more direct objectives of meeting government agency missions. A great deal of practical experience is being gained and experimentation is occurring in determining appropriate mixes of actors, activities, and information technologies to support SDI developments in different local, subnational, national, and multinational contexts. Scholarly investigators across the globe are playing a significant role in questioning and determining the extent to which social and governmental goals are being met and under what conditions.

While technological research challenges abound, most of the papers in this volume explore policy and institutional challenges to sharing georeferenced data, implementing or improving infrastructure in order to achieve desired outcomes, or enhancing the effectiveness and/or efficiency of SDI operations.

The volume begins with two articles that provide a theoretical grounding for articles that follow. The first article provides a solid review of institutional research methods that have been used previously to study SDI implementations and operational systems. After identifying theoretical and conceptual approaches used in past studies, Nama Raj Budhathoki and Zorica Nedović-Budić comment on the utility of each of the approaches in supporting SDI research and development. They propose a framework for focusing organizational and user-centric SDI research in the future in order to strengthen the conceptual base for further study and deployment of SDIs.

W. H. Erik de Man provides another solid review of the organizational-research literature as it relates to SDI development. He suggests that technical and bureaucratic issues and values seem to dominate data handling at higher levels whereas social issues and human values play a much stronger role at human levels of interaction. He suggests that a multifaceted theoretical view is needed in understanding the development of information infrastructures. He discusses information infrastructure initiatives in developing nations. While individuals are somehow attached to locality and social bonds are stronger at this level, emerging network societies will have a significant impact on societal information provision. The value of this article is perhaps in asking questions within theoretical frameworks rather than attempting to theorize or test explanations. In answer to the question of whether SDIs are special, the author seems to suggest that SDIs will perhaps achieve their ultimate success when they are no longer identifiable as such but are merely part and parcel of the overall information infrastructure to which every citizen has access.

The next two articles recommend specific theory-based approaches for the assessment of spatial data sharing. Both articles state or assume that data sharing is critical to maximizing the benefits of SDIs. Kevin McDougall, Abbas Rajabifard, and Ian Williamson explore data sharing partnerships among governmental jurisdictions by assessing their organizational contexts and identifying factors that might contribute to their success. The authors review data sharing assessment frameworks used by previous researchers, discuss the potential benefits of using a study methodology that incorporates both qualitative and quantitative assessments, and suggest circumstances or stages of study for using each form of assessment. A recommended approach is then demonstrated using a case study strategy incorporating both qualitative and quantitative methods for assessing data sharing partnerships among jurisdictions in Australia. In the next article, El-Sayed Ewis Omran, Arnold Bregt, and Joep Crompvoets focus on understanding individual and organizational behaviors in sharing spatial data. They recommend generating and testing multiple hypotheses, drawing on the theory of planned behavior and Hofstede's cultural dimensions theory.

Each of the next three articles examines a particular model for encouraging cross-jurisdictional SDI development or sharing data and services. Max Craglia and Alessandro Annoni describe INSPIRE, an emerging effort to build an SDI for Europe based on existing national and subnational initiatives and engaging users in a more structured way by organizing them in spatial data interest communities. In emphasizing partnerships and involvement of stakeholders across European jurisdictions, this new approach strives to balance stability in standards and practices and flexibility in accommodating local circumstances and change. The belief is that the cooperative approach will not only be more responsive to the direct needs of hundreds of governments, communities, and businesses but also facilitate common standards and processes across jurisdictional boundaries for more efficient and effective SDI development and maintenance.

F. Javier Zarazaga-Soria, Javier Nogueras-Iso, Miguel Á. Latre, Antonio Rodriguez, Emilio López, Pedro Vivas, and Pedro R. Muro-Medrano describe use of the INSPIRE approach in developing and maintaining an SDI that tracks border-spanning natural phenomena of interest to multiple nations. Policy clashes and institutional differences were the primary challenges inhibiting the development of common SDI services in this cross-border setting. While the technical challenges in this pilot project were substantial, different data access policies and multilingual heterogeneity were far more problematic.

Ian Williamson, Abbas Rajabifard, and Andrew Binns use the concept of an enabling platform as an organizing principle for facilitating administrative and institutional cooperation in the implementation of SDI services across governmental jurisdictions. The goal of such a collaborative approach among jurisdictions is to move from delivery of mere spatial data to delivery of specialized products and services in response to business and government needs as though from a single enterprise. The new technological capabilities will hopefully encourage jurisdictions to share not only data but also support for administration, maintenance, and expansion of new services over time.

Taking a global perspective on improving SDI implementations, Joep Crompvoets and Arnold Bregt explore the status of national clearinghouses for spatial data. Centralized government clearinghouses are often established as a means of providing network access to a nation's publicly available spatial data. The authors assess systematically the level of development and suitability of national clearinghouses throughout the world and identify critical factors for their success. The study identified 83 countries with national clearinghouses but found that many clearinghouse projects have been unable to garner sufficient continuing support for sustainability. Several possible explanations are offered. The authors argue that the shortcomings of many of the current implementations both in terms of institutional support and technological functionality could be resolved by shifting the focus from data to users and applications: providing Web services for viewing and downloading, building user-friendly interfaces with less jargon, applying International Organization for Standardization (ISO) 19115 metadata standards, and addressing similar usability and interoperability issues that would allow the immediate needs of users to be met while also supporting longer-term strategies for maintenance.

The next article proposes a generalized global approach to help enable interjurisdictional sharing of geographic data and services. Based on the wealth of SDI development experience gained in past efforts across the globe and the maturing of technologies and standards, Douglas Nebert, Carl Reed, and Roland Wagner propose the development of a universal suite of technical standards for SDIs that could be adopted by any local, subnational, or national jurisdiction and be compatible with the infrastructures of other jurisdictions. The authors point out that few current SDI implementations can seamlessly interoperate. Different SDI initiatives across the globe are implementing different international standards and using different content models for key data themes. The authors indicate that no general reference architecture currently exists for defining a framework that would enable a standards-based SDI. They identify compatible, mature geospatial standards which would allow maximum technical interoperability, and they suggest a candidate suite of standards for all SDI implementations. The authors urge the broad SDI global community to identify the requirements for this system or an alternative suite of standards and develop life cycle management procedures and documentation.

Satish Puri, Sundeep Sahay, and Yola Georgiadou seem to offer a counter but perhaps complementary position to the previous article in arguing that, while technological standardization is laudable to a degree, what is most needed for productive SDI development in much of the world is flexibility to allow diversification. They focus on the importance of metaphors in enhancing the understanding of SDI implementation and institutionalization and argue that past metaphors have emphasized top-down approaches, centralized control, and the view of information as a tradable commodity while marginalizing the roles of communities. The authors present a metaphor they believe will facilitate more diversified and productive SDIs, emphasizing the need to examine technological and organizational characteristics together rather than in mutual exclusion. They conclude that learning and skill development will often need to take precedence over improving efficiencies if SDIs are to be of greatest benefit to the populations they are intended to serve.

The next two articles present comparative studies. Tracey Lauriault and Fraser Taylor assess the spatial information and technology needs of East Timor for sustainable development and comparatively assess several operational SDIs in other nations for applicability to East Timor. The East Timorese SDI is at a very early stage of development. The political, legal, and administrative context of each nation is unique, and the authors argue that understanding and adapting to institutional constraints and opportunities will be crucial in developing a useful SDI for East Timor. In addition, overseas development agency datasets need to be coordinated and integrated in the local SDI. They conclude that the SDI models of other nations cannot be adopted by East Timor without modification and that capacity building in terms of human resource development will be critical.

The case study that follows compares the mature national SDIs of Australia and Switzerland in terms of their data and metadata management practices as well as their Web services. Like other authors, Christine Najar, Abbas Rajabifard, Ian Williamson, and Christine Giger point out that, although SDI initiatives may use the same or similar technologies and standards, comparisons are difficult due to different legal and organizational contexts. The authors develop a framework for comparing SDI initiatives using a set of clearly defined indicators intended to reflect both technical and institutional success independent of the organizational setting.

The next three articles explore legal and policy issues. Bastiaan van Loenen and Jitske de Jong assess the impact of government access policies (charging for government spatial data and restricting downstream uses) on the technical characteristics of a dataset and its direct uses. The authors hypothesized that a high-quality, expensive government spatial dataset will have strong restrictions and a low-quality government dataset will have weak restrictions. The hypothesis was readily falsified in this five-jurisdiction study. Imposition of a government cost recovery policy did not necessarily result in high-quality datasets or even better datasets than those available under more open policies. However, framing the study in this manner exposed nuanced relationships between policies and use as well as between quality and use. The authors conclude that spatial data policies may need to adapt to rather than fight or work around institutional settings framed by governmental arrangements and structures that may have been established centuries ago. In other words, institutions matter.

Katleen Janssen and Jos Dumortier examine three pieces of relatively recent legislation that will have a substantial impact on interoperability among national SDIs in Europe and on development of an SDI for Europe. They point out that the legal framework embodied in these documents incorporates certain policies that promote spatial data availability and others that hinder it. Confusion exists over which policies will or should prevail under specific circumstances, so a delineation of the proper scope of each legislative enactment is offered. The authors are particularly concerned that intellectual property rights endanger the availability of spatial data in terms of access, reuse, and sharing and may represent a considerable threat to the development of an SDI for Europe.

Mohamed Bishr, Andreas Wytzisk, and Javier Morales also address intellectual property rights but from a technological tracking and rights imposition enablement perspective. They advocate the use of a particular set of digital rights management technologies to track ownership rights in data and services available through a national SDI. This management approach rests typically on the assumption that only those agreeing to license provisions have a right to acquire and use and is often favored by private-sector companies and agencies using traditional intellectual property models. The article does not address the primary alternative technological vision in which the opportunity to acquire and use is open and users need permission only if an intended use conflicts with license provisions.

The volume ends with two articles that are perhaps the most directly related to the theme of the November 2006 GSDI-9 conference in Santiago, Chile (where all the articles were first presented)—Spatial Information: Tool for Reducing Poverty. Felicia O. Akinyemi presents the results of a survey identifying different poverty mapping studies, the types of spatial data used for poverty mapping, and the mode of usage and data sources. A primary goal of the research was to identify spatial datasets essential to poverty mapping and to determine the relevance of SDIs to poverty reduction. The author concludes that spatial data use is fast becoming a best practice for poverty assessment and that the diversity of spatial datasets in use, the large costs associated with their use, the need for consistency and accuracy in data, and the need for providing access are all pertinent poverty reduction issues that SDIs can help resolve.

In the final paper, Kwabena Asante, James Verdin, Michael Crane, Sezin Tokar, and James Rowland analyze the needs of users involved in disaster management, which includes preparedness, early warning, response, recovery, and reconstruction. While the material in this article is germane globally, natural disasters are particularly devastating in poor nations, where large populations may already be living on the edge of survival. The authors conclude that data drawn from many sources is best integrated during the preparedness phase, that national contingency plans are useful for communicating risk and initiating preparedness activities, that a dynamically linked service architecture is perhaps the most practical means of ensuring that multisector disaster managers have access to the most current information when they need it, and that SDIs in concept and practice can support all of these objectives.

REFERENCE

GSDI Newsletter. April 2006. http://gsdi.org/newsletters/GSDI/GSDInewsletterApr06.pdf

Expanding the Spatial Data Infrastructure Knowledge Base

NAMA RAJ BUDHATHOKI AND ZORICA NEDOVIĆ-BUDIĆ

UNIVERSITY OF ILLINOIS AT URBANA-CHAMPAIGN, ILLINOIS, UNITED STATES

ABSTRACT

Research on spatial data infrastructures (SDIs) is not well grounded in theory, and SDI practice often does not adequately take into account previous experiences. The purpose of this paper is to raise awareness about knowledge areas available to academics and professionals involved in studying or developing SDIs. Along with technical tools, both groups need to engage the theoretical and conceptual apparatus in their efforts to understand and address technological and organizational processes and requirements of SDIs. After briefly addressing the existing SDI literature and identifying research gaps, the paper reviews the main disciplinary areas that would contribute to institutionalization of SDIs and to ensuring their broad utility: (1) information infrastructure, (2) interorganizational collaboration-cooperation-coordination (3C), (3) intergovernmental relations, (4) action network theory, and (5) use-utility-usability (3U) of information systems. We assess their value and limitations in supporting SDI research and development. The following elements are identified as potentially contributing to the SDI conceptual framework: the mutually supporting role of SDIs, geographic information systems (GIS), and information and communication technologies (ICT) and infrastructures; the notion of an installed base and capacity building activities responsive to the local conditions and needs; consideration of political, social, economic, cultural, and institutional context; incorporation of 3C principles and opportunities; attention to intergovernmental relations and the emergence of E-governance; understanding of the networked environment of data users, producers, and managers; employing user-centered approaches; and evaluating SDI accessibility and utility. The proposed framework is comprehensive, although it excludes important but often less challenging technical topics in order to focus on organizational and user perspectives.

INTRODUCTION

A functional spatial data infrastructure (SDI) is an important asset in societal decision and policy making (Feeney 2003), effective governance (Groot 2001), citizen participation processes (McCall 2003), and private sector opportunities (Mennecke 1997). Driven by those expectations, national SDIs have grown worldwide during the last decade (Crompvoets et al. 2004; Masser 2005a; Onsrud 1998). The benefits, however, have been slow to materialize. For example, Butler et al. (2005) assert that the United States national spatial data infrastructure (NSDI) has been only partially successful after 15 years of struggle. Masser (2005a) categorizes a number of European SDIs as partially operational or nonoperational. Similarly, Crompvoets et al. (2004), in their worldwide survey of national spatial data clearinghouses, observe a declining trend of clearinghouse use. In line with these observations, Masser (2005a) cautions that "some formidable challenges lie ahead and the task of sustaining the momentum that has been built up in creating SDIs in recent years will not be easy" (p. 273).

The above cautions require close attention, particularly given the considerable amount of resources that SDIs require (i.e., on the scale of billions of dollars) (Onsrud et al. 2004; Rhind 2000). One way to secure the return on these investments is to better conceptualize and understand SDI developments and ascertain their effects. However, the SDI knowledge base is quite limited (Georgiadou et al. 2005). Georgiadou and Blakemore's (unpublished) examination of articles in seven major geographic science journals yields a disappointing finding that only 5 percent of SDI-related articles are theoretically grounded and critical. They report that most of the works are focused on either technology or applications; the conceptual domain and social and organizational ramifications have been addressed the least. While a successful SDI balances the technology and application domains, it can hardly do so without a sound theoretical foundation. Without such a knowledge base, SDI development efforts are excessively driven by either technology or application and are unlikely to become fully operational and serve the expected purposes. The conceptual knowledge and framework are crucial for informing the technological and institutional choices in a variety of circumstances and for capitalizing on the SDI promise to aid problem solving and decision making in different application realms.

In this paper, we attempt to expand the SDI theoretical base by reviewing the literature on five potentially useful knowledge areas. We first briefly identify the existing SDI research and its gaps. We then point to sources in the areas of (1) information infrastructure (II), (2) interorganizational collaboration-cooperation-coordination (3C), (3) intergovernmental relations (IGR), (4) actor network theory (ANT), and (5) use-utility-usability (3U) of information systems. We summarize the value and limitations of the reviewed knowledge areas and propose a tentative but pragmatic conceptual framework encompassing some of the key concepts. Those five fields are not comprehensively treated, and a more extensive literature review would present them more accurately and fully. Our objective is to provide information that would raise awareness of the potential that those areas bring to advancing SDI research and practice and furthering the transformation of the current worldwide SDI initiatives into functional infrastructures.

Masser (2005a) maintains that an SDI:

> . . . supports ready access to geographic information. This is achieved through the coordinated actions of nations and organizations that promote awareness and implementation of complementary policies, common standards and effective mechanism for the development and availability of interoperable digital geographic data and technologies to support decision making at all scales for multiple purposes. These actions encompass the policies, organizational remits, data, technologies, standards, delivery mechanisms, and financial and human resources necessary to ensure that those working at the (national) and regional scale are not impeded in meeting their objectives (p. 16).

This definition emphasizes the following three areas that underpin all SDIs:

1. Policy and organization (organizational, institutional, management, financial, political, and cultural issues)
2. Interoperability and sharing (backbone of SDIs)
3. Discovery, access, and use of spatial data (main purpose of SDIs)

Limited but important and encouraging seed research has been conducted in all three areas.

Policy and organization. After a decade of SDI initiation worldwide, research has begun to focus on various aspects of "second generation SDI" (Rajabifard et al. 2003). Georgiadou et al. (2005) underscore the shift from data-centric research to the notion of infrastructure; Masser (2005b) and Rajabifard et al. (2003) promote a shift from a product to a process model; Coleman et al. (2000) and Craig (2005) address human resources and leadership; Bernard and Craglia (2005) emphasize important but scarce research on the socioeconomic impact; Georgiadou and Blakemore (unpublished) sound a warning about the Western-centric and technical nature of most of the ongoing research and call for a globally relevant research program centered on the human component.

The most frequent organizational approach to SDIs is hierarchical (Rajbifard et al. 2003), with a network model as an alternative. In his evaluation of first-generation SDIs, Masser (1999) provides a generic model of national SDIs or SDI-like centers and, like most other authors, describes the growth and organization of some of the major SDI-related organizations (e.g., EUROGI, PCGIAP, Global Map; Victoria's Property Information Project) as a source of learning (Jacoby et al. 2002; Lachman et al. 2002; Masser et al. 2003). It is clear, however, that existing organizational and institutional arrangements often impede SDI advancement, and new organizational and institutional mechanisms are needed (Kok and Loenen 2005; Masser 2005b).

Interoperability and sharing. Despite the enhanced data transfer capabilities allowed by advances in information and communication technologies (ICT) and the World Wide Web in particular, sharing of spatial information is still impeded by substantial noninteroperability. This noninteroperability can be broadly classified into two categories: technical and nontechnical. According to Bishr (1998), technical interoperability has six levels: (1) network protocols, (2) hardware and operating systems, (3) spatial data files, (4) database management systems (DBMS), (5) data models, and (6) semantics. He argues that the first four items have been reasonably resolved, and research in federated database systems is expected to contribute to resolving the fifth one. The sixth one—semantics of geographic information—is addressed by a number of researchers (Bishr 1998; Fonseca et al. 2000; Harvey et al. 1999; Klien et al. 2006; Kuhn 2003; Nogueras-Iso et al. 2005; Pundt and Bishr 2002; Visser et al. 2002) and has recently benefited from a discussion of spatial ontologies (Mark et al. 2000).

In data sharing, however, nontechnical interoperability (or soft interoperability as termed by Nedović-Budić and Pinto 2001) is more challenging than the technical issues. The impediments to sharing have been identified, although the solutions to overcome them are not easily deployed (Azad and Wiggins 1995; Craig 1995; Montalvo 2003; Nedović-Budić and Pinto 1999a, 1999b; Nedović-Budić et al. 2004; Pinto and Onsrud 1995). For example, Craig (2005) argues that key individuals can make a difference in a sharing scenario; Harvey (2003) underscores trust as the most important mutual feature of the sharing entities; Nedović-Budić et al. (2004) comprehensively discuss the process and determinants of interorganizational sharing. While all these solutions are quite pragmatic and relevant to SDI policy, they are yet to be fully applied in practice.

Spatial data discovery, access, and use. Discovery of and access to spatial data are necessary initial steps in SDI use, and true SDI utility is demonstrated with a wide variety of users (Masser 2005a; Williamson 2003). The discovery of spatial data is facilitated through metadata catalogues (Craglia and Masser 2002; Craig 2005; Smith et al. 2004) and relies on metadata standards (Kim 1999). Recently, some of the metadata systems deploying a multiplicity of national and technical standards have been gradually adopting the international ISO 19115 standard, and translations have been created between different metadata standards (Nogueras-Iso et al. 2004). There are also a few preliminary assessments of the usability of the metadata standards (Fraser and Gluck 1999; Walsh et al. 2002). Several studies discuss other aspects of geoportals as gateways to SDI: Bernard et al. (2005), Maguire and Longley (2005), and Tait (2005) focus on the capabilities of second-generation geoportals to access spatial data and services; Askew et al. (2005) and Beaumont et al. (2005) describe the UK experience in building on the government's ICT investments and the difficulties in developing geoportal-related partnerships due to different levels of technological experience, goals, and expectations among the partners.

Access to spatial information is usually measured as portal hits. For example, the Geography Network receives (an encouraging) 300,000 hits by an estimated 50,000 users per day (Tait 2005). The use of spatial information seems to fall a bit behind, with some preliminary indications that contemporary SDIs do not fulfill their purpose and expectations. Crompvoets et al. (2004) report that user-unfriendly interfaces and the discipline-specific nature of metadata and clearinghouses are among the primary reasons for the declining trend in clearinghouse use. Nedović-Budić et al. (2004), in their evaluation of the use of SDIs for local planning in Victoria, Australia, and Illinois, United States, also conclude that SDIs do not effectively serve local needs. These studies reinforce the findings from a large-scale survey conducted in the United States by Tulloch and Fuld (2001) who find that using framework data in an SDI environment is challenging both technically and institutionally—technically because these data are in

various formats and of different accuracies and institutionally because the data producers are not fully prepared to share data.

SDI RESEARCH GAPS

Without claiming to be exhaustive and specific, we identify the following gaps in the current SDI literature and invite the research community to direct their future work to these general areas and many potential topics within them.

Definition and conceptualization. The many definitions of SDI (Rajabifard et al. 2003) differ in emphasis and purpose, and no clear consensus on the concept of SDI and its constituting elements and principles exists. While a multiplicity of definitions and meanings is not unusual for any phenomenon, it tends to frustrate research and development. Similarly, literature does not help much in differentiating between GIS and SDI and specifying their unique roles and relationships. For example, Bishop et al. (2000) believe that a GIS cannot be built without an SDI, whereas Georgiadou et al. (2005) argue that an SDI requires a strong GIS base. Inconsistent definitions and concept operationalizations result in ambiguous research findings and prevent comparison of studies conducted independently on the same subject (Budić 1994). In essence, they stand in the way of building a coherent body of SDI knowledge.

Models. Although the hierarchical model corresponds closely to current efforts at creating SDIs at different administrative levels, more complex horizontal and vertical interactions require further exploration and more elaborate representation. An alternative model (or models) is needed to outline SDI presence and use across all levels and organizational configurations and to accommodate all relevant participants. Public access, in particular, is a crucial component of the connectivity claimed by SDIs. While the general public is anticipated to eventually be the largest SDI user group (Dangermond 1995; McKee 2000), very few sources discuss the issue of public access and explicitly include it in SDI modeling and building attempts.

Standards. Other than the sporadic migration to ISO standards by some national SDIs, little is known about which standards are used in SDIs worldwide. Moellering (2005) started to fill this gap by reviewing metadata technical requirements and developments around the globe, including many international and national examples. Still, robust empirical work on metadata systems is lacking, for example, in terms of their matching the users' mental models, their value in assessing the fitness-for-use of the underlying data, and the complementary use of social networks in data discovery. Moreover, research on substantive standards and compliance to them in a variety of data domains is important for advancing the possibilities for transfer, sharing, and use of spatial information.

Monitoring and evaluation. Ongoing SDI research is more focused on access to spatial data than on the use and utility of the infrastructure. With utility in mind, looking at the process of SDI establishment comprehensively from conception to operation will help create a more relevant and useful infrastructure. Beyond counting portal hits, there is no clear evidence about who the users are, what they are using the information for, and how well they are served by the geoportals (Askew et al. 2005). In general, continuous monitoring and evaluation should contribute to establishing effective and valuable SDIs. Georgiadou et al. (2006)

suggest a variety of methodologically rigorous evaluation approaches suited to progressively complex foci on data, services, and E-governance. The formation of a new Spatial Data Interest Community on Monitoring and Reporting (SDIC MORE 2005) in conjunction with the implementation of the Infrastructure for Spatial Information in Europe (INSPIRE) testifies to the importance of tracking the establishment, contents, and use of SDIs. The group, however, is only beginning to identify indicators and monitoring mechanisms and procedures.

Balancing the technical and the social. We need to better understand the interaction between the technical and the nontechnical, but research efforts have been mostly limited to one or the other. In reality, the two realms interact and influence each other to give rise to a whole new set of factors, which are calibrated through a mutual adjustment between the two (Nedović-Budić 1997). Timely involvement of prospective users in the development of SDIs will contribute to enhanced usability and overall success. The diverse backgrounds and often limited skills of nonspecialists require approaches different from the ones taken for specialist users. The traditional information system development methodology of technology-centered design may work for small systems but is inadequate and too risky for SDIs. In addition, capacity building has to be included as an inherent part of SDI development (Enemark and Williamson 2004; Georgiadou and Groot 2002; Masser 2004; Williamson et al. 2003).

Politics and policy. SDIs are also susceptible to geopolitical, economic, and sociocultural issues and all the associated opportunities and threats of cyber spaces and interactions (Pickels 2004). This is particularly obvious for national SDIs, which often exhibit centralizing tendencies that run counter to federated and devolutionary system concepts. The SDI community cannot afford to overlook the relationship between the state and geographic information and thereby become a nonplayer in addressing this crucial dimension of SDI policy.

Multi- and interdisciplinary approach. SDIs draw on knowledge from many disciplines, including but not limited to sociology, cognitive science, political science, organizational studies, economics, and computer and information science (Masser 2005b). Current research, however, tends to be inward oriented, failing to reach out to other disciplines and their theories, concepts, and frameworks.

In sum, the current SDI knowledge base is not sufficient to inform development of sustainable SDIs. Therefore, in agreement with Georgiadou et al. (2005) and Masser (2005b), we direct the attention of the SDI academic and professional community toward alternative sources. The following section provides a brief overview of five key knowledge areas that can strengthen the SDI theoretical and conceptual foundation.

FIVE KNOWLEDGE AREAS

Information infrastructure. Most literature considers information infrastructure (II) in a rather narrow sense within a specified domain, for example, biology (Sepic and Kase 2002), urban planning (Langendorf 2001), academia (Begusic et al. 2003; Cramond 1999; Sepic and Kase 2002), or media (Anderson et al. 1994). Some view the Internet as II, while others equate the digitalization of libraries with II. However, the II envisioned by the former U.S. Vice President Al Gore, the U.S. Information Infrastructure Task Force (1993), and the European Union task force (Bangemann Group 1994) has much broader expectations and ramification for all sectors of society. A number of researchers also move from the domain-specific to the broad societal front and attempt to develop the general II conceptual base (Hanseth and Monteiro 1998, 2004; Monteiro 1998; Monteiro and Hanseth 1995; Star and Ruhleder 1996) (table 1). They suggest that all IIs build on their technological and social installed base and maintain that IIs are open and support any number of users and their diverse needs. These authors view information infrastructures as not only gradually expanding but also transforming, as work practices are continuously inscribed in them.

Star and Ruhleder (1996) argue that IIs cannot be independently built and maintained, but rather, they emerge through practice and get connected to other activities and structures. They criticize the highway metaphor of II as technology biased. Similarly to Borgman (2000), they view IIs as much more than the physical substrate and consider broader social relations integral to IIs. Hanseth and Monteiro (2004) suggest that some of the II characteristics may be present in certain information systems (IS), especially in interorganizational systems (IOS) or distributed information systems (DIS), and therefore, some commonalities and overlapping characteristics exist between IS and II. They state that IIs are initiated when (1) new and independent actors become involved in the development of an IOS or DIS, so that development is not controlled by one actor anymore, or

Star and Ruhleder (1996)	
Embeddedness	Infrastructure is "sunk" into (inside of) other structures, social arrangements, and technologies
Transparency	Infrastructure is transparent in use, in the sense that it does not have to be reinvented each time or assembled for each task but invisibly supports those tasks
Reach or scope	This may be either spatial or temporal—infrastructure has reach beyond a single event or one-site practice
Learned as part of membership	The taken-for-grantedness of artifacts and organizational arrangements is a *sine qua non* of membership in a community of practice. Strangers and outsiders encounter an infrastructure as a target object to be learned about. As they become members, new participants acquire a naturalized familiarity with its objects.
Links with conventions of practice	Infrastructure both shapes and is shaped by the conventions of a community of practice
Embodiment of standards	Modified by scope and often by conflicting conventions, infrastructure takes on transparency by plugging into other infrastructures and tools in a standardized fashion
Installed base	Infrastructure does not grow *de novo;* it wrestles with the "inertia of the installed base" and inherits strengths and limitations from that base
Becomes visible upon breakdown	The normally invisible quality of a working infrastructure becomes visible when the infrastructure breaks down
Hanseth and Monteiro (2004)	
Enabling	Infrastructures have a supporting or enabling function
Shared	An infrastructure is shared by a large community (collection of users and user groups)
Open	Infrastructures are open and support heterogeneous environments
Sociotechnical network	Information infrastructures are more than "pure" technology; rather, they are sociotechnical networks
Ecology of networks	Infrastructures are connected and interrelated, constituting ecologies of networks
Installed base	Infrastructures evolve by extending and improving the installed base

Table 1. Characteristics of information infrastructures.
Compiled from Star and Ruhleder 1996; Hanseth and Monteiro 2004.

(2) one of the design objectives for IOS or DIS is growth and transformation into an II (or a part of an II) in the future.

Interorganizational collaboration-cooperation-coordination. The IS literature reinforces the argument that organizational complexities increase further in interorganizational contexts and therefore require different information system development, management, and use practices (Doherty and King 2001; Lambert and Peppard 1993; Mahring et al. 2004; Suomi 1994; Williams 1997). The elements of interorganizational collaboration-cooperation-coordination (3C) are often necessary for IOS or DIS implementation and successful operation. Cooperation covers the middle ground between collaboration and coordination, with the former being least intensive and most autonomous and the latter being most intensive and least autonomous (McCann 1983).

The essential elements in studying interorganizational exchange include organizational exchange theory (Cook 1977), determinants of interorganizational relationships (including necessity, asymmetry, reciprocity, efficiency, stability, and legitimacy; Oliver 1990), and organizational interdependence (Thompson 1967). Levine and White (1969) define exchange as "any voluntary activity between two organizations which has consequences, actual or anticipated, for the realization of their respective goals or objectives" (p. 120). Exchange is usually sought with the minimum loss of organizational autonomy and power and depends on the availability of alternative resources. Thompson (1967) identifies three types of organizational interdependences: pooled, sequential, and reciprocal (in the order of increasing complexity). Kumar and van Dissel (1996) provide a typology of interorganizational systems based on type of interdependence (table 2). Meredith (1995) postulates that already existing organizational interdependence will reduce resistance to interorganizational sharing. This is particularly true for cooperative interdependence (Tjosvold 1988). However, increased

Dimension	Characteristic for the following type of interdependence		
	Pooled	Sequential	Reciprocal
Configuration			
Coordination mechanisms	Standards and rules	Standards, rules, schedules, and plans	Standards, rules, schedules, plans, and mutual adjustment
Technologies	Mediating	Long-linked	Intensive
Structurability	High	Medium	Low
Potential for conflict	Low	Medium	High
Type of IOS	Pooled information resource IOS	Value/supply-chain IOS	Networked IOS
Implementation technologies and applications	Shared databases, networks, applications, electronic markets	EDI applications, voice mail, facsimile	CAD/CASE data interchange, central repositories, desktop sharing, videoconferencing

Table 2. Organizational interdependence.
Reprinted from Kumar and van Dissel 1996, with permission of the University of Minnesota.

interdependence and need for cooperation can in some networked organizations lead to conflicts over authority, jurisdiction, and distribution of power (Ekbia and Kling 2005; Kumar and van Dissel 1996). The interdependence and greater mutual resources also tend to increase the number of decision points and thus constrain joint actions and diminish the probability of successful implementation (Aiken and Hage 1968; Pressman and Wildavsky 1984).

Finally, underlying the discussion of the value and importance of 3C to inter-organizational IS and database activity is the need to identify the motivations that would impel organizational units to get actively involved in multiparty relationships and projects. A number of factors contribute to the perceived need to seek out interorganizational geographic information relationships, whether they are voluntary or mandated (Cummings 1980). Gray (1989) refers to achievement of a shared vision and conflict resolution as the two main motivators of collaborative organizational design.

According to O'Toole and Montjoy (1984), coordination can be based on (1) authority (i.e., obligation), (2) common interest, or (3) exchange inducements based on expected or received returns.

Intergovernmental relations. As much as interorganizational systems and databases are manifestations of interorganizational relationships (Kumar and van Dissel 1996), in the public sector they also reflect models of government and intergovernmental relations (IGR). According to Cameron (2001), IGR vary along three dimensions: degree of institutionalization, extent of decision making, and level of transparency. IGR also relate directly to political and administrative decentralization (Koike and Wright 1998). For a federal context like the United States, Australia, and potentially the European Union, Agranoff (2001) proposes the pattern of intergovernmental interaction known as cooperative federalism, consisting of the following elements: federalist theory, administrative techniques, dual government structure, and context-specific cooperation. Nice and Frederickson (1995) advance a few alternative models of federalism: competitive (nation-centered, state-centered, and dual federalism), interdependent (cooperative, creative, and new federalism), and functional ("picket fence" and "bamboo fence" federalism). O'Toole (1985) differentiates between federalist models with overlapping authority, coordinative authority, and inclusive authority.

Politics are inherent in government at all levels—local, national, and international. The evolution of government toward the practice of governance[1] that is increasingly accepted worldwide more explicitly incorporates intergovernmental relations among a broader set of stakeholders and interest groups involved in decision-making processes. The increasingly participative but also politicized environment is not uncommon to collaborative alliances formed around interorganizational information systems (Kumar and van Dissel 1996). In addition to changes in institutions and the political and economic context, the intensified use of information and communication technologies (ICTs) also influences the models of governance and democratic processes (Falch 2006). For example, Radin and Romzek's (1996) comparison of Weberian and virtual bureaucracies (table 3) demonstrates how ICTs facilitate transformations from government to governance. Furthermore, Fountain's (2001) analytical framework (figure 1) relates organizational forms and institutional arrangements to the process of technology

Weberian bureaucracy	Virtual bureaucracy
Functional differentiation, precise division of labor, clear jurisdictional boundaries	Information structured using information technology rather than people; organizational structure based on information systems rather than people
Hierarchy of offices and individuals	Electronic and informal communication; teams carry out the work and make decisions
Files, written documents, staff to maintain and transmit files	Digitized files in flexible form, maintained and transmitted electronically using sensors, bar codes, transponders, handheld computers; chips record, store, analyze, and transmit data; systems staff maintain hardware, software, and telecommunications
Employees are neutral, impersonal, attached to a particular office	Employees are cross-functional, empowered; jobs limited not only by expertise but also by the extent and sophistication of computer mediation
Office system of general rules, standard operating procedures, performance programs	Rules embedded in applications and information systems; an invisible, virtual structure
Slow processing time due to batch processing, delays, lags, multiple handoffs	Rapid or real-time processing
Long cycles of feedback and adjustment	Constant monitoring and updating of feedback; more rapid or real-time adjustment possible

Table 3. Weberian and virtual bureaucracies
Reprinted from Radin and Romzek 1996, with permission of Oxford University Press.

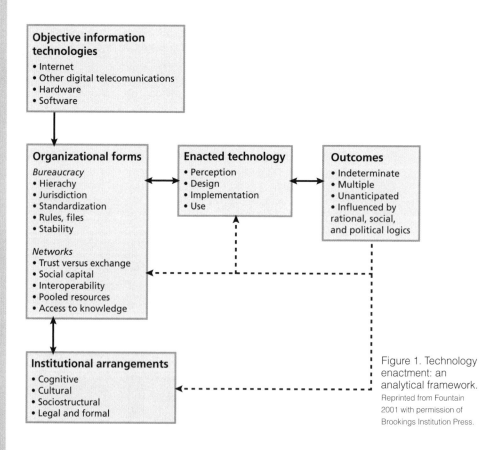

Objective information technologies
- Internet
- Other digital telecomunications
- Hardware
- Software

Organizational forms
Bureaucracy
- Hierachy
- Jurisdiction
- Standardization
- Rules, files
- Stability

Networks
- Trust versus exchange
- Social capital
- Interoperability
- Pooled resources
- Access to knowledge

Enacted technology
- Perception
- Design
- Implementation
- Use

Outcomes
- Indeterminate
- Multiple
- Unanticipated
- Influenced by rational, social, and political logics

Institutional arrangements
- Cognitive
- Cultural
- Sociostructural
- Legal and formal

Figure 1. Technology enactment: an analytical framework.
Reprinted from Fountain 2001 with permission of Brookings Institution Press.

enactment. The author suggests that different cognitive, cultural, sociostructural, and legal forms are required for hierarchical and network organizations.

Actor network theory. Actor network theory (ANT) is often used instead of conventional social theory (e.g., Giddens 1979; structuralist theory) to examine and explain the interaction between information technology and society (Hanseth et al. 2004; Monteiro and Hanseth 1995). ANT applies semiotics in explaining social phenomena and their attributes and forms as resulting from relations with other entities; in addition, all entities have to satisfy the performativity aspect of ANT, in other words, to be performed in, by, and through those relations (Law and Hassard 1999). The focus is on undoing the artificial boundaries between social and technical systems and related processes. For example, Faraj et al. (2004) employ ANT in their study of the complex interdependences that characterize the evolution of Web browsers and demonstrate that technological and human agents are inseparable in constructing new sociotechnical artifacts.

According to Callon (1986) and Mahring et al. (2004), creation of an actor network, which is also called translation, consists of four major stages: problematization, interessement (recruitment), enrollment, and mobilization (table 4). The translation process does not have to pass through all four phases and may fail at any stage. In addition to translation, there is the process of inscription of ideas in given technologies; as those technologies diffuse within specific contexts, they are assigned relevance and help achieve sociotechnical stability (Latour 1987). Another ANT phenomenon is irreversibility, which is the degree to which a network can be brought back to a state where alternative possibilities exist. Hanseth and Monteiro (1998) find that irreversibility is due to the inscription of interests into technological artifacts, whereby those individual and organizational interests customize the system and become increasingly difficult to change. In the context of changing but sometimes irreversible networks, the authors propose three actor network configurations (elements of decomposition): disconnected networks, gateways, and polyvalent networks.

Use, utility, and usability of information systems (3U). Although the terms "usability" and "usefulness" (referred to in this work as "utility") are often

Problematization	An actor initiating the process (also called focal actor) defines the identities and interests of other actors that are consistent with the interest of the focal actor. In this initial stage of building the actor network, some actors position themselves as indispensable for solving the problems defined. They define the problem and solution and also the identities and roles for other actors in the network.
Interessement (recruitment)	Convincing other actors that the interests defined by focal actors are in line with their own interest. Depending upon situation, this phase also involves creating incentives for actors so that the obstacles to bringing these actors into the network are overcome. A successful recruitment confirms the validity of problematization, locks new actors into the network, and corners the entities that are not yet co-opted.
Enrollment	The roles of the actors in the newly created network are defined. The focal actor strives to convince other actors to fully embrace the underlying ideas of the growing network and become an active part of the mission. Multilateral negotiation takes place.
Mobilization	Focal actor makes sure that all actors are acting in accordance with the underlying spirit of the network mission. The focal actor seeks continued support from all the enrolled actors in order to keep the network stable. The actors are mobilized to further stabilize and institutionalize the network.

Table 4. Actor network theory: stages of translation.
Adapted from Callon 1986 and Mahring et al. 2004.

employed interchangeably in the context of ICT systems, they are not equivalent. Blomberg et al. (1994) suggest that "usability refers to the general intelligibility of systems, particularly at the interface; usefulness means that a system's functionality actually makes sense and adds value in relation to a particular work setting" (p. 190). The concept of effective use subsumes both usability and usefulness. Effective use of ICTs, according to Gurstein (2003), is the capacity and opportunity to successfully integrate these technologies to achieve the users' self-defined or collaboratively defined goals, and it requires carriage facilities (i.e., appropriate communication infrastructure), input/output devices, tools and supports, content services, service access/provision, social facilitation (e.g., network, leadership, training), and governance. In the IS realm, DeLone and McLean (1992) suggest the amount and duration of use (e.g., number of functions performed, reports generated, charges, frequency of access) and nature and level of use as objective measures.

Although the post–World War II growth of scientific literature marked the beginning of a more systematic study of information systems, the focus of research efforts did not shift from technology to information users and their behavior until the 1980s (Wilson 1994, 2000). Consequently, the design of information systems and services started to shift from system-centered to user-centered approaches and sociotechnical designs (Eason 1988). User study is now a well-established area of information science (Bates 2005; Dervin and Nilan 1986; Dervin 1989; Foster 2004; Lamb and Kling 2003; Leckie et al. 1996; Orlikowski and Gash 1994; Savolainen 1995; Stewart and Williams 2005; Taylor 1991). Among the questions it poses are the following: How do people seek information? How is information put to use? How do information needs and activities change over time? The user-centered studies operate at two main levels of analysis: individual level (Attfield and Dowell 2003; Brashers et al. 2000; Chatman 1996; Cobbledick 1996; Ellis 1993; Savolainen 1995) and organizational level (Lamb and Kling 2003; Leckie et al. 1996; Orlikowski and Gash 1994; Taylor 1991).

In addition to individual-level studies that consider users in a more passive fashion (i.e., as relevant but not substantially influential and powerful participants), there is a prominent trend of viewing users as innovators, "sense-makers," and "domesticators" of information technologies and systems (Bruce and Hogan 1998; Dervin 1989; Griffith 1999; Stewart and Williams 2005; Williams 1997). The central tenet of domestication and its associated concept of idealization-realization of technology (Bruce 1993) is that technology gets appropriated and its meaning is constructed by situated use. By implication, designers cannot design the system; they can only invoke the design process. It is through the users' continued appropriation that an information system and services become useful.

CONCLUSIONS

This paper was motivated by the increasingly recognized failure of SDI research and practice to both utilize the existing theoretical and empirical knowledge base and develop its own conceptual framework. The majority of contributions to gray and refereed literature tend to be anecdotal, unsystematic, and isolated from the broader scientific discourse. This situation limits the development of functional and relevant SDIs worldwide. The importance of expanding the knowledge base is even more obvious when considering the magnitude and multiplicity of challenges the SDI efforts face, including politics, finance, technical capacity, human resources, and utility. In this paper we offer a substantial overview of existing SDI research, point to research gaps, and review five areas as potential major resources for strengthening the SDI conceptual base: information infrastructure, interorganizational collaboration-cooperation-coordination, intergovernmental relations, action network theory, and use-utility-usability of information systems (table 5). Figure 2 shows a tentative but pragmatic conceptual framework for SDI development.

Conceptual framework derived from the expanded SDI knowledge base. The notions of information infrastructure and of the installed base, in particular, are useful in taking a deeper look at SDIs. The concept of the installed base implies that the existing technical systems (e.g., hardware, software, and data) and organizational structures (e.g., human resources and skills, management practices, and legal arrangements) may play facilitating or constraining roles. Infrastructure openness implies that SDIs should accommodate a growing number of heterogeneous actors and artifacts. Georgiadou et al. (2005) incorporate some of these concepts in analyzing the Indian NSDI. The usefulness of the concepts, however, needs to

Knowledge area	Key premises	Value for SDI	Limitations
II	Open, transparent, standardized, and widely accessible network based on Internet and other ICT, serving a broad set of users and communities	Special type of infrastructure and the notion of the installed base	Factors, strategies, and processes for developing IIs are not elaborated or tested
3C	Interorganizational systems require 3C; they relate to interorganizational interdependences, involve complex mechanisms, and carry potential for conflict	Information sharing and exchange are fundamental to SDIs; successful 3C is necessary for SDIs to become functional and relevant	Focus on private corporations and profit maximization; difficulty in identifying viable motivators in the public sector
IGR	Models of governments and societal decision making range on a continuum from centralized to decentralized (including federalist), with different types of authority and administrative approaches	Governments at all levels are the majority stakeholders of SDIs; SDIs build upon and adjust to (as well as affect) intergovernmental settings and relationships; SDIs are an element of the envisioned virtual bureaucracy	Nongovernmental actors— private sector, academia, nonprofit organizations, and population at large (citizen associations and interest groups)— are not addressed
ANT	All phenomena take their form and attributes in relation to other entities and are "performed" in, by, and through them; membership grows through a process of translation (problematization, interessement, enrollment, and mobilization)	SDIs are often modeled as hierarchies, but they are more likely to evolve as networks and Internet-based access points to acquiring data and services; the translation process is one way of understanding and cultivating SDIs	Flexibility and uncertainty do not easily translate into implementable models; more a method for explaining and interpreting reality than for acting on it to stimulate new developments
3U	Extending traditional IS focus with sociotechnical design, user involvement and action, and evaluation	Useful in bottom-up approaches; recognizing the major role of many potential SDI users and their creativity	Developed for single systems and organizations; needs rigorous evaluation methods to apply to the evolution of SDI from data and service to E-governance

Table 5. Key premises, value, and limitations of the five knowledge areas.

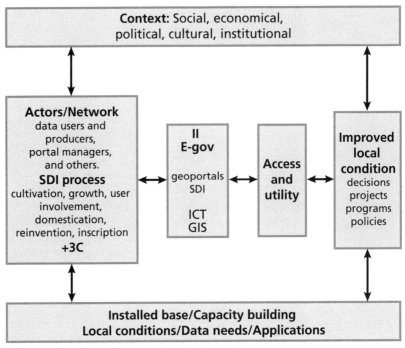

Figure 2. Proposed framework for SDI development.

be explored further. Creating SDIs with all the envisioned characteristics of a full-blown and operational infrastructure is not easy. Moreover, information infra-structures are neither created from a void nor completely designed. Rather, the process of "building" is replaced by "cultivation" of the sociotechnical installed base to gradually incorporate diverse actors in a networked environment. The cultivation approach has sufficient flexibility to accommodate local circum-stances and practices. It also turns attention to capacity building needs at all lev-els, including the so-called "interagency collaborative capacity (ICC)" (Bardach 1998), individual agency GIS capacity (Mackay et al. 2002), and citizen/user capacity (Tettey 2002).

The ideas discussed in the studies on interorganizational relationships and 3C are useful and easily applied. The majority of studies on interorganiza-tional IS are situated in the context of large corporations and employ produc-tivity and maximization of profit as success criteria (Doherty and King 2001; Johnston and Gregor 2000; Munkvold 1999; Suomi 1992, 1994; Williams 1997). Interorganizational exchange and consensus are essential factors in SDI develop-ment. The 3C concept is employed in GIS research (Azad and Wiggins 1995; Craig 2005; Harvey 2001; Nedović-Budić and Pinto 1999b; Nedović-Budić et al. 2004) but remains incompletely exploited and leaves the question of how to suc-cessfully initiate and maintain SDI coalitions among diverse stakeholders incom-pletely answered. Also, in the context of the public sector, which prevails among SDI participants, understanding intergovernmental relations and the impact on and of E-governance would also be indispensable to establishing effective SDIs.

Actor network theory offers a rich perspective on how a network of aligned interests, as well as nested smaller networks, can be created with diverse human actors and heterogeneous technical systems. ANT provides a useful theoretical toolset to investigate the coalitions required for SDIs to become functional and effective within the context of overall societal progress. Though few researchers apply actor network theory to study GIS activities (Harvey 2001; Martin 2000; Walsham and Sahay 1999), they use it within a limited organizational context and do not employ it in studying the creation of SDI networks. But more generally, we find that ANT has more facility in research than in practice. It is more helpful for observing and interpreting sociotechnical networks than for developing viable relations among targeted actors and ensuring specific outcomes of such relations.

Between usability and utility, the latter is certainly more relevant for studying large-scale infrastructures such as SDIs. The user perspective, in general, has gained widespread popularity. Gurstein's (2003) framework of effective use of information resources is applicable to SDIs. It reveals that there are other important organizational and social structures that can enable or limit SDIs. The lens of effective use thus allows us to see SDIs beyond the current paradigm of provision of and access to geospatial information. In the words of Stewart and Williams (2005, p. 2):

> Design outcomes/supplier offerings are inevitably unfinished in relation to complex, heterogeneous and evolving user requirements. Further innovation takes place as artifacts are implemented and used. To be used and useful, ICT artifacts must be 'domesticated' and become embedded in broader systems of culture and information practices. In this process artifacts are often reinvented and further elaborated.

Despite the convincing criticism of the traditional user-centered and sociotechnical approaches and their limited applicability to single systems and organizations, the proponents of more radical views have not operationalized their ideas or offered practical solutions that can be implemented in actual development projects. In huge systems like SDIs, identifying who the potential users are and how to represent them in the process of an evolving SDI remains difficult. The complexities of SDIs require further studies of use and users and continuous monitoring and evaluation. The challenges, however, should not undermine the essential importance of strong representation and active participation of users as "domesticators," "sensemakers," and "innovators" who ultimately evaluate the utility of SDIs.

The literature discussed in this paper suggests the following conceptual base: cultivation approach to SDI; focus on SDI users, access, and derived utility; capacity building in the installed base; understanding of the networking relationships and attributes of data users, producers, and managers; incorporation of 3C principles and opportunities; attention to intergovernmental relations and the emerging trends in E-governance; capitalizing on mutually interdependent and supporting roles of GIS, ICT, and II; and evaluation of SDIs in terms of their ultimate goal of improving local conditions by enabling various communities and stakeholders to get involved in decision-making processes and affect implementation of local projects, policies, and programs. Last but not least, all SDI activities and participants are situated within specific societal, cultural, and institutional contexts.

All these elements constitute the core of the proposed conceptual framework. However, the framework is only preliminary and intended to serve as a starting point for integration of the multi- and interdisciplinary knowledge base in studying and developing SDIs worldwide.

The five knowledge areas discussed in this paper are by no means sufficient or exhaustive sources for informing SDI research and practice. In fact, none of them individually offers a comprehensive knowledge base required to develop and sustain SDI networks. Knowledge areas of policy implementation, federated databases and systems development, capacity building, and public administration and finance are worth considering as well. In addition, the literature on technical concepts and models, which are also important but often less challenging, is not addressed in this paper. The five selected areas are used to illustrate the wealth of concepts and theories available and accessible to academics and professionals interested in SDIs. The expanded knowledge base provides better information for both studying and developing SDIs. By incorporating existing theoretical and empirical knowledge from other relevant fields, the SDI community will not only avoid reinventing the wheel but also be more effective in establishing SDIs and furthering scientific discourse with new insights, ideas, concepts, theories, and applications. Most importantly, the long-awaited societal benefits are more likely to emerge with SDIs that are guided by intelligence from the past as a basis for creativity and innovations for the future.

ENDNOTE

1. According to Stewart (2003), ". . .'[g]overnment' can be defined as the activity of the formal governmental system, conducted under clear procedural rules, involving statutory relationships between politicians, professionals and the public, taking place within specific territorial and administrative boundaries. It involves the exercise of powers and duties by formally elected or appointed bodies, and using public resources in a financially accountable way. 'Governance' is a much looser process often transcending geographical or administrative boundaries, conducted across public, private and voluntary/community sectors through networks and partnerships often ambiguous in their memberships, activities, relationships and accountabilities. It is a process of multistakeholder involvement, of multiple interest resolution, of compromise rather than confrontation, of negotiation rather than administrative fiat" (p. 76). In governance, transaction costs are minimized, trust maximized, and collaborative advantage extracted.

REFERENCES

Agranoff, Robert. 2001. Managing within the matrix: Do collaborative intergovernmental relations exist? *Publius* 31 (2): 31–56.

Aiken, Michael, and Jerald Hage. 1968. Organizational interdependence and intra-organizational structure. *American Sociological Review* 33 (6): 912–30.

Anderson, Rachael K., Alice Haddix, Jeannette C. McCray, and Timothy P. Wunz. 1994. Developing a health information infrastructure for Arizona. *Bulletin of the Medical Library Association* 84 (2): 396–400.

Askew, David, Sharon Evans, Ruth Matthews, and Phillipa Swanton. 2005. MAGIC: A geoportal for the English countryside. *Computers, Environment and Urban Systems* 29 (1): 71–85.

Attfield, Simon, and John Dowell. 2003. Information seeking and use by newspaper journalists. *Journal of Documentation* 59 (2): 187–204.

Azad, Bijan, and Lyna L. Wiggins. 1995. Dynamics of inter-organizational geographic data sharing: A conceptual framework for research. In *Sharing geographic information,* eds. Harlan J. Onsrud and Gerard Rushton, 22–43. New Brunswick, NJ: Center for Urban Policy Research.

Bangemann Group. 1994. *Europe and the Global Information Society.* Recommendations of the high-level group on the information highway to the Corfu European Council.

Bardach, Eugene. 1998. *Getting agencies to work together: The practice and theory of managerial craftsmanship.* Washington, DC: Brookings Institution Press.

Bates, Marcia J. 2005. An introduction to metatheories, theories, and models. In *Theories of information behavior, ASIST Monograph Series,* eds. Karen E. Fisher, Sanda Erdelez, and Lynne E. F. McKechnie, 1–24. Medford, NJ: Information Today.

Beaumont, Peter, Paul A. Longley, and David J. Maguire. 2005. Geographic information portals: A UK perspective. *Computers, Environment and Urban Systems* 29 (1): 49–69.

Begusic, Dinko, Nikola Rozic, and Hrvoje Dujmic. 2003. Development of the communication/information infrastructure at the academic institution. *Computer Communications* 26 (5): 472–76.

Bernard, Lars, and Max Craglia. 2005. SDI: From Spatial Data Infrastructure to Service Driven Infrastructure. *Position paper presented at research workshop on cross-learning between SDI and II.* International Institute for Geo-Information Science and Earth Observation (ITC). March 31–April 1. Enschede, Netherlands.

Bernard, Lars, Ioannis Kanellopoulos, Alessandro Annoni, and Paul Smits. 2005. The European geoportal: One step towards the establishment of a European Spatial Data Infrastructure. *Computers, Environment and Urban Systems* 29 (1): 15–31.

Bishop, Ian D., Francisco J. Escobar, Sadasivam Karuppannan, Ian P. Williamson, and Paul M. Yates. 2000. Spatial Data Infrastructures for cities in developing countries. *Cities* 17 (2): 85–96.

Bishr, Yaser. 1998. Overcoming the semantic and other barriers to GIS interoperability. *International Journal of Geographical Information Science* 12 (4): 299–314.

Blomberg, Jeanette, Lucy Suchman, and Randall H. Trigg. 1994. Reflections on a work-oriented design project. *Paper presented at the Participatory Design Conference (PDC'94).* October. Chapel Hill, North Carolina.

Borgman, Christine L. 2000. The premise and the promise of a Global Information Infrastructure. *Peer-Reviewed Journal of the Internet* 5 (8). http://www.firstmonday.org/issues/issue5_8/borgman/index.html (accessed January 15, 2006).

Brashers, Dale E., Judith L. Neidig, Stephen M. Haas, Linda K. Dobbs, Linda W. Cardillo, and Jane A. Russell. 2000. Communication in the management of uncertainty: The case of persons living with HIV or AIDS. *Communication Monographs* 67:63–84.

Bruce, Bertram C. 1993. Innovation and social change. In *Network-based classrooms: Promises and realities,* eds. Bertram C. Bruce, Joy K. Peyton, and Trent W. Batson, 9–32. Cambridge University Press, NY.

Bruce, Bertram C., and Maureen P. Hogan. 1998. The disappearance of technology: Toward an ecological model of literacy. In *Handbook of literacy and technology: Transformations in a post-typographic world*, eds. D. Reinking, M. McKenna, L. Labbo, and R. Kieffer, 269–81. Hillsdale, NJ: Erlbaum.

Budić, Zorica D. 1994. Implementation and management effectiveness in adoption of GIS technology in local governments. *Computers, Environment and Urban Systems* 18 (5): 295–304.

Butler, Al, Alan Voss, Dennis Goreham, and John Moeller. 2005. The national geospatial coordinating council: A dramatic new approach to build the NSDI. *GeoWorld*, October: 38–41.

Callon, Michel. 1986. Some elements of a sociology of translation: Domestication of the scallops and the fishermen of St. Brieuc Bay. In *Power, Action and Belief: A new sociology of knowledge?* ed. John Law, 196–233. London: Routledge & Kegan Paul.

Cameron, David. 2001. *The structures of intergovernmental relations*. Oxford: Blackwell Publishers.

Chatman, Elfreda A. 1996. Impoverished life world of outsiders. *Journal of the American Society for Information Science and Technology* 47 (3): 193–206.

Cobbledick, Susie. 1996. The information-seeking behavior of artists: Exploratory interviews. *Library Quarterly* 66 (4): 343–72.

Coleman, David, Richard Groot, and John McLaughlin. 2000. Human Resources issues in the emerging GDI environment. In *Geospatial data infrastructure: Concepts, cases and good practice,* 1st ed., eds. Richard Groot and John McLaughlin, 233–44. New York: Oxford University Press.

Cook, Karen S. 1977. Exchange and power in networks of inter-organizational relations. *Sociological Quarterly* 18:62–82.

Craglia, Massimo, and Ian Masser. 2002. Geographic information and the enlargement of the European Union: Four national case studies. *Journal of the Urban and Regional Information System Association* 14 (2): 43–52.

Craig, William J. 1995. Why we can't share data: Institutional inertia. In *Sharing geographic information*, eds. Harlan J. Onsrud and Gerard Rushton, 107–18. New Brunswick, NJ: Center for Urban Policy Research.

Craig, William J. 2005. White knights of Spatial Data Infrastructure: The role and motivation of key individuals. *Journal of the Urban and Regional Information System Association* 16 (2): 5–13.

Cramond, Stephen. 1999. Efforts to formalize international collaboration in scholarly information infrastructure. *Library Hi Tech* 17 (3): 272–82.

Crompvoets, Joep, Arnold Bregt, Abbas Rajabifard, and Ian Williamson. 2004. Assessing the worldwide developments of national spatial data clearinghouses. *International Journal of Geographical Information Science* 18 (7): 665–89.

Cummings, Thomas G. 1980. Inter-organization theory and organizational development. In *Systems theory for organization development,* ed. Thomas G. Cummings. New York: John Wiley & Sons.

Dangermond, Jack. 1995. Public data access. In *Sharing geographic information*, eds. Harlan J. Onsrud and Gerard Rushton, 331–39. New Brunswick, NJ: Center for Urban Policy Research.

Dawes, Sharon S. 1996. Interagency information sharing: Expected benefits, manageable risks. *Journal of Policy Analysis and Management* 15 (3): 377–94.

DeLone, William H., and Ephraim R. McLean. 1992. Information systems success: The quest for the dependent variable. *Information Systems Research* 3 (1): 6–95.

Dervin, Brenda. 1989. Users as research inventions: How research categories perpetuate inequities. *Journal of Communication* 39 (Summer): 216–32.

Dervin, Brenda, and Michael Nilan. 1986. Information needs and uses. *Annual Review of Information Science and Technology (ARIST)* 21: 3–33.

Doherty, Neil, and Malcolm King. 2001. An investigation of the factors affecting the successful treatment of organizational issues in systems development projects. *European Journal of Information Systems* 10 (3): 147–60.

Eason, Ken. 1988. *Information technology and organizational change.* Taylor & Francis, London.

Ekbia, Hamid R., and Rob Kling. 2005. Network organizations: Symmetric cooperation or multivalent negotiation? *Information Society* 21 (3): 155–68.

Ellis, David. 1993. Modeling the information-seeking patterns of academic researchers: A grounded theory approach. *Library Quarterly* 63 (4): 469–86.

Enemark, Stig, and Ian Williamson. 2004. Capacity building in land administration: A conceptual approach. *Survey Review* 39 (294): 639–50.

Falch, Morten. 2006. ICT and the future conditions for democratic governance. *Telematics and Informatics* 23 (2): 134–56.

Faraj, Samer, Dowan Kwon, and Stephanie Watts. 2004. Contested artifact: Technology sensemaking, actor networks and the shaping of the web browser. *Information Technology & People* 17 (2): 186–209.

Feeney, Mary-Ellen F. 2003. SDIs and decision support. In *Developing Spatial Data Infrastructures: From concept to reality,* eds. Ian Williamson, Abbas Rajabifard, and Mary-Ellen F. Feeney, 195–210. Boca Raton: CRC Press.

Fonseca, Fred T., Max J. Egenhofer, Clodoveu A. Davis, Jr., and Karla A. V. Borges. 2000. Ontologies and knowledge sharing in urban GIS. *Computers, Environment and Urban Systems* 24:251–71.

Foster, Allen. 2004. A nonlinear model of information-seeking behavior. *Journal of the American Society for Information Science and Technology* 55 (3): 228–37.

Fountain, Jane E. 2001. *Building the virtual state: Information technology and institutional change.* Washington, DC: Brookings Institution Press.

Fraser, Bruce, and Myke Gluck. 1999. Usability of geospatial metadata or space-time matters. *American Society for Information Science* 25:24–28.

Georgiadou, Yola, and Michael Blakemore. A journey through GIS discourses. (Unpublished.)

Georgiadou, Yola, Orlando Rodríguez-Pabón, and Kate T. Lance. 2006. SDI and E-Governance: A quest for appropriate evaluation approaches. *Journal of Urban and Regional Information Systems* 18 (2): 43–55.

Georgiadou, Yola, Satish K. Puri, and Sandeep Sahay. 2005. Towards a potential research agenda to guide the implementation of Spatial Data Infrastructures: A case study from India. *International Journal of Geographical Information Science* 19 (10): 1113–30.

Georgiadou, Yola, and Richard Groot. 2002. Policy development and capacity building for geo-information provision. A global goods perspective. In *GIS@development: The monthly magazine on geographic information science* 6 (7): 33–40.

Giddens, Anthony. 1979. *Central Problems in Social Theory: Action, Structure and Contradiction in Social Analysis.* Berkeley, CA: University of California Press.

Gray, Barbara. 1989. *Collaborating: Finding common ground for multiparty problems.* San Francisco: Jossey-Bass Publishers.

Griffith, Terri L. 1999. Technology features as triggers for sensemaking. *Academy of Management Review* 24 (3): 472–488.

Groot, Richard. 2001. Reform of government and the future performance of national surveys. *Computers, Environment and Urban Systems* 25 (4–5): 367–87.

Gurstein, Michael. 2003. Effective use: A community informatics strategy beyond the digital divide. *First Monday* 8 (12).

Hanseth, Ole, Margun Aanestad, and Marc Berg. 2004. Actor-network theory and information system. What's so Special? *Information Technology & People* 17 (2): 116–23.

Hanseth, Ole, and Eric Monteiro. 1998. Changing irreversible networks. *Paper presented at the 6th European Conference on Information Systems,* Aix-en-Provence, 4–6 (June).

Hanseth, Ole, and Eric Monteiro. 2004. *Understanding Information Infrastructure.* (Forthcoming book).

Harvey, Francis. 2001. Constructing GIS: Actor networks of collaboration. *Journal of the Urban and Regional Information System Association* 13 (1): 29–37.

Harvey, Francis. 2003. Developing geographic information infrastructures for local government: The role of trust. *Canadian Geographer* 47 (1): 28–36.

Harvey, Francis, Werner Kuhn, Hardy Pundt, Yaser Bishr, and Catharina Riedemann. 1999. Semantic interoperability: A central issue for sharing geographic information. *The Annals of Regional Science* 33 (2): 213–32.

Gore, Al (former U.S. vice president) and Information Infrastructure Task Force. 1993. *The National Information Infrastructure: Agenda for Action.* Washington, DC.

Jacoby, Steve, Jessica Smith, Lisa Ting, and Ian Williamson. 2002. Developing a common spatial data infrastructure between state and local government: An Australian case study. *International Journal of Geographical Information Science* 16 (4): 305–22.

Johnston, Robert, and Shirley Gregor. 2000. A theory of industry-level activity for understanding the adoption of interorganizational systems. *European Journal of Information Systems* 9 (4): 243–51.

Koike, Osamu, and Deil S. Wright. 1998. Five phases of IGR in Japan: Policy shifts and governance reform. *International Review of Administrative Sciences* 64:203–18.

Kim, Tschangho J. 1999. Metadata for geo-spatial data sharing: A comparative analysis. *The Annals of Regional Science* 33 (2): 171–81.

Klien, E., M. Lutz, and W. Kuhn. 2006. Ontology-based discovery of geographic information services: An application in disaster management. *Computers, Environment and Urban Systems* 30:102–23.

Kok, Bas, and Bastiaan V. Loenen. 2005. How to assess the success of National Spatial Data Infrastructures? *Computers, Environment and Urban Systems* 29 (6): 699–717.

Kuhn, Werner. 2003. Semantic reference systems. *International Journal of Geographical Information Science* 17 (5): 405–9.

Kumar, Sundeep, and Han G. van Dissel. 1996. Sustainable collaboration: Managing conflict and cooperation in interorganizational systems. *MIS Quarterly* 20:279–300.

Lachman, Beth E., Anny Wong, Debra Knopman, and Kim Gavin. 2002. *Lessons for the Global Spatial Data Infrastructure: International Case Study Analysis.* Santa Monica: Rand Co.

Langendorf, Richard. 2001. Computer-aided visualization: Possibilities for urban design, planning and management. In *Planning Support Systems: Integrating GIS and Visualization Tools,* eds. Richard Brail and Richard Klosterman, 309–59. Redlands, CA: ESRI Press.

Lamb, Roberta, and Rob Kling. 2003. Reconceptualizing users as social actors in information systems research. *MIS Quarterly* 27 (2): 197–235.

Lambert, Rob, and Joe Peppard. 1993. Information technology and new organizational forms: Destination but no road map? *Journal of Strategic Information Systems* 2 (3): 180–205.

Latour, B. 1987. *Science in Action: How to Follow Scientists and Engineers through Society.* Cambridge, MA: Harvard University Press.

Law, John, and John Hassard. 1999. *Actor network theory and after.* Oxford: Blackwell Publishers.

Leckie, Gloria J., Karen E. Pettigrew, and Christian Sylvain. 1996. Modeling the information-seeking of professionals: A general model derived from research on engineers, health care professionals and lawyers. *Library Quarterly* 66 (2): 162–93.

Levine, Sol, and Paul E. White. 1969. Exchange as a conceptual framework for the study of interorganizational relationships. In *A Sociological Reader on Complex Organizations,* ed. Amitai Etzioni. 2nd ed. New York: Holt, Rinehart and Winston.

Mackay, Ronald, Douglas Horton, Luis Dupleich, and Anders Andersen. 2002. Evaluating organizational capacity development. *The Canadian Journal of Program Evaluation* 17 (2): 121–50.

Maguire, David J., and Paul A. Longley. 2005. The emergence of geoportals and their role in Spatial Data Infrastructures. *Computers, Environment and Urban Systems* 29 (1): 3–14.

Mahring, Magnus, Jonny Holmstrom, Mark Keil, and Ramiro Montealgre. 2004. Trojan actor-network and swift translation: Bringing actor-network theory to IT project escalation studies. *Information Technology & People* 17 (2): 210–38.

Mark, David, Max Egenhofer, Stephen Hirtle, and Barry Smith. 2000. Ontological foundations for Geographic Information Science. University Consortium for Geographic Information Science (UCGIS). 2000 Research White Papers. http://www.ucgis.org/priorities/research/research_white/2000%20Papers/emerging/ontology_new.pdf (accessed June 25, 2006).

Martin, Eugene W. 2000. Actor-networks and implementation: Examples from conservation GIS in Ecuador. *International Journal of Geographical Information Science* 14 (8): 715–38.

Masser, Ian. 1999. All shapes and sizes: The first generation of National Spatial Data Infrastructures. *International Journal of Geographical Information Science* 13 (1): 67–84.

Masser, Ian. 2004. Capacity Building for Spatial Data Infrastructure Development. *Keynote presentation at 7th International Seminar on GIS for developing countries (GISDECO).* May 10–12. Johor, Malaysia.

Masser, Ian. 2005a. *GIS Worlds: Creating Spatial Data Infrastructures,* 1st ed. Redlands, CA: ESRI Press.

Masser, Ian. 2005b. Some Priorities for SDI Related Research. *Paper presented at the Global Spatial Data Infrastructure* 8. Cairo, Egypt. April 16–21.

Masser, Ian, Santiago Borrero, and Peter Holland. 2003. Regional SDIs. In *Developing Spatial Data Infrastructures: From concept to reality,* 1st ed., eds. Ian Williamson, Abbas Rajabifard, and Mary-Ellen F. Feeney, 59–77. Boca Raton: CRC Press.

McCall, Michael K. 2003. Seeking good governance in participatory-GIS: A review of processes and governance dimensions in applying GIS to participatory spatial planning. *Habitat International* 27 (4): 549–73.

McCann, Joseph E. 1983. Design guidelines for social problem-solving interventions. *Journal of Applied Behavioral Science* 19 (2): 177–92.

McKee, Lance. 2000. Who wants GDI? In *Geospatial Data Infrastructure: Concepts, cases and good practice,* 1st ed., eds. Richard Groot and John McLaughlin, 13–24. Oxford University Press, New York.

Mennecke, Brian E. 1997. Understanding the role of geographic information technologies in business: Applications and research directions. *Journal of Geographic Information and Decision Analysis* 1 (1): 44–68.

Meredith, Paul H. 1995. Distributed GIS: If its time is now, why is it resisted? In *Sharing geographic information,* eds. Harlan J. Onsrud and Gerard Rushton, 7–21. New Brunswick, NJ: Center for Urban Policy Research.

Moellering, Harold. 2005. *World spatial metadata standards.* Oxford: Elsevier Science.

Montalvo, Uta Wehn de. 2003. In Search of rigorous models for policy-oriented research: A behavioral approach to spatial data sharing. *The Journal of Urban and Regional Information Systems* 15:19–28.

Monteiro, Eric. 1998. Scaling information infrastructure: The case of next-generation IP in the Internet. *The Information Society* 14:229–45.

Monteiro, Eric, and Ole Hanseth. 1995. Social shaping of information infrastructure: On being specific about the technology. In *Information Technology and Changes in Organizational Work,* eds. Wanda J. Orlikowski, Goeff Walsham, Matthew R. Jones, and Janice I. DeGross, 325–43. Norwell, MA: Kluwer Academic Publishers.

Munkvold, Bjorn E. 1999. Challenges of IT implementation for supporting collaboration in distributed organizations. *European Journal of Information Systems* 8 (4): 260–72.

Nedović-Budić, Zorica, Mary-Ellen F. Feeney, Abbas Rajabifard, and Ian Williamson. 2004. Are SDIs serving the needs of local planning? Case study of Victoria, Australia and Illinois, USA. *Computers, Environment and Urban Systems* 28 (4): 329–51.

Nedović-Budić, Zorica, and Jeffrey K. Pinto. 1999a. Interorganizational GIS: Issues and prospects. *The Annals of Regional Science* 33 (2): 183–95.

Nedović-Budić, Zorica, and Jeffrey K. Pinto. 1999b. Understanding interorganizational GIS activities: A conceptual framework. *Journal of the Urban and Regional Information System Association* 11 (1): 53–64.

Nedović-Budić, Zorica. 1997. GIS technology and organizational context: Interaction and adaptation. In *Geographic Information Research: Bridging the Atlantic,* eds. Massimo Craglia and Helen Couclelis, 165–84. London: Taylor & Francis.

Nedović-Budić, Zorica, and Jeffrey K. Pinto. 2001. Organizational (Soft) GIS interoperability: Lessons from the U.S. *International Journal of Applied Earth Observation and Geoinformation* 3 (3): 290–98.

Nedović-Budić, Zorica, Jeffrey K. Pinto, and Lisa Warnecke. 2004. GIS database development and exchange: Interaction mechanisms and motivations. *Journal of the Urban and Regional Information System Association* 16 (1): 15–29.

Nice, David D., and Patricia Frederickson. 1995. *The politics of intergovernmental relations*, 2nd ed. Chicago: Nelson-Hall.

Nogueras-Iso, J., F. J. Zarazaga-Soria, R. Bejar, P. J. Alvarez, and P. R. Muro-Medrano. 2005. OGC Catalog Services: A key element for the development of Spatial Data Infrastructures. *Computers & Geosciences* 31 (2): 199–209.

Nogueras-Iso, J., F. J. Zarazaga-Soria, J. Lacasta, R. Bejar, and P. R. Muro-Medrano. 2004. Metadata standard interoperability: Application in the geographic information domain. *Computers, Environment and Urban Systems* 28 (6): 611–34.

Oliver, Christine. 1990. Determinants of inter-organizational relationships: Integration and future direction. *Academy of Management Review* 15 (2): 241–65.

Onsrud, Harlan J., Barbara Poore, Robert Rugg, Richard Taupier, and Lyna Wiggins. 2004. The future of the Spatial Information Infrastructure. In *A Research Agenda for Geographic Information Science,* eds. Robert B. McMaster and E. Lynn Usery, 225–55. Boca Raton: CRC Press.

Onsrud, Harlan J. 1998. Compiled Responses by Questions for Selected Questions: Survey of National Spatial Data Infrastructures. http://www.spatial.maine.edu/~onsrud/gsdi/Selected.html (accessed December 3, 2005).

Orlikowski, Wanda J., and Debra C. Gash. 1994. Technological frames: Making sense of information technology in organizations. *ACM Transactions in Information Systems* 12 (2): 174–207.

O'Toole, Laurence J., Jr., ed. 1985. *American intergovernmental relations: Foundations, perspectives, and issues.* Washington, DC: CQ Press.

O'Toole, Laurence J., Jr., and Robert S. Montjoy. 1984. Interorganizational policy implementation: A theoretical perspective. *Public Administration Review* 84 (6): 491–503.

Pickles, John. 2004. *A History of spaces: Cartographic reason, mapping and the geo-coded world.* London: Routledge.

Pinto, Jeffrey K., and Harlan J. Onsrud. 1995. Sharing geographic information across organizational boundaries: A research framework. In *Sharing geographic information,* eds. Harlan J. Onsrud and Gerard Rushton, 44–64. New Brunswick, NJ: Center for Urban Policy Research.

Pressman, J., and A. Wildavsky. 1984. *Implementation,* 3rd ed. Berkeley, CA: University of California Press.

Pundt, Hardy, and Yaser Bishr. 2002. Domain ontologies for data sharing: An example from environmental monitoring using _eld GIS. *Computers & Geosciences* 28:95–102.

Radin, Beryl A., and Barbara S. Romzek. 1996. Accountability expectations in an intergovernmental arena: The national rural development partnership. *Publius* 26 (2): 59–82.

Rajabifard, Abbas. 2003. SDI diffusion: A Regional case study with relevance to other levels. In *Developing Spatial Data Infrastructures: From concept to reality,* 1st ed., eds. Ian Williamson, Abbas Rajabifard, and Mary-Ellen F. Feeney, 78–94: Boca Raton: CRC Press.

Robey, D., and C. A. Sales. 1994. Designing Organizations. *Journal of Management Information Systems* 10 (4): 183–211.

Rajabifard, Abbas, Mary-Ellen F. Feeney, and Ian Williamson. 2003. Spatial Data Infrastructures: Concept, nature and SDI hierarchy. In *Developing Spatial Data Infrastructures: From concept to reality,* 1st ed., eds. Ian Williamson, Abbas Rajabifard, and Mary-Ellen F. Feeney, 17–40. Boca Raton: CRC Press.

Rajabifard, Abbas, Mary-Ellen F. Feeney, Ian Williamson, and Ian Masser. 2003. National SDI initiatives. In *Developing Spatial Data Infrastructures: From concept to reality,* 1st ed., eds. Ian Williamson, Abbas Rajabifard, and Mary-Ellen F. Feeney, 95–109. Boca Raton: CRC Press.

Rhind, David. 2000. Funding an NGDI. In *Geospatial data infrastructure: Concepts, cases and good practice,* 1st ed., eds. Richard Groot and John McLaughlin, 39–56. New York: Oxford University Press.

Savolainen, Reijo. 1995. Everyday life information seeking: Approaching information seeking in the context of "way of life." *Library and Information Science Research* 17 (3): 259–94.

Spatial Data Interest Community on Monitoring and Reporting (SDIC MORE). 2005. Spatial Applications Division of Leuven (SADL), Katholieke Universiteit Leuven and Laboratorio di Sistemi Informativi Territoriali ed Ambientali (LABSITA), University of Rome La Sapienza. http://www.sdic-more.org (accessed April 11, 2006).

Sepic, Ron, and Kate Kase. 2002. The national biological information infrastructure as an e-government tool. *Government Information Quarterly* 19 (4): 407–24.

Smith, Jessica, William Mackaness, Allison Kealy, and Ian Williamson. 2004. Spatial Data Infrastructure requirements for mobile location based journey planning. *Transactions in GIS* 8 (1): 23–44.

Star, Susan L., and Karen Ruhleder. 1996. Steps toward an ecology of infrastructure: Design and access for large information spaces. *Journal of Information Systems Research* 7 (1): 111–34.

Stewart, Murray. 2003. Towards collaborative capacity. In *Urban transformation and urban governance: Shaping the competitive city of the future,* ed. Boddy Martin, 76–89. Bristol, UK: The Policy Press.

Stewart, James, and Robin Williams. 2005. The wrong trousers? Beyond the design fallacy: Social learning and the user. In *User involvement in innovation processes: Strategies and limitations from a socio-technical perspective,* ed. Harald Rohracher, 9–35. Munich: Profil-Verlag.

Suomi, Reima. 1992. On the concept of inter-organizational information systems. *Journal of Strategic Information Systems* 1 (2): 93–100.

Suomi, Reima. 1994. What to take into account when building an inter-organizational information system. *Information Processing & Management* 30 (1): 151–59.

Taylor, Robert S. 1991. Information use environments. *Progress in Communication Sciences* 10:217–55.

Tait, Michael G. 2005. Implementing geoportals: Applications of distributed GIS. *Computers, Environment and Urban Systems* 29 (1): 33–47.

Tettey, Wisdom J. 2002. ICT, local government capacity building, and civic engagement: An evaluation of the sample initiative in Ghana. *Perspectives on Global Development and Technology* 1 (2): 165–92.

Thompson, James D. 1967. *Organizations in action.* New York: McGraw-Hill.

Tjosvold, Dean. 1988. Cooperative and competitive dynamics within and between organizational units. *Human Relations* 41 (6): 425–36.

Tulloch, David, and Jennifer Fuld. 2001. Exploring county-level production of framework data: Analysis of the national framework data survey. *The Journal of Urban and Regional Information Systems* 13 (2): 11–21.

Visser, U., H. Stuckenschmidt, G. Schuster, and T. Vogele. 2002. Ontologies for geographic information processing. *Computers & Geosciences* 28:103–17.

Walsh, Kuuipo A., Cherri M. Pancake, Dawn J. Wright, Sally Haerer, and F. J. Hanus. 2002. "Humane" interfaces to improve the usability of data clearinghouses. Paper presented at the GIScience 2002 conference.

Walsham, Geoff, and Sandeep Sahay. 1999. GIS for district-level administration in India: Problems and opportunities. *MIS Quarterly* 23 (1): 39–66.

Williams, R. 1997. The social shaping of information and communications technologies. In *The Social Shaping of information superhighways: European and American roads to the information society,* eds. H. Kubicek, W. H. Dutton, and R. Williams, 299–338. New York: St. Martin's Press.

Williams, Trevor. 1997. Interorganisational information systems: Issues affecting interorganisational cooperation. *Journal of Strategic Information Systems* 6 (3): 231–50.

Williamson, Ian. 2003. SDIs: Setting the scene. In *Developing Spatial Data Infrastructures: From concept to reality,* 1st ed., eds. Ian Williamson, Abbas Rajabifard, and Mary-Ellen F. Feeney, 3–16. Boca Raton: CRC Press.

Williamson, Ian P., Abbas Rajabifard, and Enemark Stig. 2003. Capacity building for SDIs. *Proceedings of 16th United Nations Regional Cartographic Conference for Asia Pacific.* July 14–18. Okinawa, Japan.

Wilson, Thomas D. 1994. Information needs and uses: Fifty years of progress. In *Fifty years of information progress,* ed. B. C. Vickery, 15–51. *A Journal of Documentation Review.*

Wilson, Thomas D. 2000. Human information behavior. *Informing Science* 3 (2): 49–55.

Are Spatial Data Infrastructures Special?

W. H. ERIK DE MAN

DEPARTMENT OF URBAN AND REGIONAL PLANNING AND GEO-INFORMATION MANAGEMENT,
INTERNATIONAL INSTITUTE FOR GEO-INFORMATION SCIENCE AND EARTH OBSERVATION (ITC),
ENSCHEDE, THE NETHERLANDS

ABSTRACT

Understanding SDIs and their effects will benefit from considering commonalities with other information infrastructures rather than from focusing only on what is special to them. Information infrastructures are embedded within social systems. They are multifaceted and face similar challenges and dilemmas in their development, adoption, and application. Evaluation of information infrastructures has to go beyond performance indicators only. A major dilemma is how to provide for the needed stability and sustainability of the information infrastructure without ignoring unstable and conflicting environmental conditions. How to find a balance between static structures and dynamic structuring? The article proposes that the main tenet of the institutional perspective—being valued and trusted—will be helpful in dealing with this dilemma. Specifically, Orlikowski's duality-of-technology concept and actor network theory together may provide for this. Although SDIs may not be characteristically different from other information infrastructures, the article argues that the concept of space does matter. Development, adoption, and application of any information infrastructure will be different at different spatial (or geographical) levels because of the different social contexts. Technical and bureaucratic issues and values seem to dominate data handling at higher levels, whereas social issues and human values play a pronounced role at local levels of human interaction. Information infrastructures encompassing different geographical and administrative levels may face the problem of having to deal with different and sometimes conflicting institutional arrangements in different social systems. This challenges the hierarchical information infrastructure model, where higher-level infrastructures can be subdivided into lower-level ones and are made up of them. They may not add up. The article briefly discusses development, adoption, and application of information infrastructures in developing countries. Advanced information infrastructures imposed on developing countries from outside easily erode existing institutional arrangements. Finally, the article discusses some implications for the geospatial information community: what lies beyond SDIs?

The ever-growing attention within the geospatial information community to spatial data infrastructures (SDIs) invites the question of whether SDIs are distinct from other kinds of information infrastructures, like those for health care and telemedicine or digital libraries. To be sure, SDIs are special in that they apply specialised tools and concepts for handling spatial data and in that their implementation and use require understanding of basic geographic and cartographic principles (Obermeyer and Pinto 1994, pp. 52–70). The article argues, however, that understanding SDIs and their effects will benefit more from considering commonalities with other kinds of information infrastructures than from focusing on what is special to them only. This leads to the question, "Are the notions of space and spatial relevant to information infrastructures in general?" The article argues that the development, adoption, and application of any information infrastructure are different at different spatial (or geographical) levels because of the different social contexts. This challenges the hierarchical SDI model (Rajabifard et al. 2003, pp. 28–37). The predominantly social and institutional focus of these arguments also sets the scene for discussing the relevance of SDIs and other information infrastructures for developing countries, as the "promises" of SDIs are often magnified for these countries. The article argues that developing countries are generally poor in public institutions, and SDIs and other information infrastructures should reinforce existing institutional capacity rather than jeopardize it. Finally, the article discusses some implications of these arguments for the geospatial information community: what is the future beyond SDIs?

From the outset it must be clear that the article is both exploratory and speculative in addressing differences between SDIs and other information infrastructures. This is almost unavoidable at this stage of inquiry. The approach for understanding the concept of SDI is twofold. First, SDIs are (information) infrastructures. It follows that the general properties of infrastructures apply to SDIs. This speculation is not without risk. Therefore—and this is the second line of inquiry in understanding SDIs—the article also aims at identifying and formulating relevant questions on the development, adoption, and applied use of SDIs rather than providing firm guidelines. At this initial stage, the article presents these questions, challenges, and dilemmas in a more-or-less theoretical framework. This approach implies that the article is likely to ask more questions regarding implementation and use of SDIs than it may possibly answer. No matter how unsatisfactory this may be, asking relevant questions and indicating possible challenges and dilemmas will contribute more to the understanding of SDIs than will answers to irrelevant questions. Thus, the article is squarely rooted in the interpretative tradition of inquiry (Verstehen).

Before exploring these issues, we have to clarify the concept of SDI. In the literature, SDI is described in various ways, reflecting its multifaceted character. For example, Crompvoets et al. (2004, p. 665) view SDIs as being about facilitation and coordination of the exchange and sharing of spatial data between stakeholders in the spatial data community. Williamson (2003) speaks of an SDI both as an initiative (p. 3) and as a concept (p. 4). Nevertheless, he also speaks in terms of "building" SDIs (p. 4). Groot and McLaughlin (2000, p. 4) use the term ("geospatial data infrastructure" in their terminology) for certain "activities." Clearly more than just artefacts, SDIs aim at facilitating and coordinating exchange, sharing, accessibility, and use of spatial data; they encompass networked spatial databases and data handling facilities and are complexes of interacting institutional, organisational, technological, human, and economic resources. Rajabifard et al. (2002, pp. 14–17) recognise an emerging approach oriented toward management of information assets instead of linkage of available databases only. They consider this a shift from a product-based approach to a process-based approach.

Understanding of the concept of SDI, however, is hampered by literature that is replete with the "promise" of promoting sustainable development and good governance. This promise carries the risk of overselling the idea. The so-called SDI Cookbook, for example, emphasises the vital role of geographic information in decisions at the local, regional, and global levels. It mentions crime management, business development, flood mitigation, environmental restoration, community land use assessment, and disaster recovery as just a few areas in which decision makers are benefiting from geographic information and the associated infrastructure (i.e., SDI) that supports information discovery, access, and use of this information in the decision-making process (Nebert 2004, p. 6). Rajabifard et al. (2002, p. 11) notice an increasing recognition in the literature that communities investing in spatial information systems benefit from "development of a spatial information marketplace, social stability, reduced resource disputes, improved environmental management, and improvement of land administrative systems." However, notwithstanding the conceptual and theoretical developments, operational implementation of SDIs appears unruly and sometimes even problematic. (For some critical observations, see Crompvoets et al. 2004.) Masser (2005, pp. 258–261) suggests that the claims of many countries of being involved in some form of SDI development should be treated with caution—there may be an element of wishful thinking in some of them. He also stresses the need to rigorously examine claims that SDIs will promote economic growth, better government, and improved environmental sustainability, and that more attention should be given to possible negative effects. This is not a trivial issue, because what we call success depends largely on the parameters of success, as Mol (2002, p. 235) reminds us.

The article first argues that the obvious differences between SDIs and other information infrastructures are not fundamental, and many commonalities exist. Different infrastructures face similar challenges and dilemmas in their development, adoption, and application. It follows that the understanding of specific SDI initiatives is multifaceted and has to go beyond objective-oriented performance indicators only. Subsequently, the article argues that information infrastructures are conditioned by the social systems in which they are embedded but may contribute to their structuring as well (the duality of information infrastructures). Conditions for stability and sustainability and for change and flexibility of information infrastructures may be found in the contextual structures. Implementation of SDIs and other information infrastructures follows an unruly process of negotiation and alignment between heterogeneous actors rather than following a neat and well-planned development trajectory. This view draws attention to (potential) opponents and allies of the information infrastructure initiative in addition to its proponents. After stressing the commonalities between SDIs and other information infrastructures, the article addresses the importance of the notion of space for information infrastructures in general. The article argues that the social settings of information infrastructures differ between different spatial (or geographical) levels and speculates that information infrastructures are differently developed, adopted, and applied at the different levels. This challenges the hierarchical model of information infrastructures, where higher-level entities are made up of lower parts; they may not add up. Next, the article discusses briefly how these arguments relate to the development, adoption, and application of information infrastructures in developing countries. The final section brings together the main arguments and propositions about the role of space in information infrastructure development, adoption, and application. It also discusses some implications of these arguments for the geospatial information community: what if SDIs lose their distinctiveness and become part of a general information infrastructure supporting decision making at different governance levels?

ARE SDIs DIFFERENT FROM OTHER INFORMATION INFRASTRUCTURES?

SDIs rely both on spatial data and on the technology and concepts to handle these data. Hence, they may be different from other information infrastructures like those for health care and telemedicine or digital libraries. But are these differences fundamental? Before addressing this question, the concept of information infrastructure in general needs to be clarified first. Star (1999, pp. 380–82) defines an (information) infrastructure as having the following properties:

- **Embeddedness.** Infrastructure is sunk into and inside of other structures, social arrangements, and technologies.

- **Transparency.** Infrastructure is transparent to use, in the sense that it does not have to be reinvented each time or assembled for each task but invisibly supports those tasks.

- **Reach of scope.** Infrastructure has reach beyond a single event or one-site practice.

- **Learned as part of membership.** The taken-for-grantedness of artefacts and organisational arrangements is a *sine qua non* of membership in a so-called community of practice.

- **Links with conventions of practice.** Infrastructure both shapes and is shaped by the conventions of a community of practice.

- **Embodiment of standards.** Infrastructure takes on transparency by plugging into other infrastructures and tools in a standardised fashion.

- **Built on an installed base.** Infrastructure does not grow *de novo*; it wrestles with the inertia of the installed base and inherits strengths and limitations from that base.

- **Becomes visible upon breakdown.** The normally invisible quality of a working infrastructure becomes visible when the infrastructure breaks down.

- **Is fixed in modular increments, not all at once or globally.** Because an infrastructure is big, layered, and complex, and because it means different things locally, it is never changed from above. Changes take time and negotiation, and adjustments with other aspects of the systems are involved.

This definition not only emphasises the multifaceted character of the information infrastructure concept but also suggests that information infrastructures in practice may be quite different from their conceptual ideal. For example, in reviewing the literature on specific SDI projects, Masser (2005, p. 17) concludes that

SDIs cannot be realized without coordinated action on the part of governments. Would this apparently top-down tendency jeopardise the bottom-up and organic modes of development as conceptualised by Star? It seems, however, that SDIs can also be built from the bottom up and develop organically (McLaughlin and Groot 2000, p. 273; McDougall et al. 2002; and Harvey 2003), while many other types of information infrastructures may be developed from the top down as well. SDIs are diverse, as are other information infrastructures. The question, however, is whether SDIs are fundamentally different from other information infrastructures. The likely answer is no—SDIs are in many respects similar to other information infrastructures and face similar challenges and dilemmas in their implementation and use.

SDIs are not fundamentally different in technology, implementation, and use. The last three decades or so have witnessed impressive development of specialised and complex tools for handling spatial data. But these technical developments alone hardly justify claims that spatial or geographic information technology (GIT) is a specialised area, distinct from mainstream information technology. Reeve and Petch (1999, pp. 177–85) assume that the convergence of computing toward open systems and interoperability may lead to the disappearance of GIS-specific hardware and software.

Are SDIs different from other information infrastructures in their implementation? McLaughlin and Groot (2000, p. 273) view the absence of a master architect as essential to SDIs. Instead, SDIs emerge as almost organic webs of partnerships and relationships evolving purposefully within a given jurisdiction. This holds for other kinds of information infrastructures as well, at least conceptually. They are not designed by blueprint but, as Dahlbom and Mathiassen (1993, p. 128) put it, by "cultivating a process rather than designing a product" (see also Aanestad 2002). Because of their nature, information infrastructures rely on communication of data between a variety of data providers and data users. This requires development and diffusion of standards. Monteiro and Hanseth (1995) consider this the distinguishing characteristic of information infrastructures (see also Star 1999). The need for standards in successful implementation of SDIs is also recognised in the literature (Crosswell 2000). Moreover, the multitude of stakeholders involved in any information infrastructure clearly creates the need for alignment of different expectations and other interests. In the literature, this need has been recognised for information infrastructures in general (Monteiro and Hanseth 1995, Aanestad and Hanseth 2000) and for the special case of SDIs (Martin 2000; Harvey 2001). It seems that SDIs and other information infrastructures are not fundamentally different in their implementation. Whether each specific case adheres to these conceptual characteristics, however, is a different matter.

Some see a distinguishing factor of GIT (and therefore of SDIs) in application and use and argue that this requires understanding of basic geographic and cartographic principles (Obermeyer and Pinto 1994, pp. 52–70). Substantive knowledge of the application domain, however, seems to be conditional for any information infrastructure to be effective.

SDIs and other information infrastructures face similar challenges. The literature reviewed so far suggests similarities between SDIs and other information infrastructures, and the similarities are multifaceted. The various common facets facilitate understanding of SDIs and of other information infrastructures from different theoretical perspectives (De Man 2006, pp. 332–36). This section briefly outlines some of the salient perspectives on the concept of information infrastructure and speculates about major challenges in their development, adoption, and application: exclusion (access denial), fragmentation, technocracy (technocentricity), isolation (from use), and discontinuity (obsolescence).

Information infrastructures are about communication and sharing of data and information. Communication is what they have in common with language. In his sociopolitical study of language, De Swaan (2001, pp. 102–6) noticed cases where language is perverted by elites from a means of communication into a means of exclusion and continuous domination of the underprivileged. The article assumes that information infrastructures face similar conditions and barriers for communication and exchange, notably power positions. Powerful actors may prevent others from having full and direct access to these infrastructures. Moreover, computerised information infrastructures tend to favour the relatively easy official and formalised data and records over the more difficult informal data capturing local knowledge. Such information infrastructures would increasingly reflect and transmit biased images of reality.

Information infrastructures may also be regarded as particular cases of networked infrastructures in general. As a consequence, they all would have "network externalities," where all users benefit when a new user joins the network because of the ability to communicate with more actors (Monteiro and Hanseth 1995; North 1990, p. 7). But like other infrastructures, information infrastructures would also have potentially fragmenting, discriminating, and exclusionary effects. For example, in their study on networked infrastructures, Graham and Marvin (2001) discuss technological mobility and the urban condition and how modern infrastructure developments generally help sustain the fragmentation of the social and material fabric of these spatial entities (p. 33). Often this leads to exclusion and segregation in a struggle between different social groups and interests.

Information infrastructures in general encompass both technical and social elements and may therefore be regarded as sociotechnical systems. The question of whether technology is primarily technical or social has been extensively dealt with in the literature under the rubrics of social construction of technology, actor network theory, and duality of technology. From the perspective of social construction of technology, technology is primarily a product of human action under prevailing social conditions (Bijker 1995; Bijker and Law 1992; Latour and Woolgar 1979). The actor network theory (ANT) views the process of developing networked assemblies as interplay between heterogeneous actors—technical and nontechnical elements tied together in actor networks (Callon 1980 and 1986; Latour 1999; Law 1992, 2000; Law and Mol 2002). Because (most) actors pursue their own interests, actor networks develop in a process of negotiating and aligning various expectations and interests. ANT, however, is not a unified body of concepts (Aanestad 2003, p. 6). Literature on ANT includes references to information technology (Aanestad 2002, 2003; Monteiro and Hanseth 1995) and GIS (Harvey 2000, 2001; Martin 2000). From the perspective of duality of technology, technology is the product of human action and also provides structure for human action (Orlikowski 1992, p. 406). This perspective builds upon Giddens' "structuration" theory (1984): repetitive, patterned, and routine behaviour develops into societal structures (notably institutions), while these structures, in turn, shape behaviour. We return to these perspectives in the next section.

Information infrastructures are supposed to support a wide group of stakeholders in the communication and sharing of data and information. However, some actors may perceive information as power and are reluctant to share this power with other actors. As a result, the ability of information infrastructures to provide data and information for the common good can be significantly curtailed—self-interest may be at odds with common interests. Hardin (1968) argues in his seminal article on "the tragedy of the commons" that users of common-pool resources (commons) are caught in an inevitable process that leads to the destruction of the resources on which they depend. One way of coping with this dilemma is to strengthen central control. But this would mean that the very essence of commonality might be jeopardised. Managing the commons might also be dealt with in the tradition of "coping" with tragedies of the commons (Ostrom 1990, 1999, 2000, 2005). This perspective provides a repertoire of concepts and approaches that may help identify critical factors for success and failure of information infrastructures. For example, the notion of coproduction by various actors would draw attention to the potential of synergy within information infrastructures. The notion of polycentrism may help in viewing information infrastructures as complex adaptive systems and in identifying conditions for sustainability. This perspective suggests that SDIs and other information infrastructures need a broader scope of analysis than the one limited to narrowly defined economic issues such as monopoly, markets, and privatisation (Onsrud 1998).

Information infrastructures generally operate within unstable environments. Therefore, the ability to adapt may be critical to their success and sustainability. Adaptation to evolving circumstances, in turn, requires not only the ability to learn but also the ability to learn how to learn, in the sense of "double-loop" learning (Argyris and Schön 1978, pp. 7–29). Learning involves adaptation to internal as well as external circumstances (environments). Morgan (1997, pp. 86–90) suggests that learning organisations must develop capacities to scan and anticipate changes in their environment; to question, challenge, and change operating norms and assumptions; and to allow appropriate strategic organisations to emerge. He sees bureaucratisation as a major barrier to double-loop learning. This situated learning is essentially a social process and comes largely from day-to-day practice. Or, as Lave and Wenger (1991) suggest, situated learning involves a process of engagement in a "community of practice" (Lesser and Storck 2001).

Once implemented, SDIs and other information infrastructures may develop institutional properties in communicating, connecting, and sharing abilities between various stakeholders. They appear to become part of the social structures in which they are embedded. Institutionalisation of information infrastructures means that these infrastructures are valued within a group of stakeholders and that they are considered relevant to perceived social needs. The institutionalisation perspective may help in understanding conditions for sustainability of information infrastructures and addressing the problems of obsolescence and irrelevance. (For institutionalisation of geographic information technologies, see De Man 2000, 2003. For institutionalisation of information technology in general, see Orlikowski 1992.) We return to the issue of institutionalisation of information infrastructures in some detail in the next section.

Challenges, key concepts, common perspectives, and dilemmas. Table 1 summarises the major challenges, key concepts, and relevant common perspectives for understanding information infrastructures as outlined above. The challenges call for a multifaceted view. This approach is not meant, however, to replace emerging evaluation systems (Steudler 2003, pp. 237–45). Rather, it shows that evaluation of SDI initiatives has to go beyond objective-oriented performance indicators only, as important as they may be. A multifaceted view does not mean an all-encompassing and exhaustive (evaluation) framework. Finally, the challenges also suggest that development of SDI initiatives and other information infrastructures is not a linear process. Infrequent and unruly processes generally tend to be convulsive and revolve around dilemmas, as Argyris and Schön (1974, pp. 30–34, 99–102, 114–20) point out.

DUALITY OF INFORMATION INFRASTRUCTURES: VALUE AND TRUST

A central tenet of this article is that SDIs, like other information infrastructures (or infrastructures in general for that matter), are embedded within their own social systems. What are these social systems and how do (information) infrastructures relate to them? And what are the critical conditions for their implementation and use? For SDIs, stakeholders in the spatial data community are the immediate social context because, as we have seen, SDIs are about facilitation and coordination of the exchange and sharing of spatial data (Crompvoets et al. 2004, p. 665). In exploring the social context for the development, adoption, and application of information infrastructures, the argument revolves around the dilemma

Challenges	Key concepts	Common perspectives
Exclusion (access denial)	• Power • Communication • Coproduction • Commons	• Political sociology of language (De Swaan 2001) • Coping with the tragedy of the commons (Ostrom 1990, 1999, 2000, 2006)
Fragmentation	• Power • Networking • Network externalities • Network collapse	• Networked infrastructures (Monteiro and Hanseth 1995) • Splintering urbanism (Graham and Marvin 2001) • Cultivating networks (Dahlbom and Mathiassen 1993, Aanestad and Hanseth 2000, Aanestad 2002)
Technocracy (technocentricity)	• Power • Emergent properties • Mutual negotiations (translations) between heterogeneous actors	• Social construction of technology (Bijker 1995, Bijker and Law 1992, Latour and Woolgar 1979) • Actor network theory (Callon 1980 and 1986, Law 1992, Law 2000, Law and Mol 2002, Latour 1999, Martin 2000, Harvey 2000 and 2001, Aanestad and Hanseth 2000, Aanestad 2002) • Duality of technology (Orlikowski 1992)
Isolation (from use)	• Social and situational learning • Adaptation	• Community of practice (Lave and Wenger 1991) • Organizational learning (Argyris and Schön 1978, Morgan 1997)
Discontinuity (obsolescence)	• Institutionalisation • Embeddedness • Structuration	• Structuration of society (Giddens 1984) • Duality of technology (Orlikowski 1992) • Institutionalisation of geographic information technologies (De Man 2000) • Cultivating networks (Dahlbom and Mathiassen 1993, Aanestad and Hanseth 2000, Aanestad 2002)

Table 1. Major challenges, key concepts, and common perspectives for understanding information infrastructures.

of how to provide for the needed stability and durability (or sustainability) while not ignoring the unstable and conflicting environmental conditions—in other words, how to find a balance between static structures (with their inherent danger of freezing) and dynamic structuring and a balance between standardised global solutions and local contexts (Ciborra 1998, pp. 8–12; Ciborra 2005, p. 261; Hanseth et al. 1996; Rolland and Monteiro 2002; Hanseth et al. 2006).

Institutional context of information infrastructures. What are the social conditions for implementation and use of information infrastructures? This question is essentially about conditions for exchange and sharing of data and information within a social system, which are social practices. Social structure enables or facilitates social practice and at the same time is reproduced by (patterned) social practice and action. Giddens calls this the "duality of structure," and it is one of the main propositions of his "structuration" theory (1984, pp. 19, 25–28). The most deeply embedded structural properties are institutions (Giddens 1984, p. 17), which are the cornerstones of trust in society (Zijderveld 2000, p. 73). March and Olson (1989, p. 162) also stress the dual relationship between institutions and their environment. Institutions not only respond to their environments but also create them at the same time. North views institutions as the rules of the games in a society—the humanly devised constraints that shape human interaction (North 1990, pp. 1, 6, 12, 62–64). Their major role is to reduce uncertainty and transaction costs. Robertson (1982, p. 93) views an institution as a stable cluster of values, norms, statuses, roles, and groups that develop around a basic social need. Zijderveld (2000, pp. 30–33) emphasises the process which produces institutions: institutionalisation. Individual and subjective behaviour develops into a collective and objective pattern, which in turn has influence on subsequent individual and subjective actions, thoughts, and feelings. This reciprocity (or duality) is similar to Giddens' "structuration of institutions" (1984, p. xxi). Finally, institutionalisation is a process of feedback (Buckley 1967, p. 137). Institutions, therefore, may become stronger or weaker over time or may even decay (Broom et al. 1981, pp. 18, 320–479.) Similarly, Perri 6 (2003, p. 399) argues that viable sets of institutions are not necessarily statically stable but have the capability of being sustained within their environment. Building on the work of Mary Douglas (e.g., 1978) and Thompson et al. (1990), Perri 6 proposes that institutions are under pressure from the different solidarities that emerge from two basic factors in organising social life: social regulation and social integration. Institutional viability, then, is a settlement of rival pressures for institutional similarity and tolerance of dissimilarity (p. 411).

Returning to the question of how practices of exchange and sharing of data and information develop within social systems, the following propositions are made. Exchange and sharing of data and information are social practices that develop around shared needs within groups of actors (stakeholders). With repetition, these social practices tend to become collective and objective patterns that, in turn, influence subsequent exchange and sharing of data and information. Gradually, these structural properties of the social system tend to become institutionalised.

Duality of information infrastructures. The next question is about the role of information infrastructures in the institutionalisation of exchange and sharing of data and information. Application of information technology, and thus of information infrastructures, is supposed to influence human action, notably, the exchange and sharing of data and information and the use of that information in decision making. This requires that the technology in use be valued and trusted by those concerned (De Man 2000, p. 145). In this sense, information infrastructures may be regarded as parts of social structures and would share their characteristic of duality. Orlikowski (1992, p. 406), following Giddens' structuration theory, understands technology as both the product of human action under prevailing structural properties within social systems (socially constructed) and as having structural properties by itself—once applied, technology tends to become reified and institutionalised. As a consequence, technology as facilitating or constraining human action would both be conditioned by those institutional properties and be shaping them. The impact of duality of technology, however, will also depend on prevailing cultural conditions. Some cultures are more active and innovative, allowing for change, whereas other cultures tend to avoid uncertainty (Etzioni 1968; Hofstede 1980, 1997; De Man and Van den Toorn 2002.) The duality of technology thesis leads to the following propositions. Any information infrastructure not only is embedded within a specific social system but also tends to become institutionalised over time when effectively addressing a collectively perceived need in the exchange and sharing of data and information. The institutional property of an information infrastructure would have a strong influence on the continued exchange and sharing of data and information within the host social system. This institutional duality also provides for sustainability in the development, adoption, and application of information infrastructures.

Information infrastructures are negotiated. The other side of the duality coin assumes that information infrastructures are assembled, developed, and applied and may have some autonomy with respect to the prevailing institutional properties of their host social system. (To be sure, technology tends to become institutionalised eventually, but initially this may not be the case.) The processes of developing information infrastructures themselves can be viewed from the perspective of actor network theory (ANT) as ongoing processes of negotiating and aligning the various expectations and interests of a wide variety of heterogeneous actors, both human and nonhuman. ANT recognises that these processes not only involve proponents of the information infrastructure initiative but may also have opponents at some stage. Alliances may be brought into collaboration but generally only temporarily. ANT also recognises that these negotiations do not necessarily lead to stable and durable situations; the resulting information infrastructure will "never be finished," so to speak. In short, ANT recognises changing contextual conditions and the need for flexibility in development, adoption, and application of information infrastructures. Martin (2000, pp. 733–35) suggests, on the basis of his empirical study of environmental geographic information systems (GIS) in Ecuador, some conditions for stability and sustainability in the development of GIS. It seems that stakeholder involvement (participation), collaboration, and trust are important conditions. These are also major conditions for institutionalisation (De Man 2000). This leads to the proposition that ANT and the tenet of duality of technology (structuration theory) together provide scope for understanding development, adoption, and application of effective, viable, and thus sustainable information infrastructures. This scope recognises the complex, contentious, unruly, and multifaceted character of these processes. Sahay and Walsham (1996, 1997) come to a similar conclusion in their analysis of developing GIS in an Indian government agency. They combined the theoretical perspective of the role of social context (structuring) in the implementation of GIS with that of the process of implementation itself. Finally, being valued and trusted can be viewed as major conditions for effective implementation of any information infrastructure.

The notion of space is not irrelevant to information infrastructures in general. To what extent would space matter for development, adoption, and application of an information infrastructure beyond the specialised tools and concepts for handling spatial data and the required understanding of basic geographic principles to apply geographic information technology effectively?

Different conceptions of space at different spatial levels. Human activity takes place in space. Space surrounds us and is omnipresent (Tversky 2003, 2004). Space is more than georeferenced location. Space may be understood as subjectively conceived of by individuals. Objects in space are not just objects. Their meaning to us depends on what they offer, provide, or furnish, or do not. In the ecological psychology of J. J. Gibson, the "affordances" of our environment are what matters (Smith and Mark 2001). Consequently, environment (or space) and activities (or behaviour) are intertwined. Smith (2001) speaks of "physical–behavioural units," recurrent types of settings, which serve as the environment for human activities. They are complexes of times, places, actions, and things. Individuals likely conceive of physical-behavioural units differently to the extent that they are more or less intimately connected to their behavioural conditions. (For example, my physical–behavioural unit is different when I work around my home than when I fly from, say, Amsterdam to Manila.) These different conceptions of space are somewhat similar to Tversky's "multiplicity of mental spaces": space of the body, space around the body, and space of navigation (Tversky 2001, 2003). Tversky observes that the human mind conceives of some large spaces as integrated wholes rather than piecemeal as space is intimately experienced (Tversky 2003, p. 72). These different conceptions of space are akin to generalisations normally associated with enlargement of geographic scale (or reduction of cartographic scale).

On the other hand, human responses to problematic environmental conditions are hardly specialized and fragmented (or compartmentalized) at the local level, as this is generally the case at higher levels of physical–behavioural units. For example, management of various risks—like flooding, earthquakes, health hazards, and unemployment—seems to be integrated into livelihood at the community level (Heijmans 2004, p. 120), whereas these various risks are generally dealt with separately through specialised agencies at the national level. (The distinction between fragmented and integrated ways of dealing with problematic issues has also been made by Fred Riggs [1962, 1964] in his theory of "prismatic society.")

This brief reflection suggests that space tends to be more integrated into wholes at higher spatial levels whereas behavioural conditions tend to be more specialised and fragmented at those levels. Moreover, physical–behavioural units tend to be less intimately connected to human and social conditions and values at higher geographical and administrative levels than at the local level. The NIMBY (not in my backyard) syndrome illustrates the kind of social and emotional values that may come with lower-level physical–behavioural units.

Social setting of space. Space is generally shared with others. Copresence is fundamental to social life because it provides for social encounters (Giddens 1984, pp. xxiv, 4, 70–81). In other words, space (understood in terms of physical–behavioural units) is a setting for social life. The intensity of social encounters and of social life in general, can be characterised by the degree of social capital, which refers to connections among individuals—social networks and the norms of reciprocity and trustworthiness that arise from them (Putnam 2000, p. 19). Social capital develops the "I" into the "we" (Putnam 1995, p. 67). Put differently, social capital refers to the norms and networks that enable people to act collectively (Woolcock et al. 2000, p. 226). This implies that the members of the network have common interests and share a common understanding of issues facing them. It follows that social capital is a resource for structuring individual behaviour into social practice, in other words, for social life.

Assuming that physical–behavioural units are the setting for social life, it also follows that social capital is a resource for structuring these physical–behavioural units. Moreover, and to the extent that social capital may have a different intensity and therefore a different impact at different spatial (or geographical) levels, space will be differently structured and conceived of at these different levels. This does not mean, however, that these different social settings of space are neatly separated. To the contrary, different settings may easily overlap because individuals are generally involved in multiple and different social settings.

Information infrastructures are different at different spatial levels. What are the possible implications of these propositions about the spatially differentiated social context for the development, adoption, and application of information infrastructures? From the outset it must be clear, however, that prevailing social conditions may support or hamper development of (information) infrastructures but will not determine them. Therefore, the answer to the question about the impact of social context may at best inform the development, adoption, and application of information infrastructures about possible opportunities and threats regarding their content, role, and degree of complexity.

First, the handling of data, whether spatial or nonspatial, tends to be dominated by technical and bureaucratic values and issues at higher spatial levels, whereas social and human values and issues seem to play a more pronounced role at local levels of physical–behavioural units. As a consequence, the desired content of information infrastructures will be differently conceived of at different spatial levels. Particularly at these lower levels, much of the information is informal: beliefs, values, expectations, and other interests. Therefore, the selection of core or framework data for SDI (Groot et al. 2000, pp. 4–6) may well turn out to be more contentious in multilevel SDIs than initially anticipated. How do these data, directly or indirectly, address commonly perceived problems in the exchange and sharing of spatial data across the different spatial levels? Of course, the arguments so far by no means imply that the spatial level is the only determining factor for data handling. Masser (1980), for instance, points at different planning environments reflecting planning for growth, on the one hand, and planning for decline, on the other—for example, the opening of new primary schools and the closure of existing primary schools. Growth tends to emphasise technical considerations, whereas decline tends to revolve around political and procedural questions (*ibid.,* pp. 44–46). It seems that differentiated (planning) environments have different effects on the required information and, consequently, on data handling.

Second, information infrastructures play different societal roles at different spatial levels. At higher administrative levels one generally deals with issues *about* lower-level physical–behavioural units and the concerned actors rather than *with* them. (For example, governments set conditions for the behaviour of their citizens.) But information infrastructures may also play a role in the shaping and integration of social systems. As Giddens points out (1984, pp. 143, 191, 377), social systems can be "stretched" across time and space, on the basis of mechanisms of social and system integration. Rose (1999) argues that modern information and communication technologies are such mechanisms, able to integrate social systems both spatially and temporally. Information infrastructures may contribute to the shaping (or structuring) of social systems, but this will likely be different at different spatial levels for the aforementioned reasons. The initiative of establishing an infrastructure for spatial information in Europe (INSPIRE) provides an interesting case to illustrate this point. Whose problems in exchange and sharing of spatial data and information are addressed? At which levels—Europe, member nation, local, or "Euregio" (cross-border)—are these problems commonly perceived? Are these problems commonly perceived by functionally different stakeholders across these levels: Commission, Council, Parliament, professional communities, business, citizens, and so on? A related question is whether SDIs for different territorial units at the same spatial level are similar. One can argue that this is not necessarily the case. It would all depend on similarities in societal decision making (governance) between these units. For example, and all other things being equal, one may expect more similarity in the development, adoption, and application of information infrastructures at the subnational level within unitary states than in federal forms of statehood. Unitary states tend to have more similar policies among their administrative regions.

Third, one can argue that the degree of required complexity in development, adoption, and application of information infrastructures tends to be considerably higher at local levels because of the complex worldviews and structural properties at these levels. This would necessitate specialised approaches to capture such local, indigenous knowledge adequately. A case in point is the category of community-based mapping approaches, either computerised or not, like participatory GIS, community mapping, and sketch mapping (Pickels 1995, pp. 9–11; Harris et al. 1995; and Craig et al. 2002). Moreover, information infrastructures at local levels may have to compete with existing, intricate, and often informal institutional arrangements for information sharing and communication. This competition might be a major reason that developing a regional information infrastructure for health care in the relatively densely populated Southern Norway proved to be more difficult than a similar development for the larger but sparsely populated Northern Norway (Aanestad 2005). Implementation and use of information infrastructures covering different geographical and administrative levels may be even more problematic because they have to deal with different and sometimes conflicting institutional properties of the various social systems involved.

Finally, information infrastructures at different geographical and administrative levels may have to complement each other. For the above reasons, higher-level information infrastructures cannot simply be subdivided into lower-level ones and are not made up of them. This challenges the hierarchical structure as the only possible, or most desirable, model for the information infrastructure concept (Rajabifard et al. 2003, pp. 28–37). In addition, information infrastructures at lower levels are likely to be complemented with informal information like local, indigenous knowledge in order to be effective. The need for complementarity can direct and inform the debate on how information infrastructure initiatives meet demands at various administrative levels (as in the INSPIRE initiative). For instance, how could SDIs meet demands for different jurisdictions at local, state, national, regional, or even global levels (Rajabifard et al. 2002, pp. 17–21); how to involve local governments in developing national SDIs (Harvey and Tulloch 2003; McDougall et al. 2002); and how to address challenges of growing networks in telemedicine (Aanestad 2002, pp. 39–43).

IMPLICATIONS OF SPACE FOR INFORMATION INFRASTRUCTURES IN DEVELOPING COUNTRIES

Are SDIs and other information infrastructures relevant to developing countries? The literature is replete with "promises." What are the salient conditions?

Developing countries are poor in public institutions. Caiden and Wildavsky (1974, pp. 46, 52–53) have argued that developing countries are poor for more reasons than lack of money. Their poverty extends to information, trained manpower, and public institutions. In addition, developing countries lack functional redundancy as reserve and security, but above all as facilitator of change. It would follow that the implementation and use of information infrastructures must take into account how existing potentials and (public) institutions can be preserved, if not strengthened, rather than just being innovative and modern. Successful introduction and application of information technology require the consideration of specific and critical conditions, technical as well as social (Krishna and Walsham 2005, p. 129), and must not be oriented exclusively toward advanced systems (Chilundo and Aanestad 2003). Existing arrangements and capacities for communication and information sharing also need to be considered, in terms of how they can be improved instead of being wasted or jeopardised. Of particular interest are "counter networks" that develop when people, regions, and sectors for whatever reason are excluded from formal networks and that are based on local practices (Mosse and Sahay 2003, p. 38). Probably, the conditions for implementation and use of information infrastructures in developing countries are not fundamentally different from those in more advanced countries. It seems, however, that (post-)industrialised economies are generally in better condition than transitional economies to take the risk of mistakes and failure in the implementation and use of technologically advanced information infrastructures.

Different institutional properties at different spatial levels. Most developing countries are characterised by different institutional arrangements at different geographical and administrative levels. For example, many local communities have developed local institutional arrangements for addressing dilemmas in common resource use and collective problem solving (Ostrom 1999). Often, these grassroots arrangements are intricate, fused, personalised, and informal in remote and rural communities, whereas they are more formal, specialised, and focused ("diffracted") in or near bigger cities (Riggs 1962, 1964) and at higher levels of governance. Moreover, Riggs argues in his theory of prismatic society (1962, 1964) that societies may develop from traditional and relatively fused ways of conducting affairs to more or less modern and diffracted ways. Existing local institutional arrangements may be at risk when formal arrangements are introduced from outside. For example, when resources previously controlled by local communities have been nationalised, state control has usually proven to be less effective and efficient, if not disastrous, than control by those directly affected (Ostrom 1999, p. 495). In addition, local institutional arrangements are generally based on personal acquaintances. This is likely the case for communication and information sharing. Mosse and Sahay (2003, p. 47) found in their study on health information systems in Mozambique that communication practices were seen as informal, locally specific, and taking place in an improvised manner but with a degree of harmony that obscured the chaos within. In their study on the introduction of GIS by the Indian Ministry of Environment and Forests for forest management, Barett et al. (2001, pp. 18–19) conclude that while there is a need for increased levels of technology-related trust, there is a preference for continued face-to-face interactions. But they also concluded (p. 19) that traditional and local forms of knowledge systems are vulnerable to revision and change, because of the application of advanced technology in the form of satellite data and GIS. Moreover, sensitive information such as deforestation figures was inadvertently revealed by these applications. Heeks (2002, p. 10) also addresses this dilemma. He considers the mismatch between "hard" and rational design assumptions on the one hand and "soft" societal reality on the other as one of the major reasons for failure in the implementation of information systems in developing countries. At the same time, however, successful information systems are supposed to bring some positive change in their environment (*ibid.*, p. 6). Therefore, a major challenge for (advanced) information infrastructures is to improve existing capacity in communication and information sharing instead of jeopardising or wasting it.

CONCLUSIONS AND REFLECTIONS: BEYOND SDIs

Central to the article is the question of whether SDIs are fundamentally different from other information infrastructures. Arguments were given for why they are not. The article addressed this question from a conceptual point of view and took an exploratory and speculative approach. The goal of the article is to inform the development, adoption, and application of specific information infrastructure initiatives by presenting questions, challenges, and dilemmas in a more-or-less theoretical framework rather than to provide explanatory theories. The main conclusions and implications are discussed below.

SDIs and other information infrastructures are not fundamentally different. The concept of space does not differentiate between SDIs and other information infrastructures but rather between development, adoption, and application of any information infrastructure at different spatial levels. Information infrastructures are different in content, role, and degree of complexity at these different spatial levels.

SDIs and other kinds of information infrastructure are multifaceted and face similar challenges and dilemmas. This suggests that understanding specific information infrastructure initiatives requires a multifaceted view but not necessarily an all-encompassing and exhaustive (evaluation) framework.

Information infrastructures have to deal with the major dilemma of how to provide for the needed stability and sustainability in the development, adoption, and application process and not ignore potentially unstable and conflicting environmental conditions. The article suggests that the perspective of negotiated development of the actor network theory together with the institutional perspective of Giddens' structuration theory (and Orlikowski's duality of technology thesis) provides scope for informing development, adoption, and application of effective and sustainable information infrastructure initiatives.

Finally, the article suggests that development, adoption, and application of information infrastructures need to be complementary in that information infrastructures at different spatial levels may have to complement each other. In addition, information infrastructures at lower levels are likely to be complemented with informal information, like local, indigenous knowledge.

Information infrastructure beyond space and locality? The central argument in the article hinges on two related assumptions. First, space (as physical–behavioural units) tends to be more intimately connected to human and social conditions at the local level. Second, social bonds are stronger at this level. But what about the fact that "people live their lives individually in their own networks that stretch across territorially defined boundaries" (Hajer 2003, p. 88)? Would this not jeopardise the argument? For sure, in the unusual cases where individuals are neither behaviourally nor socially attached to a locality, the relevance of the argument vanishes. But under those circumstances the meaning of territorial authority vanishes as well. Therefore, we may assume that individuals are somehow attached to a locality, however weak this attachment may be. Nevertheless, it can be expected that emerging network societies will have a significant impact on societal information provision, specifically on development, adoption, and application of information infrastructures.

Beyond SDIs? What are the possible implications for the spatial data community at large if SDIs lose their distinctiveness and become part of a general information infrastructure? Theoretical developments in governance (Hajer and Wagenaar 2003) support the proposition that over time effective information infrastructures tend to become an integral and characteristic part of the institutional governance system in which they are implicated. This will pose an enormous challenge for practice and research in the geoinformation science community and the spatial data community.

Governance is multifaceted. It is a pluricentric, networked process involving risk and uncertainty and reflecting ideal and empirical realities (Van Kersbergen and Van Waarden 2004, pp. 151–52). Governance is a societal and deliberative practice. It follows that effective information infrastructures as an integral part of governance facilitate deliberation and learning between heterogeneous actors. This functionality goes beyond the provision of processed data only. Development, adoption, and application of information infrastructure becomes a social process, involving the questions of how to communicate, debate, and integrate different perceptions of reality and how to integrate local and indigenous knowledge with administrative knowledge at different levels. Understanding "what happens on the ground" requires truly interdisciplinary research between the geoinformation science community and the social sciences (including sociologists and scholars of public administration). Instead of simply bringing together technical sciences and social sciences with their respective paradigms, this approach aims at developing a genuine sociotechnical science in its own right beyond the realm of traditional positivism. Moreover, research and practice will also become less distinct. The contours of this sociotechnical research agenda can already be found in the literature. Both in public administration and information technology one finds new vocabularies to describe emerging fields, as current terminology appears inadequate to describe what happens. For example, Hajer and Wagenaar (2003, pp. 1–25) refer to governance, institutional capacity, networks, complexity, trust, deliberation, and interdependence as a new vocabulary for describing developments in governing the public domain. Likewise, Ciborra (1998) speaks of care, hospitality, and cultivation in describing developments in information systems thinking. Contemporary geoinformation science may find it difficult to accommodate this learning-by-doing.

ACKNOWLEDGMENTS

I thank Ian Masser, Arnold Bregt, Joep Crompvoets, Abbas Rajabifard, Yola Georgiadou, Emile Dopheide, Lyande Eelderink, Corazon Mendoza, and anonymous reviewers for their encouragement, comments, and suggestions on earlier versions of this article.

REFERENCES

Aanestad, M. 2002. *Cultivating networks: Implementing surgical telemedicine.* Doctoral thesis. Oslo, Norway: University of Oslo.

Aanestad, M. 2003. The camera as an actor: Design-in-use of telemedicine infrastructure in surgery. *Computer Supported Cooperative Work* 12:1–20.

Aanestad, M. 2005. Building information infrastructures in the public sector. *Position paper at the First Research Workshop on cross-learning on Spatial Data Infrastructures (SDI) and Information Infrastructures (II).* ITC. Enschede, The Netherlands. March 31–April 1.

Aanestad, M., and O. Hanseth. 2000. Implementing open network technologies in complex work practices: A case from telemedicine. In *Organisational and social perspectives on information technologies,* eds. R. Baskerville, J. Stage, and J. I. DeGross, 355–69. Dordrecht: Kluwer.

Argyris, C., and D. A. Schön. 1974. *Theory in practice: Increasing professional effectiveness.* San Fransisco, CA: Jossey-Bass.

Argyris, C., and D. A. Schön. 1978. *Organizational learning: A theory of action perspective.* Reading, MA: Addison-Wesley.

Barett, M., S. Sahay, and G. Walsham. 2001. Information technology and social transformation: GIS for forestry management in India. *The Information Society* 17 (1): 5–20.

Bijker, W. E. 1995. *Of bicycles, bakelites, and bulbs: Towards a theory of sociotechnical change.* Cambridge, MA: The MIT Press.

Bijker, W. E., and J. Law, eds. 1992. *Shaping technology/building society: Studies in sociotechnical change.* Cambridge, MA: The MIT Press.

Broom, L., P. Selznick, and D. Broom-Darroch. 1981. *Sociology,* 7th ed. New York: Harper and Row.

Buckley, W. 1967. *Sociology and modern systems theory.* Englewood Cliffs, NJ: Prentice Hall.

Caiden, N., and A. Wildavsky. 1974. *Planning and budgeting in poor countries.* New York: John Wiley & Sons.

Callon, M. 1980. Struggles and negotiations to define what is problematic and what is not: The sociology of translation. In *The social process of scientific investigation: Sociology of the Sciences Yearbook,* eds. K. D. Knorr, R. Krohn, and R. D. Whitley, 197–219. Dordrecht and Boston, MA: Reidel.

Callon, M. 1986. Some elements of a sociology of translation: domestication of the scallops and the fishermen of Saint Brieuc Bay. In Power, action and belief: A new Sociology of Knowledge? *Sociological Review Monograph* 32, ed. J. Law, 196–233. London: Routledge.

Chilundo, B., and M. Aanestad. 2003. Vertical or integrated health programmes? *In-progress paper presented at the Joint WG 8.2 & 9.4 IFIP Working Conference IS Perspectives and Challenges in the Context of Globalization.* Athens University of Economics and Business. June 15–17, 2003.

Ciborra, C. U. 1998. Crisis and foundations: An inquiry into the nature and limits of models and methods in the information systems discipline. *Journal of Strategic Information Systems* 7:5–16.

Ciborra, C. 2005. Interpreting e-government and development: Efficiency, transparency or governance at a distance? *Information Technology & People* 18 (3): 260–79.

Craig, J., T. M. Harris, and D. Weiner, eds. 2002. *Community participation and geographic information systems.* London: Taylor & Francis.

Crompvoets, J., A. Bregt, A. Rajabifard, and I. Williamson. 2004. Assessing the worldwide developments of national spatial data clearinghouses. *Int. Journal of GIS* 18 (7): 665–89.

Crosswel, P. L. 2000. The role of standards in support of GDI. In *Geospatial data infrastructure: Concepts, cases and good practice,* eds. R. Groot and J. McLaughlin, 57–83. Oxford: Oxford University Press.

Dahlbom, B., and L. Mathiassen. 1993. *Computers in context: The philosophy and practice of systems design.* Cambridge, MA: Blackwell.

De Man, W. H. E. 2000. Institutionalization of geographic information technologies: Unifying concept? *Cartography and GIS* 27 (2): 139–51.

De Man, W. H. E. 2003. Cultural and institutional conditions for using geographic information: Access and participation. *URISA Journal* 15 (APA2): 23–27.

De Man, W. H. E. 2006. Understanding SDI: Complexity and institutionalization. *Int. Journal of GIS* 20 (3): 329–43.

De Man, W. H. E., and W. Van den Toorn. 2002. Culture and the adoption and use of GIS within organisations. Int. *Journal of Appl. Earth Observation and Geoinformatics* 4:51–63.

De Swaan, A. 2001. *Words of the world: The global language system.* Cambridge/Oxford: Polity Press/Blackwell Publishers.

Douglas, M. 1978. *Cultural Bias*. Occasional Paper No. 35. London: Royal Anthropological Institute.

Etzioni, A. 1968. *The active society*. New York: Free Press.

Giddens, A. 1984. *The constitution of society: Outline of the theory of structuration*. Berkeley, CA: University of California Press.

Graham, S., and S. Marvin. 2001. *Splintering urbanism: Networked infrastructures, technological mobilities and the urban condition*. London: Routledge.

Groot, R., and J. McLaughlin, eds. 2000. *Geospatial data infrastructure: Concepts, cases and good practice*. Oxford: Oxford University Press.

Hajer, M. 2003. A frame in the fields: Policymaking and the reinvention of politics. In *Deliberative policy analysis: Understanding governance in the network society*, eds. M. A. Hajer and H. Wagenaar, 88–110. Cambridge, UK: Cambridge University Press.

Hajer, M. A., and H. Wagenaar, eds. 2003. *Deliberative policy analysis: Understanding governance in the network society*. Cambridge, UK: Cambridge University Press.

Hanseth, O., E. Monteiro, and M. Hatling. 1996. Developing information infrastructure standards: The tension between standardisation and flexibility. *Science, Technology and Human Values* 21 (4): 407–26.

Hanseth, O., E. Jacucci, M. Grisot, and M. Aenestad. 2006. Reflexive standardization: Side-effects and complexity in standard making. *Management Information Systems Quarterly* 30 (special issue August): 563–81.

Hardin G. 1968. The tragedy of the commons. *Science* 162:1243–48.

Harris, T. M., D. Weiner, T. A. Warner, and R. Levin. 1995. Chapter 9: Pursuing social goals through participatory geographic information systems. In *Ground truth: The social implication of geographic information systems*, ed. J. Pickels, 196–222. New York: The Guilford Press.

Harvey, F. 2000. The social construction of geographic information systems. *Int. Journal of GIS* 14 (8): 711–13.

Harvey, F. 2001. Constructing GIS: Actor networks of collaboration. *URISA Journal* 13 (1): 29–37.

Harvey, F. 2003. Developing geographic information infrastructures for local government: The role of trust. *The Canadian Geographer* 47 (1): 28–36.

Harvey, F., and D. Tulloch. 2003. *Building the NSDI at the base: Establishing best sharing and coordination practices among Local Government*. Project report. Minneapolis and New Brunswick: University of Minnesota and State University of New Jersey.

Heeks, R. 2002. Failure, success and improvisation of information systems projects in developing countries. Working Paper 11, Development Informatics, Manchester: Inst. For Development Policy and Management, University of Manchester.

Heijmans, A. 2004. From vulnerability to empowerment. In *Mapping vulnerability: Disasters, development and people*, eds. G. Bankoff, G. Frerks, and D. Hilhorst, 115–27. London: Earthscan.

Hofstede, G. 1980. *Culture's consequences: International differences in work-related values*. Beverly Hills, CA: Sage.

Hofstede, G. 1997. *Cultures and organizations: Software of the mind*. London: McGraw-Hill.

Krishna, S., and G. Walsham. 2005. Implementing public information systems in developing countries: Learning from a success story. *Information Technology for Development* 11 (2): 123–40.

Latour, B. 1999. On recalling ANT. In *Actor network theory and after*, eds. J. Law and J. Hassard, 15–25. Oxford: Blackwell Publishers.

Latour, B., and S. Woolgar. 1979. *Laboratory life: The social construction of scientific facts*. Beverly Hills, CA: Sage.

Lave, J., and E. Wenger. 1991. *Situated learning: Legitimate peripheral participation*. Cambridge: Cambridge University Press.

Law, J. 1992. Notes on the theory of the actor network: Ordering, strategy and heterogeneity. *Systems Practice* 5:179–393.

Law, J. 2000. *Networks, relations, cyborgs: On the social study of technology*. http://www.lancs.ac.uk/fss/sociology/papers/law-networks-relations-cyborgs.pdf.

Law, J., and A. Mol. 2002. *And if the global were small and non-coherent?* Centre for Science Studies at the Department of Sociology, Lancaster University, Lancaster, United Kingdom. http://www.comp.lancs.ac.uk/sociology096jl.html (accessed January 28, 2002).

Lesser, E. L., and J. Storck. 2001. Communities of practice and organizational performance. *IBM Systems Journal* 40 (4): 831–41.

March, J., and J. Olsen. 1989. *Rediscovering institutions: The organizational basis of politics*. New York: The Free Press.

Martin, E. W. 2000. Actor-networks and implementation: Examples from Conservation GIS in Ecuador. *Inter. Journal of GIS* 15 (8): 715–38.

Masser, I. 1980. The limits to planning. *Town Planning Review* 51 (1): 39–49.

Masser, I. 2005. *GIS Worlds: Spatial data infrastructures*. Redlands, CA: ESRI Press.

McDougall, K., A. Rajabifard, and I. Williamson. 2002. From little things big things grow: Building the SDI from Local Government up. Paper presented at the Joint AURISA and Institution of Surveyors Conference. Adelaide, South Australia. November 25–30, 2002.

McLaughlin, J., and R. Groot. 2000. Advancing the GDI concept. In *Geospatial data infrastructure: Concepts, cases and good practice*, eds. R. Groot and J. McLaughlin, 269–75. Oxford: Oxford University Press.

Mol, A. 2002. Cutting surgeons, walking patients: Some complexities involved in comparing. In *Complexities: Social studies of knowledge practices*, eds. J. Law and A. Mol, 218–57. Durham: Duke University Press.

Monteiro, E., and O. Hanseth. 1995. Social shaping of information infrastructure: On being specific about technology. In *Information technology and changes in organisational work*, eds. W. Orlikowski, G. Walsham, M. Jones, and J. DeGross, 325–43. London: Chapman and Hall.

Morgan, G. 1997. *Images of organization*. Thousands Oaks, CA: Sage Publications.

Mosse, E. L., and S. Sahay. 2003. Counter networks, communication and health information systems: A case study from Mozambique. In *Organizational information systems in the context of globalization*, eds. Korpela, M., R. Montealegre, and A. Poulymenakou, 35–51. Boston: Kluwer.

Nebert, D. D., ed. 2004. *Developing spatial data infrastructures: The SDI Cookbook*. Global Spatial Data Infrastructure Association. http://www.gsdi.org/docs2004/Cookbook/cookbookV2.0.pdf.

North, D. 1990. *Institutions, institutional change, and economic performance*. London: Cambridge University Press.

Obermeyer, N. J., and J. K. Pinto. 1994. *Managing geographic information systems*. New York: The Guilford Press.

Onsrud, H. J. 1998. The tragedy of the information commons. In *Policy issues in modern cartography*, ed. D. R. Fraser Taylor, 141–58. Oxford: Pergamon.

Orlikowski, W. J. 1992. The duality of technology: Rethinking the concept of technology in organizations. *Organization Science* 3 (3): 398–427.

Ostrom, E. 1990. *Governing the commons: The evolution of institutions for collective action*. London: Cambridge University Press.

Ostrom, E. 1999. Coping with tragedies of the commons. *Annual Review of Political Science* 2:493–535.

Ostrom, E. 2000. The danger of self-evident truths. *PS: Political Science & Politics* 33 (1): 33–44.

Ostrom, E. 2005. *Understanding institutional diversity*. Princeton, NJ: Princeton University Press.

Pickels, J. 1995. Chapter 1: Representation in an electronic age: Geography, GIS, and democracy. In *Ground truth: The social implication of geographic information systems*, ed. J. Pickels, 1–30. New York: The Guilford Press.

Putnam, R. D. 1995. Bowling alone: America's declining social capital. *Journal of Democracy* 6 (1): 65–78.

Putnam, R. D. 2000. *Bowling alone: The collapse and revival of American community*. New York: Simon and Schuster.

Rajabifard, A., M.-E. F. Feeney, and I. Williamson. 2002. Future directions for SDI development. *Int. Journal of Appl. Earth Observation and Geoinformatics* 4:11–22.

Rajabifard, A., M.-E. F. Feeney, and I. Williamson. 2003. Chapter 2: Spatial data infrastructures: Concepts, nature and SDI hierarchy. In *Developing spatial data infrastructures: From concept to reality*, eds. I. Williamson, A. Rajabifard, and M.-E. F. Feeney, 17–40. London: Taylor & Francis.

Reeve, D., and J. Petch. 1999. *GIS organisations and people: a socio-technical approach*. London: Taylor & Francis.

Riggs, F. W. 1962. The "sala" model: An ecological approach to the study of comparative administration. *Philippine Journal of Public Administration* 6:3–16.

Riggs, F. W. 1964. *Administration in developing countries: The theory of Prismatic Society*. Boston, MA: Houghton Mifflin.

Robertson, I. 1982. *Sociology*. New York: Worth Publishers.

Rolland, K. H., and E. Monteiro. 2002. Balancing the local and the global in infrastructural information systems. *The Information Society* 18:87–100.

Rose, J. 1999. Towards a structurational theory of IS: Theory development and case study illustrations. In *Proceedings of the 7th European Conference on Information Systems*, eds. J. Pries-Heje, C. Ciborra, K. Kautz, J. Valdor, E. Christiaanse, D. Avison, and C. Heje. Copenhagen: Copenhagen Business School.

Sahay, S., and G. Walsham. 1996. Implementation of GIS in India: Organizational issues and implications. *Int. Journal of GIS* 10 (4): 385–404.

Sahay, S., and G. Walsham. 1997. Social structure and managerial agency in India. *Organization Studies* 18 (3): 415–44.

Smith, B. 2001. Objects and their environments: From Aristotle to ecological ontology. In *The life and motion of socio-economic units (GISDATA 8),* eds. A. Frank, J. Raper, and J. P. Cheylan, 79–97. London: Taylor & Francis.

Smith, B., and D. Mark. 2001. Geographical categories: An ontological investigation. *Intern. Journal of GIS* 15 (7): 591–612.

Star, S. L. 1999. The ethnography of infrastructure. *American Behavioral Scientist* 43 (3): 377–91.

Steudler, D. 2003. Chapter 14: Developing evaluation and performance indicators for SDIs. In *Developing spatial data infrastructures: From concept to reality,* eds. I. Williamson, A. Rajabifard, and M.-E. F. Feeney, 235–46. London: Taylor & Francis.

Thompson, M., R. Ellis, and A. Wildavsky. 1990. *Cultural theory.* Boulder, CO: Westview Press.

Tversky, B. 2001. Multiple mental spaces. In *Visual and spatial reasoning in design,* eds. J. S. Gero, B. Tversky, and T. Purcell, 3–13. Sydney, Australia: Key Centre of Design Computing and Cognition.

Tversky, B. 2003. Structures of mental spaces: How people think about space. *Environment and Behavior* 35 (1): 66–80.

Tversky, B. 2004. Narratives of space, time and life. *Mind & Language* 19 (4): 380–92.

Tversky, B., and P. U. Lee. 1998. How space structures language. In *Spatial cognition,* eds. C. Freksa, C. Habel, and K. F. Wender, 157–76. Berlin: Springer.

Van Kersbergen, K., and F. Van Waarden. 2004. 'Governance' as a bridge between disciplines: Cross-disciplinary inspiration regarding shifts in governance and problems of governability, accountability and legitimacy. *European Journal of Political Research* 43:143–71.

Williamson, I. 2003. Chapter 1: SDIs: Setting the scene. In *Developing spatial data infrastructures: From concept to reality,* eds. I. Williamson, A. Rajabifard, and M.-E. F. Feeney, 3–16. London: Taylor & Francis.

Woolcock, M., and D. Narayan. 2000. Social capital: Implications for development theory, research and policy. *The World Bank Research Observer* 15 (2): 225–49.

Zijderveld, A. C. 2000. *The institutional imperative: The interface of institutions and networks.* Amsterdam: Amsterdam University Press.

6, Perri. 2003. Institutional viability: A neo-Durkheimian theory. *Innovation: the European Journal of Social Science Research* 16 (4): 395–415.

A Mixed-Method Approach for Evaluating Spatial Data Sharing Partnerships for Spatial Data Infrastructure Development

KEVIN MCDOUGALL,[1,2] ABBAS RAJABIFARD,[2] AND IAN P. WILLIAMSON[2]

UNIVERSITY OF SOUTHERN QUEENSLAND, TOOWOOMBA,[1] AND UNIVERSITY OF MELBOURNE, VICTORIA,[2] AUSTRALIA

ABSTRACT

In recent years interjurisdictional partnerships have emerged as an important mechanism for establishing an environment conducive to data sharing and hence the facilitation of SDI development. However, unless the partnership arrangements are carefully designed and managed to meet the business objectives of each partner, it is unlikely that they will be successful or sustainable in the longer term. The purpose of this paper is to focus on the methodological approaches and relevant issues for researching these new data sharing partnerships and their relationships to SDI development. This paper proposes a research methodology for investigating both the organisational context of data sharing partnerships and the factors that contribute to the success of interjurisdictional data sharing initiatives. The paper examines past research and theory in spatial data sharing and examines the characteristics of a number of existing data sharing models and frameworks. The use of a mixed-method approach to evaluate local-state government partnerships in Australia is described. Finally, the validation of the mixed-method approach and its generalisation to other SDI and data sharing initiatives is discussed.

Spatial information plays an important role in many social, economic, and political decisions. Governments, business, and the general community rely on spatial information for practical decision making on a daily basis (Onsrud and Rushton 1995). In emergency services and disaster management the value of accurate and relevant information such as address, vehicular access, location of services, property ownership, climate, and topography is crucial for directing and managing response efforts. However, rarely do all of these datasets reside within the one organisation or jurisdiction, and hence cooperation and data sharing amongst these organisations is essential. Although there is a history of good cooperation between local, state, and national jurisdictions during disaster management, at other times the sharing of data has been problematic.

With local government being a custodian of a number of strategic spatial datasets, it has a crucial role to play in the development of the state and national spatial data infrastructures (SDIs), which rely heavily on the vertical integration of spatial data from the lower levels of government (Harvey 2000). In recent years, a number of cooperative partnerships between local and state governments have emerged. These partnerships are relatively new arrangements that have been established to facilitate the improved sharing of spatial data and to realise the full potential of a spatial data infrastructure (National Research Council 1994). However, in order to achieve maximum benefit from such arrangements, it is important to understand the factors that contribute to the successful and sustainable operation of these partnerships.

Organisational, technical, legal, and economic issues continue to impede the integration of spatial information in heterogeneous data sharing environments (Masser 1998; Masser and Campbell 1994; Nedović-Budić and Pinto 2001; Onsrud and Rushton 1995). Although research has identified that these inter-organisational issues remain a priority, there have been few systematic evaluations of the mechanisms and factors that facilitate the interorganisational efforts (Nedović-Budić and Pinto 2001). In particular, the vertical integration of multiple levels of data across multiple levels of government continues to be a major impediment to a fully robust national SDI (Harvey et al. 1999). Masser (2005) identifies a pressing need for more research on the nature of data sharing in a multilevel SDI environment, particularly with respect to the organisational issues.

Partnerships are considered to be essential for SDI development because they provide a mechanism to allow organisations to work together to achieve SDI goals and share implementation responsibilities and the eventual partnership benefits (Wehn de Montalvo 2001). Experiences in several countries have identified a number of problems with establishing partnerships at every level of government. These problems include the poor structure of partnerships, lack of awareness of partnership benefits, poorly defined responsibilities of each partner, fear of losing control of data, limited funding, and lack of a buy-in (Wehn de Montalvo 2001, 2003b). Although many issues have been identified, the key problem of "how to package these research insights into a coherent and effective program or set of guidelines" remains (Nedović-Budić and Pinto 2001). Kevany (1995) also identifies as one of the most important areas of future research the establishment of a set of factors (values) for both successful and unsuccessful data sharing environments which can be applied to future initiatives.

The importance of partnerships and collaboration has been promoted and reported by the National Mapping Committee of the National Research Council (National Research Council 1994, 2001) and the Geodata Alliance (Johnson et al. 2001) through documented success stories and identification of key success factors. However, these documents also suggest that more rigorous efforts need to be pursued to improve our understanding of collaborative initiatives. A better understanding of the existing jurisdictional partnership arrangements could assist in the development of a more universal and successful model for collaboration. The benefits from such a model should lead to the improved development of spatial data infrastructures at all levels, which in turn should positively impact all sectors of the government, business, and community.

This paper will firstly review a variety of existing data sharing models and frameworks with respect to their characteristics, strengths, and limitations. The mixed-method research approach is then described as a suitable method for examining existing data sharing partnerships. This methodology will then be examined in the context of evaluating local-state government data sharing partnerships for SDI development. Finally, the utility of this approach and its validity will be discussed.

DATA SHARING MODELS AND FRAMEWORKS

The sharing of spatial data is not new; however, in recent times the importance of spatial data sharing as a mechanism for building and sustaining the development of spatial data infrastructures has been highlighted (National Research Council 1994). Several contributions have been made to the understanding of data sharing within and across organisations, including the willingness of organisations to share their data. These contributions range in complexity and detail, but it is useful to review a number of these models and frameworks to gain a better understanding of existing theory and practice.

One of the early efforts to describe a classification framework for data sharing was undertaken by Calkins and Weatherbe (1995). The four primary components of their taxonomy included (1) characteristics of the organisation, (2) data, (3) exchange, and (4) constraints and impediments. Kevany (1995) proposed a more detailed structure to measure the effectiveness of data sharing. This structure is based on the author's experience across a range of projects, particularly at the county and city levels in the United States. Thirty factors that influence data sharing were identified under the following nine broad areas: sharing classes, project environment, need for shared data, opportunity to share data, willingness to share, incentive to share, impediments to sharing, technical capability for sharing, and resources for sharing.

Data sharing can also be viewed in terms of antecedents and consequences. A framework proposed by Obermeyer and Pinto (1994) and Pinto and Onsrud (1995) includes a number of antecedents—such as incentives, superordinate goals, accessibility, quality of relationships, bureaucratisation, and resource scarcity—which precede the process of data sharing. The effects of these events and factors then mediate a range of data sharing consequences such as efficiency, effectiveness, and enhanced decision making. Azad and Wiggins (1995) proposed a typology based on interorganisational relations (IOR) and dynamics. The authors argue that spatial data sharing across many agencies is fundamentally an organisational affair and that the concept of organisational autonomy is a critical issue in data sharing.

Another framework which examines organisational data sharing is put forward by Nedović-Budić and Pinto (1999) and draws on the Kevany model (1995), which was largely experience based. The conceptual framework draws on a broader literature base to derive four theoretical constructs: interorganisational context, motivation, coordination mechanisms, and outcomes. The theoretical foundations of this framework provide a very useful basis for further development and assessment of spatial data sharing initiatives. Wehn de Montalvo (2002) suggests that sharing, by its very nature, is a human behaviour and therefore it should be explored from a human behavioural perspective. The author used the theory of "planned behaviour" as an organising framework for investigating the willingness to share spatial data. The model maps the process of data sharing using belief structures and the predictive power of intentional behaviour. Table 1 summarises the various models and frameworks proposed by different authors.

The data sharing models in table 1 rely on a range of theoretical and experiential approaches. Increasingly, the importance of organisational and behavioural issues is recognised, and there is growing support for empirical models. The recent assessments of these models and theories (Nedović-Budić and Pinto 1999; Wehn de Montalvo 2003a) have identified the advantages of utilising both qualitative and quantitative approaches to better understand and evaluate data sharing arrangements. To understand the issues associated with data sharing within

Model/framework	Characteristics	Strengths	Limitations
Calkins and Weatherbe (1995)	Taxonomy based on characteristics of organisation, data, exchange process, and constraints/impediments	Framework recognises organisational issues and nature of exchange	Limited with respect to motivations, policy, and capacity of organisations
Kevany (1995)	Factor- and measure-based model	Very comprehensive list of factors that can be rated based on existing exchanges	Based on personal experience and not supported by theoretical foundations
Obermeyer and Pinto (1994), Pinto and Onsrud (1995)	Conceptual model based on antecedents and consequences	Based on exchange and organisational theory; basis for further research	Mainly conceptual and has limited depth or justification of factors
Azad and Wiggins (1995)	Typology based on IOR and dynamics	Attempts to classify organisation dynamics and behaviour (Oliver 1990)	Lack of justification for the initial premise that data sharing leads to the loss of autonomy and independence and lack of empirical evidence
Nedović-Budić and Pinto (1999)	Based on the theoretical constructs of context, motivation, mechanisms, and outcomes	Broad theoretical basis supported by quantitative validation in later studies	May not predict potential willingness to share data
Wehn de Montalvo (2003)	Based on theory of planned behaviour	Strong theoretical basis that is strengthened by a mixed-method approach	Model is predictive (by design) and may not be directly applicable to the analysis of existing initiatives

Table 1. Data sharing models.

the context of a data sharing partnership, the following research questions need to be addressed:

1. How can our understanding of existing interjurisdictional data sharing models be utilised to improve their operation and sustainability in the context of SDI development?

2. How can these partnership models be more rigorously described and classified?

3. What are the motivations for, and barriers to, the participation of governments in spatial data sharing partnerships?

4. What are the factors that contribute to the success of these data sharing partnerships?

5. Can these factors be used to identify the capacity of each partner to successfully participate in these partnerships?

6. Can a generic framework or model be developed to guide future spatial data sharing partnerships?

The first and second questions are primarily qualitative in nature and seek to explain the nature of interjurisdictional partnerships. The next three questions are more quantitative and seek to identify and measure a number of issues or factors. The final question requires the blending of both qualitative and quantitative approaches to better guide the development of a generic framework or model. To investigate these questions more fully, we propose a mixed-method approach which integrates both qualitative and quantitative strategies. The theory of mixed methods is discussed in detail below to demonstrate its applicability to the classification and evaluation of spatial data sharing partnerships.

MIXED-METHOD APPROACH

The debate over the benefits of qualitative versus quantitative methods continues, with the proponents in each camp vigorously defending the benefits and rigor of each approach (Tashakkori and Teddlie 2003). New methods in theory and practice such as participatory approaches, advocacy perspectives, critical appraisal, and pragmatic ideas have continued to emerge (Lincoln and Guba 2000). However, in recent times researchers have begun to reexamine these previously isolated strategies (Creswell 2003). The field of mixed methods has developed as a pragmatic approach to utilise the strengths of both qualitative and quantitative methods.

Mixed-method research is not new but a logical extension of the current reexamination and exploration of new practices. As Creswell (2003) puts it,

> Mixed methods research has come of age. To include only quantitative or qualitative methods falls short of the major approaches being used today in the social and human sciences. . . . The situation today is less quantitative versus qualitative and more how research practices lie somewhere on the continuum between the two. . . . The best that can be said is studies tend to be more quantitative or qualitative in nature.

The definitions for qualitative and quantitative methods vary with individual researchers (Thomas 2003). Mixed-method design can incorporate techniques from both the qualitative and the quantitative research traditions in a unique approach to answer research questions that could not be answered in another way (Tashakkori and Teddlie 2003). However, the mixed-method approach differs from qualitative and quantitative research paradigms (Brannen 1992) and can provide a number of advantages. Teddlie and Tashakkori (2003) identify three reasons that mixed-method research may be superior to single-approach designs:

1. Mixed-method research can answer research questions that other methodologies cannot

2. Mixed-method research provides better (stronger) inferences

3. Mixed methods provide the opportunity for presenting a greater diversity of views

The above reasons provide a sound basis for justifying the application of the mixed-method approach to SDI partnership research. Firstly, the mixed-method approach not only enabled the exploration and description of existing partnership arrangements, particularly the "why" and "how" of the arrangements, but also facilitated the measurement or quantification of the value of these arrangements. The research questions identified previously are also difficult to answer through any single approach. A case study approach was deemed suitable for addressing the "why" and "how" questions. However, to evaluate large multi-participant data sharing partnerships, a quantitative approach was considered more appropriate.

Secondly, the weaknesses of a single approach are minimised through the complementary utilisation of other methods. The qualitative case study approach provided the opportunity to investigate the organisational aspects of the partnerships in greater depth, while a quantitative survey of a larger number of partnership participants facilitated a greater breadth of views. Finally, the opportunity to investigate and present a greater diversity of views was considered important in validating the research findings. This was valuable because it led to the reexamination of the conceptual framework and underlying assumptions of each of the two methods (Teddlie and Tashakkori 2003). The diversity and divergence of perspectives between different levels of jurisdictions such as state and local governments is well known. Importantly, this reflects the reality of the relationships and hence the health of the partnership arrangements.

An important consideration in using a mixed-method approach is the way in which the qualitative and quantitative methods are combined (Brannen 1992). The two strategies can be combined in three ways according to Bryman (1998):

1. Preeminence of quantitative over qualitative

2. Preeminence of qualitative over quantitative

3. Qualitative and quantitative are given equal weight

In the first approach, the qualitative work may be undertaken prior to the main quantitative study and may be used as a basis for hypothesis testing, developing the research instrument, or clarifying quantitative data. The qualitative work may be performed at an early stage but can also be revisited later. In the second approach, the quantitative study can be conducted before the main study or at the end of the main study. It can provide background data to contextualise small intensive studies, test hypotheses derived through qualitative methods, or provide a basis for sampling and comparison. The final approach provides equal weighting to each method. The two studies are considered separate but linked and can be performed simultaneously or consecutively. The processes may be linked at various stages in the research process and then integrated to formulate final outcomes.

Priority, implementation timing, stage of integration, and theoretical perspectives can assist in classifying the mixed-method approach. Creswell et al. (2003) propose six design types based on these four criteria. These design types can be used to assist researchers in identifying the most suitable mixed-method approach for a particular study, particularly when and how to integrate the two methods. The design types proposed by Creswell et al. are classified primarily as either sequential or concurrent. For the sequential design, the order of the quantitative and qualitative studies may be dictated by the research problem and whether a more exploratory or explanatory approach is required. Alternatively, the two studies could be conducted concurrently, with the results of each study being interpreted during the analysis stage.

The mixed-method approach is not without problems, and care must be taken in the integration and interpretation phases of the research (Bryman 1992). However, when properly balanced and guided by an understanding of the research purposes and problems, the mixed-method approach is a powerful research strategy. To more clearly illustrate the mixed-method approach, we examine its use for the classification and evaluation of local-state government spatial data sharing partnerships from a methodological perspective.

USE OF THE MIXED-METHOD APPROACH TO ASSESS DATA SHARING PARTNERSHIPS IN AUSTRALIA

Local government is a rich source of accurate and detailed spatial information, which is utilised not only at the local level but also increasingly at other levels of government. In countries that have a system of federated states, such as Australia, the building of state and national SDIs increasingly relies on the involvement of local government. Although institutional problems still present some of the greatest challenges in building multijurisdictional SDIs, the technical and physical capacity of the smaller jurisdictions can affect their ability to participate with larger, bigger-budget jurisdictions.

The mixed-method research design illustrated in figure 1 consists of a four-stage process which culminates in the synthesis and development of a new model for local-state government SDI partnerships. This design draws together a generalised design framework for case study approaches proposed by Yin (1994), Onsrud et al. (1992), Lee (1989), and Williamson and Fourie (1998). The three-stage process of Williamson and Fourie (1998) is expanded to include quantitative methods used to identify and measure the impact and effectiveness of the data sharing partnership models. The design also includes the integration of both qualitative and quantitative results and a process of model validation.

A number of mixed-method design frameworks have emerged in recent times (Creswell et al. 2003; Johnson and Onwuegbuzie 2004; Nedović-Budić unpublished; Tashakkori and Teddlie 1998; Wehn de Montalvo 2003a). The design in figure 1 starts with the identification of research questions and proceeds to organisational case studies, a quantitative survey, and synthesis of results. The four stages are discussed in detail below.

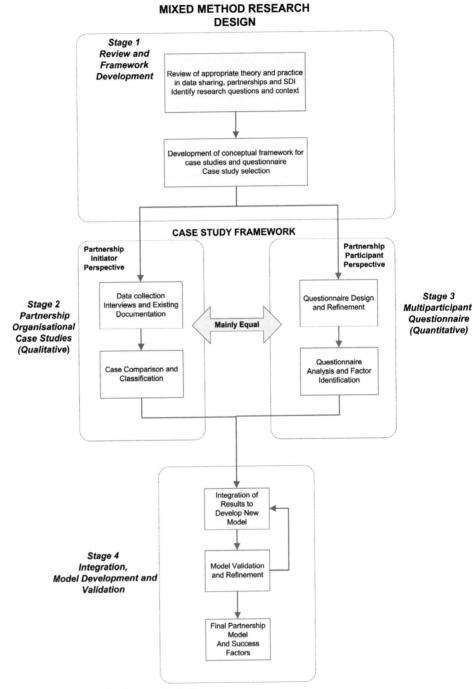

Figure 1. Mixed-method research design.

Stage 1: Review of theory and framework development. The first stage of the research provides the foundation for development of a suitable conceptual framework for the initial data collection and assessment. For the organisational case studies of the state governments, the conceptual framework was developed from organisational and collaboration theories. A variety of researchers (Child et al. 2005; Gray 1985; Mulford and Rogers 1982; Oliver 1990; Prefontaine et al. 2003) have identified a number of important dimensions of collaboration including the collaborative environment, the determinants for collaboration, the collaborative process, and the performance of collaborative initiatives. The theory enabled the development of a basic framework for exploring the initiation, development, and operation of the state government partnerships. One of the primary purposes was to investigate the contribution of data sharing partnerships to SDI development at local and state levels. Therefore, the conceptual framework for the local-government questionnaires was developed around the SDI elements identified by a range of authors (Coleman and McLaughlin 1998; Groot 1997; National Research Council 1993; Rajabifard and Williamson 2001). These components include data, people, standards, institutional framework/policies, and technology/access arrangements.

Case study selection. The case studies investigated existing data sharing partnerships between state and local governments in Australia which had been established to facilitate the sharing of property-related data. The three Australian states of Queensland, Victoria, and Tasmania were chosen for the study. The states were selected on the basis of already established data sharing arrangements and a variety of characteristics including geographic area, population, and the number of local governments. The state of Queensland is the second largest state in Australia by area. Its capital city of Brisbane represents one of the largest local government jurisdictions in the world. Queensland also has a relatively large number of local governments, 125 in total, including many in remote rural communities with very small populations.

At the other end of the spectrum, the state of Tasmania is a compact island state that has only 29 local governments and approximately half a million people. It provided a contrasting study of a smaller jurisdiction both in area and in the number of partnership participants. The third case selected was the state of Victoria, with 79 local governments. Victoria is one of the most populated states in Australia and is also well advanced in its partnership arrangements. It falls in between the other two states in geographic area and number of local governments. The 3 states represent almost 50 percent of Australia's population, approximately 35 percent of the total number of local governments, and about 25 percent of the land area. The states represent contrasting mixtures of local governments, geography, and institutional arrangements.

Stage 2: Organisational case studies of partnerships (qualitative component). A key objective of the qualitative component of the case studies was to examine the organisational frameworks of each of the state-government-initiated partnerships. A structured case study methodology as recommended by Yin (1994) was utilised. An SDI framework consisting of the key areas of policy, data, people, access arrangements, and technology/standards provided the basis for the investigation.

Case study data collection. For this qualitative component, the methods of data collection focused on two primary forms of evidence: interviews and existing documentation. A semistructured interview technique was utilised to collect data from staff within each state government agency that was charged with the management of the partnership arrangement. The interviews covered the following topics:

- Organisation overview and role of partnership
- Historical developments within the partnership
- Existing policy arrangements
- Understanding of the data and data sharing processes
- Operational and resource aspects of the partnership
- Organisational and institutional arrangements
- Barriers and issues (legal, technical, economic, institutional)

The people interviewed included the partnership initiators, partnership managers, and staff involved in various data sharing activities.

The other key source of evidence for the case studies consisted of historical documentation which had been in existence since the design and development of the partnerships. The documentation varied from state to state but included some of the following:

- Initial proposal documents for the partnership
- Descriptive documentation such as that available on Web sites
- Examples of individual partnership agreements
- Internal review documents of the arrangements
- External consultancy reports
- Conference and journal papers describing the arrangements

In the evaluation of each of the documents, care was taken to recognise the strengths and weaknesses of the various forms of documentation, particularly with respect to any bias. In case studies, one of the most important uses for documentation is to corroborate and augment evidence from other sources to minimise possible bias.

CASE STUDY COMPARISON AND CLASSIFICATION

An important objective of the research was to compare and classify the different partnership arrangements in existence. Basic comparators included:

- Length of partnership
- Extent of data sharing
- Quantification of resources
- Communication mechanisms and frequency
- Number of partners
- Geographic extent
- Environmental context

To further explore the nature and sustainability of the SDI partnerships in comparison to partnerships operating in other disciplines, a typology for classifying the partnership models was developed. The typology included the following dimensions:

- Nature of partnership
- Partnership goals
- Negotiation processes
- Resource or funding model
- Governance model
- Project management
- Performance measurement
- Maturity and organisational learning

Stage 3: Multiparticipant questionnaire (quantitative component). In order to assess the motivating factors, constraints, and effectiveness of local-state government data sharing partnerships, a questionnaire was administered to the local governments in the three states. The purpose of the questionnaire was to assess a range of factors that might influence the success or failure of the data sharing partnerships, particularly from a local government perspective. The questionnaire was constructed around the existing knowledge of SDI frameworks, especially the participants' understanding of policies, data holdings, people, access arrangements, and standards/technology. In addition to the SDI framework, the questionnaire investigated the organisational setting, partnerships, and collaborations and the participants' perspectives on the existing partnership arrangements.

The questionnaire consisted of eight sections:

1. The **Organisation** section quantified the size of the local government in terms of the number of properties and staff and provided an assessment of its general ICT capacity including specific GIS and spatial data capacities.

2. The **Policy on Use of Spatial Data** section explored existing policies within the local government for access and pricing of spatial information including issues of legal liability, copyright, and privacy.

3. The **Accessing Spatial Data** section examined the organisation's arrangements for accessing and pricing of spatial information from the perspectives of both internal and external users.

4. The **About Spatial Data** section examined the sources of spatial data, the key providers, and the status of their data holdings.

5. The **Spatial Data Standards and Integration** section investigated the use of standards and the degree of integration of the organisation's spatial data systems with other core systems, providing an indication of the level of maturity of spatial information systems within the organisation.

6. The **About People** section explored the human resources of the organisation including staff turnover and access to training.

7. The **Partnerships and Collaboration** section explored the perceived strength of the organisation's relationship with a range of organisations, the barriers to collaboration, the drivers for collaboration, and the types of existing collaborations.

8. The **Specific Data Sharing Partnerships** section examined the organisation's specific attitudes toward and experiences with an existing SDI partnership.

For the majority of questions the responses were standardised and categorised on a five-point Likert scale. Some questions asked for numeric data such as number of staff or land parcels. Participants could also provide comments on each area of the questionnaire. A draft questionnaire was distributed to three local governments to check for terminology and understanding of the questions. The questionnaire was then converted into a Web form to enable digital collection of the data and facilitate a higher return rate. The Web-based questionnaire was then tested internally and externally by two local governments to ensure that the URL provided was accessible and also that responses were being recorded on the Web server.

QUESTIONNAIRE DISTRIBUTION AND ANALYSIS

The distribution of the questionnaire was undertaken after consultation with each of the state agencies. The questionnaire sought responses from local governments in a number of areas that could reflect poorly on the state government agency, so a degree of sensitivity was required. Privacy of customer or partner information also became an issue. Under state and federal government privacy legislation, permission must be sought from individuals before their contact details can be disclosed. This became a significant issue, as it was critical that the questionnaire be sent to the correct partnership contact person rather than randomly targeted local government staff. The privacy issue was addressed by the state government agency making the initial contact with the local government agency and seeking their consent to be involved with the study. Once consent was obtained, the details were passed on to the researcher. The questionnaire response rate was 56 percent, which was considered extremely satisfactory, given the diversity of local governments being investigated.

The data from the questionnaires was automatically recorded into an Excel spreadsheet via the Web server. This process was extremely effective, as it eliminated encoding and transcription errors and facilitated direct transfer to the analysis software (SPSS). Initial descriptive statistics identified a number of early trends in the responses from the different state jurisdictions, particularly in the areas of information policy and outcomes delivered through the partnerships. Factor analysis was then utilised to identify clusters of variables (components), which were then correlated with the outcome variables using a regression model. Through this modelling, components which had contributed significantly to the success of the partnership outcomes were identified.

Stage 4: Integration, model development, and validation. After the completion of the case studies and questionnaire analysis the results were integrated to develop a new data sharing partnership model. The case study results assisted in clarifying the initial conceptual framework and typology of the existing partnerships in each of the three state government jurisdictions. The descriptive and comparative analysis enabled a clearer understanding of the organisational structures, policy objectives and goals, partnership structure, progress and outcomes, resource requirements, and sustainability. The perspectives gained from these cases assisted in answering some of the research questions related to how and why the spatial data sharing initiatives were put in place and identified some of the major issues related to their implementation. Importantly, it should be noted that the descriptive case studies primarily provided the perspective of the partnership initiator and manager rather than partnership participants.

The development of a generic model required the perspectives of local governments for a more balanced view of the success of the data sharing arrangements. The results of the questionnaire identified the capacities and motivations of local governments to participate in data sharing partnerships. The quantitative analysis enabled these factors to be identified and modelled against partnership outcomes.

Interjurisdictional (local and state) partnerships inevitably create challenges for each level of government. The research found that state–local government data sharing partnerships differ in a number of ways from other interjurisdictional data sharing. Firstly, for a comprehensive solution to data sharing between state and local governments the partnership arrangements need to be established on a one-to-many basis. The qualitative case studies showed that a systemised approach to partnership negotiation, data licensing, data maintenance, partner communication, data exchange, and project management is critical to the success of these endeavours.

Table 2 identifies some of the differences among the three state jurisdictions. Both the Victorian and the Tasmanian data sharing partnerships from the outset had appropriate resources, clear goals, and strong leadership. However, the Queensland partnership struggled to gain the support of local governments because of poor initial funding and a restrictive policy framework that limited the local governments in conducting their business activities using state government data.

The findings of the state government level investigations were supported by the quantitative statistics of the local government survey (figure 2). The areas of

Collaborative stage	Victorian Property Information Project (PIP)	Queensland Property Location Index (PLI) Project	Land Information System Tasmania (LIST)
Establishment and direction setting Goal setting Negotiation Agreements	A clear common goal for the project. Well-managed process of negotiation and development of policy and institutional structures.	Business case for the project was limited. Goals unclear, and policy framework worked against data-sharing agreements.	High-level strategy and clear overall goals. Policy and negotiation strategy well-structured. Agreements very detailed.
Operation and maintenance Project management Maintenance Resources Communication	Project management has been good since inception, maintenance infrastructure developed progressively, some resource limitations. Communication with stakeholders and partners has been positive.	Poor institutional arrangements led to resource limitations and poor project support. Culture of interjurisdictional sharing emerging only now. Confused channels of communication due to dispersed organisational structure.	LIST started with strong overall leadership and project support. Project generally had strong resources and was technology focused. Issues of local government communication and data maintenance now starting to emerge.
Governance Governance structures Reporting Performance management	Early project efforts focused on negotiation and data exchange. Performance management now part of the process. Improved governance arrangements emerging.	There appears to have been little performance management or reporting. No governance structure in place which includes the key stakeholders.	Initial governance and reporting structures were appropriate, but as project matures new governance models are required.

Table 2. Qualitative assessments of the performance of state partnerships.

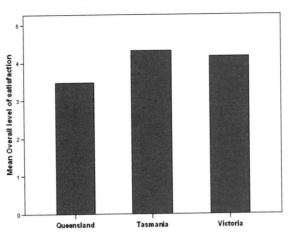

Figure 2. Levels of satisfaction reported by local governments (Likert scale).

weakness in the partnership processes identified at the state government level were reflected in the overall level of satisfaction in the local government survey. Areas such as policy formulation at the state government level have a strong influence on the corresponding policy developments at the local level. Clear partnership goals, continuous and open communication, and adequate funding also have a strong influence on partnership outcomes.

DISCUSSION AND CONCLUSIONS

The research methodology described above builds on similar models proposed by Yin (1994), Onsrud et al. (1992), Lee (1989), and Williamson and Fourie (1998) for case study approaches with the addition of quantitative methods. The mixed-method approach has already been utilised successfully by a number of researchers studying spatial data sharing (e.g., by Wehn de Montalvo [2003a] for assessing the willingness to share spatial data and by Nedović-Budić [unpublished] for assessing adoption of GIS technology). However, the possible utility and validity of the approach deserve further comment.

Qualitative approaches such as case studies have often been viewed as inferior to quantitative approaches, suitable primarily for stand-alone descriptions of phenomena or as exploratory research preliminary to the real research of generating hypotheses and testing them statistically (Benbasat 1984). Although such comments on earlier studies were common, rigorous (Yin 1994) and scientific (Lee 1989) case study frameworks now exist.

For research reported in this article, the case study method was selected as the primary qualitative strategy for examining a number of spatial-data-sharing partnership models in different jurisdictions, particularly from an organisational perspective. The case study approach was deemed suitable for examining these partnership models for several reasons. Firstly, data sharing partnership models can be studied in their natural settings and provide the opportunity to learn from state-of-the-art approaches and practice (Benbasat et al. 1987; Maxwell 1996). Secondly, the case study approach allows the asking of the "how" and "why" research questions and investigation of the nature and complexity of spatial data sharing partnerships (Benbasat et al. 1987; Yin 1994). Thirdly, the case study approach can provide a suitable framework for analysis and classification of partnership models (Lee 1989; Yin 1994). Finally, the case study approach provides a high level of data currency as well as data integrity (Bonoma 1985).

The incorporation of the quantitative dimension with the use of a questionnaire strengthens the case study approach by facilitating efficient inclusion of a large number of participant perspectives and comprehensive and quick analysis of this data. It can also assist in identification of key factors, correlations, and possible trends for developing an improved partnership model.

In the study the qualitative and quantitative components were generally completed concurrently. The qualitative organisational cases were evolving during the course of the study, with some periodic updates of the organisational environment. The questionnaires were completed over a six- to nine-month period and reviewed as the need arose. The evidence from each component was given equal weight, although this was often difficult to confirm. Finally, the integration of the two strategies was achieved at the analysis stage. This process facilitated the corroboration of results and confirmation of the importance of issues.

The triangulation of existing theory, case studies, and survey results informs the final model (figure 3). The internal validity of the model should, in theory, be superior to each of the singular approaches. However, care must always be exercised in early conceptual development and design, as in addition to the potential for complementarity, the risk of conflicting results exists.

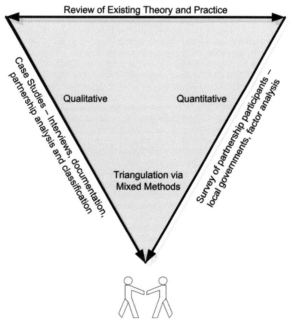

Validity of Methods Through Triangulation

Review of Existing Theory and Practice

Case Studies – Interviews, documentation, partnership analysis and classification

Qualitative

Quantitative

Survey of partnership participants – local governments, factor analysis

Triangulation via Mixed Methods

Improved Partnership Model and Success Factors

Figure 3. Method triangulation model.

The difficulty in generalising the findings from the small number of cases being analysed is often identified as the weakness of the case study approach. By undertaking a more wide-ranging survey of partnership participants, the findings of the case studies were strengthened.

The purpose of this paper was to examine the methodological approaches and issues which arise in researching spatial data sharing partnerships and their relationships to SDI development. As partnerships continue to emerge, it is important to understand their success and contribution to building SDIs. In the past, discrete research approaches and models have provided valuable starting points for measuring and classifying data sharing efforts. However, a mixed-method approach provides a useful strategy to build on the existing theory and to more rigorously evaluate the results of these partnership efforts.

ACKNOWLEDGMENTS

We acknowledge the support of the state and local governments in Queensland, Victoria, and Tasmania, the University of Southern Queensland, and the members of the Centre for SDI and Land Administration in the Department of Geomatics at the University of Melbourne. (However, the views expressed in the paper do not necessarily reflect the views of these groups.)

REFERENCES

Azad, B., and L. L. Wiggins. 1995. Dynamics of inter-organizational data sharing: A conceptual framework for research. In *Sharing geographic information*. H. J. Onsrud and G. Rushton eds., 22–43. Centre for Urban Policy Research, New Brunswick, NJ.

Benbasat, I., D. K. Goldstein, and M. Mead. 1987. The case research strategy in studies of information systems. *MIS Quarterly* 11 (3): 369–89.

Benbasat, I., ed. 1984. *The information systems research challenge* 2. Experimental research methods, Harvard Business School Press, Boston.

Bonoma, T. V. 1985. Case research in marketing: Opportunities, problems, and a process. *Journal of Marketing Research* 22 (2): 199–208.

Brannen, J. 1992. Combining qualitative and quantitative approaches: An overview. In *Mixing Methods: qualitative and quantitative research*, ed. J. Brannen. Avebury, Aldershot, England. 3–37.

Bryman, A. 1992. Quantitative and qualitative research: Further reflections on their integration. In *Mixing Methods: Qualitative and quantitative research*, ed. J. Brannen, 57–78. Avebury, Aldershot, England.

Bryman, A. 1998. *Quantity and quality in social research*. Unwin Hyman, London.

Calkins, H. W., and R. Weatherbe. 1995. Taxonomy on spatial data sharing. In *Sharing geographic information*, eds. H.J. Onsrud and G. Rushton, 67–75. Centre for Urban Policy Research, New Brunswick, NJ.

Child, J., D. Faulkner, and S. B. Tallman. 2005. *Cooperative strategy*. Oxford University Press, Oxford.

Coleman, D. J., and J. McLaughlin. 1998. Defining global geospatial data infrastructure (GGDI): Components, stakeholders and interfaces. *Geomatica* 52 (2): 129–43.

Creswell, J. W. 2003. *Research design: Qualitative, quantitative, and mixed method approaches,* 2nd ed. Sage Publications, Thousand Oaks, CA.

Creswell, J. W., V. L. Plano Clark, M. L. Gutmann, and W. E. Hanson. 2003. Advanced mixed methods research designs. In *Handbook of mixed methods in social and behavioral research*, eds. A. Tashakkori and C. Teddlie, 209–40. Thousand Oaks, CA: Sage Publications.

Gray, B. 1985. Conditions facilitating interorganizational collaboration. *Human Relations* 38 (10): 911–36.

Groot, R. 1997. Spatial data infrastructure (SDI) for sustainable land management. *ITC Journal* 1997 (3/4): 287–93.

Harvey, F. J. 2000. Potentials and pitfalls for vertical integration for the NSDI. Final report of a survey of local government perspectives.

Harvey, F. J., B. P. Buttenfield, and S. C. Lambert. 1999. Integrating geodata infrastructures from the ground up. *Photogrammetric Engineering and Remote Sensing* 65 (11): 1287–91.

Johnson, R., Z. Nedovic-Budic, and K. Covert. 2001. Lessons from practice: A guidebook to organizing and sustaining geodata collaboratives. Geodata Alliance, June 20, 2002.

Johnson, R. B., and A. J. Onwuegbuzie. 2004. Mixed methods research: A research paradigm whose time has come. *Educational Researcher* 33 (7): 14–26.

Kevany, M. J. 1995. A proposed structure for observing data sharing. In *Sharing geographic information,* eds. H. J. Onsrud and G. Rushton. Centre for Urban Policy Research, New Brunswick, NJ. 76–100.

Lee, A. S. 1989. A scientific methodology for MIS case studies. *MIS Quarterly* 8 (1): 33–50.

Lincoln, Y. S., and E. G. Guba. 2000. Paradigmatic controversies, contradictions, and emerging confluences. In *Handbook of qualitative research,* 2nd ed., eds. N. K. Denzin and Y. S. Lincoln, 163–88. Thousand Oaks, CA: Sage Publications.

Masser, I. 1998. *Governments and geographic information.* London: Taylor & Francis.

Masser, I. 2005. *GIS worlds: Creating spatial data infrastructures.* Redlands, CA: ESRI Press.

Masser, I., and H. Campbell. 1994. Information sharing and the implementation of GIS: Some key issues. In *Innovations in GIS* 1:217–27, ed. M. F. Worboys. London: Taylor & Francis.

Maxwell, J. A. 1996. *Qualitative research design: an iterative approach* 41. Applied social research methods series. Thousand Oaks, CA: Sage Publications.

Mulford, C. L., and D. L. Rogers. 1982. Chapter 2 Definitions and Models. In *Interorganizational coordination: Theory, research and implementation,* eds. D. A. Rogers and D. A. Whettons, 9–31. Iowa State University Press, Ames, Iowa.

National Research Council. 1993. *Toward a coordinated spatial data infrastructure for the nation.* Washington, DC: National Academy Press.

National Research Council. 1994. *Promoting the national spatial data infrastructure through partnerships.* Washington, DC: National Academy Press.

National Research Council. 2001. *National spatial data infrastructure partnership programs: Rethinking the focus.* Washington, DC: National Academy Press.

Nedović-Budić, Z. Exploring the utility of mixed method research in GIS. (In progress.)

Nedović-Budić, Z., and J.K. Pinto. 1999. Understanding interorganizational GIS activities: a conceptual framework. *Journal of Urban and Regional Information Systems Association* 11 (1): 53–64.

Nedović-Budić, Z., and J. K. Pinto. 2001. Organizational (soft) GIS interoperability: Lessons from the U.S. *International Journal of Applied Earth Observation and Geoinformation* 3 (3): 290–98.

Obermeyer, N. J., and J. K. Pinto. 1994. *Managing geographic information systems.* New York: The Guildford Press.

Oliver, C. 1990. Determinants of inter-organizational relationships: integration and future directions. *Academy of Management Review* 15 (2): 241–65.

Onsrud, H. J., and G. Rushton. 1995. Sharing geographic information: An introduction. In *Sharing geographic information.* H. J. Onsrud and G. Rushton eds. Centre for Urban Policy Research, New Brunswick, New Jersey. xiii–xviii.

Onsrud, H. J., J. K. Pinto, and B. Azad. 1992. Case study research methods for geographic information systems. *URISA Journal* 1 (4): 32–43.

Pinto, J. K., and H. J. Onsrud. 1995. Sharing geographic information across organisational boundaries: A research framework. In *Sharing geographic information,* eds. H. J. Onsrud and G. Rushton, 45–64. New Brunswick, NJ: Centre for Urban Policy Research.

Prefontaine, L., L. Ricard, H. Sicotte, D. Turcotte, and S. S. Dawes. 2003. New models of collaboration for public service delivery: Critical success factors of collaboration for public service delivery. CTG. http://www.ctg.albany.edu (May 1, 2006).

Rajabifard, A., and I. P. Williamson. 2001. Spatial data infrastructures: Concept, SDI hierarchy and future directions. Proceedings of Geomatics'80, Tehran, Iran.

Tashakkori, A., and C. Teddlie. 1998. *Mixed methodology: Combining qualitative and quantitative approaches* 46. Applied social research methods series. Thousand Oaks, CA: Sage Publications.

Tashakkori, A., and C. Teddlie, eds. 2003. *Handbook of mixed methods in social and behavioral research.* Thousand Oaks, CA: Sage Publications.

Teddlie, C., and A. Tashakkori. 2003. Major issues and controversies in the use of mixed methods in the social and behavioral sciences. In *Handbook of mixed methods in social and behavioral research,* eds. A. Tashakkori and C. Teddlie, 3–50. Thousand Oaks, CA: Sage Publications.

Thomas, R. M. 2003. *Blending qualitative and quantitative research methods in theses and dissertations.* Thousand Oaks, CA: Corwin Press.

Wehn de Montalvo, U. 2001. Strategies for SDI implementation: A survey of national experiences. Proceedings of 5th Global Spatial Data Infrastructure Conference, Cartagena de Indias, Colombia, May 21–25, 2001.

Wehn de Montalvo, U. 2002. Mapping the determinants of spatial data sharing. Proceedings of 8th EC-GI and GIS Workshop, Dublin, Ireland, July 3–5, 2002.

Wehn de Montalvo, U. 2003a. In search of rigorous models for policy-orientated research: A behavioural approach to spatial data sharing. *URISA Journal* 15 (1): 19–28.

Wehn de Montalvo, U. 2003b. *Mapping the determinants of spatial data sharing.* Aldershot, England: Ashgate Publishing Ltd.

Williamson, I. P., and C. Fourie. 1998. Using the case study methodology for cadastral reform. *Geomatica* 52 (3): 283–95.

Yin, R. K. 1994. *Case study research: Design and methods,* 2nd ed., vol. 5. Applied social research methods series. Thousand Oaks, CA: Sage Publications.

Spatial Data Sharing: A Cross-Cultural Conceptual Model

EL-SAYED EWIS OMRAN, ARNOLD BREGT, AND JOEP CROMPVOETS

CENTRE FOR GEO-INFORMATION, WAGENINGEN UNIVERSITY, WAGENINGEN, THE NETHERLANDS

ABSTRACT

In order to make full use of spatial data infrastructures, spatial data sharing (SDS) is essential. Various authors indicate that the attitudes both of individuals and of organizations toward SDS are quite often problematic. For various reasons and motivations, SDS is far from optimal. However, research on individual and organizational behavior in SDS is in its infancy and presents a challenge for new theory development. The objective of this paper is to shed light on the interaction between individual and organizational SDS behaviors and their social and cultural aspects. A new theoretical model is proposed. This model integrates concepts from multiple theories: theory of planned behavior, culture (grid-group) theory, and Hofstede's cultural dimensions. The relationships within the model are formulated in 23 hypotheses. The hypotheses have not yet been tested. Knowledge about relationships among individuals and organizations derived from the emerging model may provide insights into the attitudes of individuals and organizations toward SDS.

Many countries are developing spatial data infrastructures (SDIs) in order to better manage their spatial datasets (Rajabifard and Williamson 2004) for supporting various applications. The development of these datasets is often done with little coordination among various organizations, and as a consequence duplication of effort and wasting of resources occur (Warnecke et al. 1998; Wehn de Montalvo 2003a; Omran et al. 2006). In order to reduce this duplication, spatial data sharing (SDS) is essential. In many instances individuals and organizations are unwilling to share data across and within organizations. SDS behavior is strongly related to sociocultural context. Understanding and changing individual and organizational behaviors could be the key to improving spatial data sharing.

Individual spatial data sharing behavior has not received adequate attention in either research or practice. Even when social issues are considered, the focus is mainly on people as participants in the implementation process (Eason 1993), political issues (Buchanan 1993), or better design of decision support tools (Medyckyj-Scott and Hearnshaw 1993) rather than on psychological factors related to data sharing. Based on sociocultural theories, personal factors that strongly influence the individual decision to share data include attitudes, experiences, self-confidence, empathy, fatalism, motivation, behavior, trust, ability to cope with uncertainty, and incentives. In our assessment, the influences of these factors on SDS have not been sufficiently investigated. The current study was motivated by the question, "What factors influence individual SDS behavior?"

Another issue germane to spatial data sharing is the question of organizational resistance to sharing data. Resistance to share data may be due to a lack of motivation. Organizations are motivated by organizational needs and capabilities (Calkins and Weatherbe 1995), the advantages of synergisms (Craig 1995), and appeals to professionalism and common goals (Obermeyer 1995). These common or "superordinate" objectives are among the noneconomic reasons for sharing (Tjosvold 1988; Pinto and Onsrud 1995). Appropriate organizational motivation is required for data sharing; incentives can also motivate the organizations to share their data. The current study was also motivated by the question, "What factors influence organizational SDS behavior?"

To answer these two questions, Tayeb (1988) proposed two lines of research. The first line is institutionalism, which deals with structural aspects of organizations. The second line is "ideationalism," which focuses on the intentions, attitudes, and values of organization members. The relationship between individual and organizational behaviors and data sharing is very complex (Dueker and Vrana 1995).

Many sociocultural theories (e.g., theory of planned behavior and culture theory) can be used to characterize individual and organizational behaviors and describe relationships between them. Hofstede (1991, 2001) and Hofstede and Hofstede (2005) argue that five dimensions can be used to classify societies according to their culture: power distance, uncertainty avoidance, individualism/collectivism, masculine/feminine, and long-term/short-term orientation. Power distance (PD) represents the extent of adherence to formal authority and the degree to which less powerful members will accept unequal distribution of power. This dimension addresses how a society handles inequalities among people. Uncertainty avoidance (UNA) refers to how much people feel threatened by ambiguity, as well as the felt importance of rules and standards. This dimension addresses how a society reacts to the fact that the future is unknown, for example, whether it tries to control the future or lets it happen. Power distance and uncertainty avoidance have consequences for the way people build their institutions and organizations. Individualism/collectivism refers to the basic level of behavior regulation. It refers to the degree of interdependence a society maintains among individuals. In an individualistic society, the ties between individuals are loose. In a collectivist society people integrate into strong, cohesive groups and tend to do what is best for the group. Masculine cultures emphasize work and material accomplishments. In contrast, feminine cultures put human relationships at the forefront, and work is seen as a way to support the more important things in life. A long-term orientation (LTO) means that people are more concerned with the long-term effects of their decision. A short-term orientation (STO) tends toward consumption and maintaining materialistic status.

Although Hofstede made a major contribution to the study of organizations within a cultural setting, he did not empirically investigate the relationships between the five dimensions and the attitudes and behaviors of individuals and organizations. So, it is important to discern in what ways individuals and organizations are influenced by Hofstede's dimensions. How does national culture influence the attitudes of individuals and organizations toward SDS?

Although the bulk of the literature focuses on technical aspects of spatial data sharing, the emphasis of this paper is on individual and organizational aspects. The objective of this paper is to develop a conceptual model that describes the willingness of individuals and organizations to share spatial data. Data sharing by individuals and organizations in a sociocultural context serves as a starting point. The approach is to ground the assessment of variables in well-accepted theories. The innovative aspect of the model is the integration of different theories and concepts. Such a model increases our insight into the SDS behaviors of individuals and organizations and might potentially be used to explain differences between societies and organizations.

After an overview of SDS concepts and gaps, we propose an SDS model, describe the theoretical foundation and hypothesis development, and discuss the merits of the model.

SPATIAL DATA SHARING: CONCEPTS AND GAPS

Spatial data sharing is generally considered problematic. A considerable number of SDS relationships have failed to meet their founders' expectations. Porter (1987) and Park and Ungson (1997) report that the failure rate in interorganizational relationships is approximately 50 percent. Organizations, however, continue to form these relationships, and as a result failures are expected to continue or even increase (Miles and Snow 1992).

Calkins et al. (1991) present factors that could influence institutional data sharing: bureaucratic procedures, cooperation, organizational structure, corporate culture, and political environment. Kevany (1995) explores factors that may create a sharing environment and identifies opportunities, incentives, impediments, and resources as the main factors that influence SDS. Pinto and Onsrud (1995) state that, under conditions of resource scarcity, organizations tend to be driven by the desire to maintain some form of control over other organizations.

As risks increase, so does the need for trust. Trust is mostly connected to risks and risk taking (Mayer et al. 1995; Coulter and Coulter 2002) and influences both individuals and organizations (Doney and Cannon 1997).

Most of the SDS frameworks in the literature are based on the authors' experiences with data sharing. An exception is the work done by Wehn de Montalvo (2001; 2003a,b), who proposed a model of SDS perceptions and practices in South Africa from a social psychological perspective. Also, Nedović-Budić et al. (2004) proposed a model that includes the motivation behind sharing. These two examples move towards a more widely grounded theoretical approach to SDS. However, if we consider all the literature on SDS, the following research gaps are still observed:

1. No comprehensive theory-based framework for analyzing relevant factors exists.
2. The relationships between factors have not been adequately investigated.
3. The proposed experimental frameworks have not been verified.
4. Sociocultural aspects of SDS have not been adequately considered.
5. No systematic analysis of SDS between individuals and organizations has been performed.

The literature identifies uncertainty, incentives, resource scarcity, autonomy, rules, and similar factors within particular sociocultural settings as explaining, predicting, or modeling SDS. However, the integration of such factors in an overall model is missing, and little is known about the influences of these factors on the reasons that individuals and organizations are willing or not willing to share data. Sociocultural perspectives provide a useful point of departure for exploring this issue.

PROPOSED SPATIAL DATA SHARING MODEL

Interactions among and between individuals and organizations are a complex phenomena, and SDS behaviors across contexts cannot be described by a single theory. Our proposed model integrates insights from three theories: theory of planned behavior (Ajzen 1991), culture (grid-group) theory (Douglas 1970; Thompson et al. 1990), and Hofstede's (1980) culture dimensions. These theories are strong candidates for developing a more generalizable approach to assessment of SDS because they have already been investigated and identified by other researchers as having relevancy in this domain. These theories have received strong empirical support in the social sciences, having been widely applied and tested with considerable proven explanatory and predictive value for the behaviors of individuals, organizations, and even countries. We expect that these theories can also be used for modeling spatial data sharing, both for individuals and for organizations.

Overall model. Figure 1 shows the main components of the proposed SDS model. SDS is influenced by individual and organizational behaviors. Individual behavior (micro level) is analyzed by employing the major concepts of the theory of planned behavior (TPB). Organizational behavior (macro level) is studied by using the culture (grid-group) theory. The individual and the organizational levels are linked within the model in two ways: by the cultural dimensions

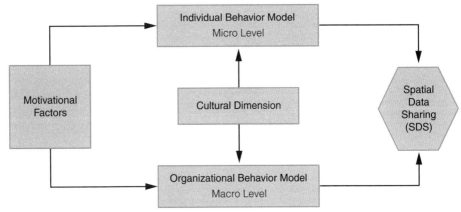

Figure 1. Main components of the SDS model.

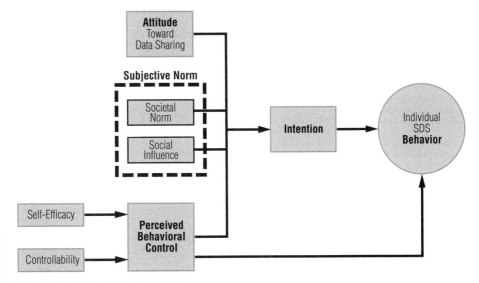

Figure 2. Individual-behavior submodel.

of Hofstede and by motivational factors derived from literature. Nakata and Sivakumar (2001) argue that Hofstede's cultural dimensions serve as the most powerful culture theory for social research. In addition, there are potential motivational factors (trust, uncertainty, incentives, resource scarcity, rules, and autonomy) that affect individual and organizational SDS behaviors. We argue that cultural dimensions in combination with motivational factors could be used as a link between the two submodels described below.

Individual-behavior submodel (micro level). The individual submodel is based mainly on TPB (figure 2). Ajzen (1991) and Ford et al. (2003) indicate that TPB has been developed with individuals as units of analysis. Ajzen (1991) argues that a central factor in TPB is the intention of individuals to demonstrate a particular behavior. The intention of individuals to engage in SDS is closely linked to actual behavior. Ajzen (1988, 1991) proposes that intentions are assumed to capture the motivational factors that influence a behavior. The stronger the intention for a particular behavior, the more likely is the behavior itself. At the level of the individual, we measure willingness to share spatial data. Ajzen (1985, 1988, 1991) argues that the behavioral, normative, and control beliefs are influenced by a wide variety of cultural, personal, and situational factors.

The intention of each individual is based on the attitude, subjective norm (SN), and perceived behavior control (PBC) relative to data sharing. In order to predict the spatial data sharing intention of an individual, we need to predict these three underlying factors. Attitude is defined as the degree of positive or negative value for SDS. Subjective norm is defined as the social pressure for sharing felt by the individuals. Subjective norm is based on societal norm and social influence. Societal norm refers to norms of the larger societal community, while social influence reflects opinions from family, friends, and peers. PBC is the extent to which the individual controls the sharing procedures for a particular spatial dataset. PBC is influenced by the individual's judgment of his own capabilities (self-efficacy) and by his confidence in the data sharing process (controllability). By understanding and estimating these three factors, we can assess an individual's intention for SDS.

Organizational-behavior submodel (macro level). The organizational submodel is based on culture theory. Thompson et al. (1990) propose that any organizational setting consists of two dimensions: grid (action) and group (identity) (figure 3). Adapting the theory to SDS requires specific definitions of the grid and group concepts. "Grid" refers to the degree of individual freedom in SDS and rules of authority that limit how people behave toward one another. In cultures with strong grids, everyone has a well-defined place in his or her organization. Institutions classify individuals and restrict their transactions. Moving away from a strong grid, dependence decreases and autonomy, control, and competition open up (Douglas 1978). This paves the way for freedom of transactions. "Group" refers to the degree to which individuals are member of groups or networks (social boundedness). The more an individual is incorporated into bounded units, the more his choice is subject to group determination (Douglas 1978). In combination, these two key dimensions can produce four organizational settings for SDS—hierarchy, egalitarianism, individualism, and fatalism (always potentially present in any group or organization).

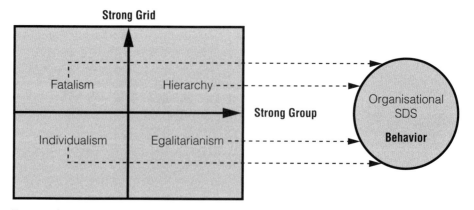

Figure 3. Organisational-behavior submodel.

Linkage between the individual- and organizational-behavior submodels. The proposed model combines Hofstede's cultural dimensions with the motivational factors to link the two submodels.

First, Hofstede's cultural dimensions play an important role in combining the individual and organizational submodels. For the individual-behavior model, we assume that cultural dimensions influence the values and weights of the predictors for intention (attitude, SN, and PBC). Ajzen (1991), Straub et al. (1997), and Ford et al. (2003) expect that national culture influences the weighting of the predictors of intention in TPB. For example, in a culture that is more individualistic, the effects of subjective norms are low and the effects of attitude and perceived behavioral control are high (i.e., the individual's own opinions are more important). Likewise, in the organizational behavior model, egalitarianism within organizations is expected in such a culture.

Second, motivational factors (e.g., trust) can influence individual and organizational behaviors. For example, Weick et al. (1999) argue that the relationships between individuals and organizations based on trust are characterized by strong ties. These strong ties lead to a more cooperative attitude towards spatial data sharing. Another important reason for adding motivational factors is that the cultural dimensions of Hofstede probably do not explain all relations. The exact relationship between Hofstede's cultural dimensions and motivational factors on the one hand and the variables in the model on the other has not yet been empirically tested. Hypotheses on the nature of these relations are discussed below.

THEORETICAL FOUNDATION AND HYPOTHESIS DEVELOPMENT

The proposed model and the hypotheses are presented in figure 4. The theoretical foundation and hypothesis development are presented in the next part based on Hofstede's cultural dimensions and motivational factors.

Hofstede's cultural dimensions. The individualism/collectivism dimension represents a continuum. Hofstede and Hofstede (2005) explain that in an individualist society people are expected to look after themselves. In contrast, a collectivist society finds people integrated into strong, cohesive groups. Hofstede and Bond (1988) demonstrate that collectivistic societies have strong relations within in-groups. In-group relations focus on maintaining harmony (Bond and Smith 1996). Once collectivistic societies have established a positive attitude toward data sharing, they tend to internalize it and take it into their in-group circle. Pavlou and Chai (2002) found that the relationship between attitude and transaction intention is stronger in collectivist societies than in individualist societies. Thus, we would expect that a higher level of collectivism leads to a more positive attitude towards SDS. **Hypothesis 1:** The positive relationship between attitude and the intention for SDS is stronger in collectivist cultures than in individualist cultures.

The intentions of people to engage in data sharing are a function of societal norms and social influence. Hofstede (1991) argues that members of individualistic societies prefer self-sufficiency while those in collectivistic cultures acknowledge their interdependent nature and obligations to the group. Hofstede and Hofstede (2005) indicate that an individualist culture is one in which the ties between individuals are loose. **Hypothesis 2:** The positive relationship between

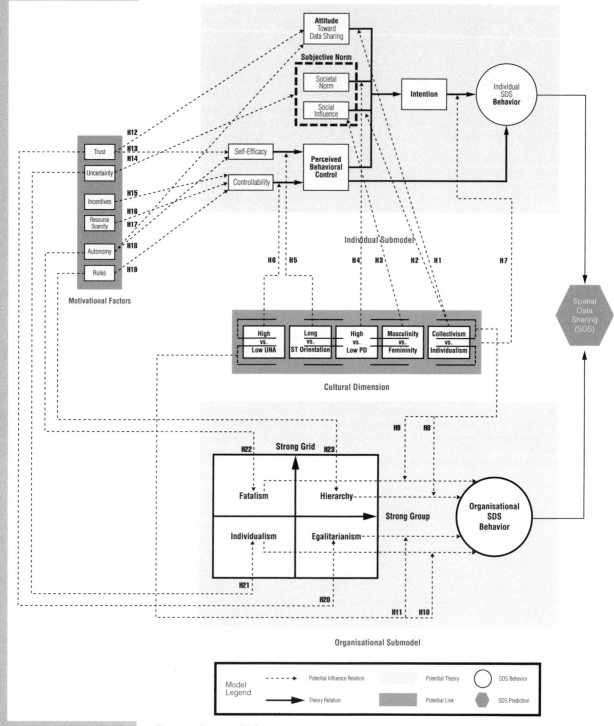

Figure 4. Detailed SDS model.

social norms and the intention for SDS is stronger in collectivist cultures than in individualist cultures.

The cultural dimension of masculinity/femininity relates to one's self-concept: who am I, and what is my task in life? A society is called masculine when emotional gender roles are clearly distinct. In feminine cultures, emotional gender roles overlap (Hofstede and Hofstede 2005). We see the influence of cultural masculinity in the emphasis on competitiveness and SDS success. In highly masculine environments, individuals are driven toward cooperation and innovation in order to prove their worthiness. This creative energy can be expected to result in higher levels of SDS. Chiasson and Lovato (2001) report that a subjective (social) norm is a significant antecedent of the intention for information system adoption. The higher the level of cultural masculinity, the higher the intention for SDS. **Hypothesis 3:** The positive relationship between social norms and the intention for SDS is stronger in masculine cultures.

The second relevant cultural dimension is power distance (PD), which is the extent to which people accept a hierarchical system with an unequal power distribution. In cultures high in power distance, SDS decisions are made by superiors without consulting their subordinates, and employees fear disagreements with their superiors (Hofstede 1980, Hofstede and Hofstede 2005). Superiors tend to be autocratic, and subordinates willingly do as they are told (Hofstede 1991). Thus, PD is closely related to societal influence. Cultures higher in PD are likely to impede SDS. Lower-level employees tend to wait for instructions. In contrast, cultures low in power distance have a more cooperative relationship between superiors and subordinates. Pavlou and Chai (2002) found that the relationship between subjective norm and online transaction intention is stronger in cultures with high power distance. Thus, high PD can be expected to result in lower levels of SDS. **Hypothesis 4:** The negative relationship between societal influence and the intention for SDS is stronger in cultures with high power distance.

Ajzen suggests (1991) that PBC reflects beliefs regarding access to resources and opportunities required to facilitate a behavior and emphasizes (2002) that PBC denotes a subjective degree of control over a behavior (e.g., the perceived ease or difficulty of sharing data). Mathieson (1991) showed that behavioral control influences the intention to use an information system. A positive relationship between control and intentions was found by Taylor and Todd (1995) for users in a computer resource center. Pavlou (2002) found the same results for e-commerce behavior. High PBC should have a positive effect on SDS intentions—since individuals do not fear opportunistic behavior from bosses—and is likely to reduce barriers to SDS.

According to Hofstede and Hofstede (2005), LTO plays an important role in day-to-day decisions, giving people more control over their actions. Cultures with LTO focus on future rewards. Pavlou and Chai (2002) found that the positive relationship between perceived behavior control and transaction intention is stronger in societies characterized by long- versus short-term orientation. Therefore, an LTO environment would foster the intention for SDS. The higher the level of LTO, the higher the intention for SDS. **Hypothesis 5:** The positive relationship between PBC (self-efficacy) and the intention for SDS is stronger in long-term-oriented cultures.

Uncertainty avoidance (UNA) is "related to anxiety, need for security and dependence upon experts" (Hofstede 1980). Under conditions of high levels of uncertainty, individuals avoid unfamiliar situations and tend to develop a conservative attitude. A culture that is high in uncertainty avoidance would exhibit a rule orientation and employment stability. In such a society, change and innovation are not valued. SDS would not be sought or welcomed. As a result, individuals are likely to have no incentive to share spatial data. Individuals feel that "what is different is dangerous." **Hypothesis 6:** The negative relationship between PBC (controllability) and the intention for SDS is stronger in cultures characterized by high uncertainty.

All of the above cultural dimensions influence an individual's intention for SDS. Ajzen (1988, 1991) assumed that intention captures the motivational factors that influence behavior, which indicate how much effort individuals plan to exert to perform the behavior. Cultures high in individualism are likely to value personal time and personal accomplishments, whereas cultures high in collectivism value group integration more than individual desires. Collectivist cultures believe that it is best for the individual if the group is cohesive (Hofstede 1980; Hofstede and Hofstede 2005). In addition, cultures with high PD are likely to impede SDS by weakening the two-way communication between individuals that is necessary for high levels of SDS. In high-PD cultures, employees tend to wait for instructions from managers, who do not welcome innovative ideas about data sharing from below. In contrast, low-PD cultures allow for a more participative and egalitarian relationship between superiors and subordinates. Karahanna et al. (1999) found that the high intention of top management, and supervisors significantly influenced adoption of technology. So, the stronger the intention to engage in an SDS, the more likely should be its achievement. **Hypothesis 7:** The positive relationship between intention and SDS behavior is stronger in cultures high in collectivism, masculinity, and LTO and low in PD and uncertainty avoidance.

Thompson et al. (1990) propose that any organizational setting falls into one of four types: hierarchy (strong grid/strong group), egalitarianism (strong group/ weak grid), individualism (weak group/weak grid), and fatalism (strong grid/ weak group). In a hierarchy, an individual has strong binding internal regulations and strong group boundaries. In individualism, members have a loose personal network and no strong binding to any group. An egalitarian organization is a closed sectarian community that has elaborate rules for keeping individuals equal (Rayner 1988); because of strong boundaries between groups, members have no external contacts other than in or via the group. In fatalism, individuals have fewer social resources for participation, and the isolation creates dependency on others (Gross and Rayner 1985). SDS behavior depends on organizational culture. **Hypothesis 8:** The negative relationship between hierarchical organizations and SDS is stronger in cultures low in collectivism, masculinity, and LTO and high in PD and UNA. **Hypothesis 9:** The negative relationship between fatalistic organizations and SDS is stronger in cultures low in collectivism, masculinity, and LTO and high in PD and UNA. **Hypothesis 10:** The positive relationship between individualistic organizations and SDS is stronger in cultures low in individualism, femininity, PD, and UNA and high in LTO. **Hypothesis 11:** The positive relationship between egalitarian organizations and SDS is stronger in cultures low in individualism, femininity, PD, and UNA and high in LTO.

Motivational factors. Trust in data sharing is a behavioral belief that directly influences attitude, and it indirectly affects behavioral intentions for SDS. The relationship between trust and attitude can be explored by viewing trust from the perspective of TPB as a behavioral belief (Pavlou 2002). Trust is related to positive feelings, beliefs, and attitudes (McKnight and Chervany 2002; Adobor 2005, 2006). Trust creates positive feelings towards SDS. Moreover, trust in SDS creates confidence in the behavior of another party. Trust does not directly influence control through self-efficacy (SE), but it can be a facilitating condition. Bandura (1986) defines SE as individual judgment of a person's capabilities to perform a behavior. Self-efficacy beliefs could influence choice of activities, effort expended, as well as thought patterns and emotional reactions (Bandura 1982, 1991). The concept of SE can be applied to an individual's judgment of his capabilities to engage in SDS. Trust gives the individuals perceptual resources (trust beliefs) to gain control over their activities. A belief that a person will behave in accordance with expectations is likely to increase SDS behavior. **Hypothesis 12:** Trust positively influences favorable attitude toward SDS. **Hypothesis 13:** Trust positively influences perceived behavioral control for SDS.

According to Hofstede (1980), some cultures foster greater uncertainty in people than others do. Societal rules, rituals, religious orientations, and technologies are cultural forces that shape an individual's response to uncertainty. The more uncertain the task, the harder it is to schedule work activities in advance and the greater the reliance on ad hoc arrangements. Smith (1973) points out that social influence plays a role as people seek to reduce uncertainty. Oliver (1990) and Pfeffer and Salancik (1978) argue that individuals and organizations try to establish relationships in order to achieve stability. **Hypothesis 14:** Uncertainty positively influences subjective norms for the intention for SDS.

SDS is encouraged where an incentive for sharing exists. This argument captures the question frequently asked before a person makes a commitment: "What's in it for me?" (Pinto and Onsrud 1995). From this perspective, an organization or its key members must expect a payment or some other incentives for the establishment of an SDS relationship. Craig (1995) sees "institutional inertia" as a major problem. If everyone is focused on the mission and mandates of the agency, there may be no incentives for activities like sharing data. So, the willingness of an organization to participate in SDS is directly related to the perceived reward (e.g., money, access to data, and so forth). Economic exchange relationships between organizations can stimulate SDS. **Hypothesis 15:** Incentives for individuals have a positive influence on SDS.

Ajzen (2002) defined controllability as individual judgment about the availability of resources and opportunities to perform the behavior. Resource scarcity motivates individuals and organizations to cooperate with one another. When resources are scarce and organizations are unable to generate them, the organizations are more likely to establish ties with each other (Molnar 1978). Pfeffer and Salancik (1978) argue that resource scarcity prompts organizations to attempt to exert power, influence, or control over organizations that possess the required scarce resources. Thus, perceived resource scarcity is likely to influence the intention for SDS in a positive way. **Hypothesis 16:** Perceived resource scarcity has a positive influence on the intention for SDS.

Any decision to engage in SDS influences the autonomy of the stakeholders. Organizational reluctance to share data due to loss of autonomy and control over information sources and organizational power is widely acknowledged (Azad and Wiggins 1995; Meredith 1995; Provan 1982). Spatial data can be viewed as a form of power. Individuals and organizations are less likely to share their data if they are losing power in the relationship. **Hypothesis 17:** Autonomy negatively influences attitudes towards SDS. **Hypothesis 18:** Autonomy negatively influences perceived behavioral control for SDS.

Enhancement of organizational legitimacy has been cited as a motivation for organizations to cooperate. Galbraith and Nathanson (1978) demonstrate that rules and procedures are central to any interorganizational cooperation. McCann and Galbraith (1981) also discuss rules and procedures as techniques for coordinating activities, controlling behavior, and maintaining organizational structure. Ruekert and Walker (1987) report that written or formalized rules and procedures have a significant positive relationship with the perceived effectiveness of organizational relations. **Hypothesis 19:** Organizational rules positively influence perceived behavioral control for SDS.

Organizational trust is "the subjective belief with which a population of organizations performs transactions according to their confident expectations" (McKnight and Chervany 2002; Bhattacharya et al. 1998; Doney and Cannon 1997). Trust is a driver for cooperation (Morgan and Hunt 1994; Adobor 2005, 2006) and contributes to organizational performance by enabling people to share valuable information with each other (Mayer et al. 1995; Kramer and Tyler 1996). Tulloch and Harvey (2006) argue that institutions share data with people they know and trust. The groups have strong boundaries between them, and individuals have no external contacts other than in or via the group (egalitarian structure). **Hypothesis 20:** Trust positively influences egalitarian organizations to share spatial data.

Organizations have different objectives when they participate in interorganizational relationships, and these relationships can therefore take different forms (Bensaou and Venkatraman 1995; Grandori 1997). Uncertainty can affect organizational relationships by keeping institutions small and stimulating organizational individualism. Individualistic organizations have loose personal networks, without strong binding to any group. Bradley and Nolan (1998) argue that the high pace of change has pressured organizations to cooperate more and demands more rapid information sharing. **Hypothesis 21:** Uncertainty positively influences individualistic organizations to share spatial data.

Autonomy limits relations between organizations (fatalism). Fatalists operate in isolation, and as a consequence they have a more negative attitude towards data sharing (Gross and Rayner 1985). Organizational reluctance to share data due to a fear of losing autonomy and control over information sources is widely acknowledged (Pinto and Azad 1994; Meredith 1995). **Hypothesis 22:** Autonomy negatively influences fatalistic organizations in sharing spatial data.

It is important to distinguish between the concept of bureaucratic control and the effects of bureaucracy on SDS. With strong bureaucratic control, organizations tend to become protective and to actually inhibit the flow of information across organizational borders. However, bureaucracy overall may have a positive effect on the sharing of information. Deshpande and Zaltman (1987) and Moenaert and Souder (1990) suggest that increased formalization produces a more harmonious influence on the development of cooperation and information sharing. **Hypothesis 23:** Organizational rules positively influence hierarchical organizations to share spatial data.

DISCUSSION

Many decisions are based on spatial data. The development and maintenance of these data have become large cost components in the use of technology to address today's problems. Billions of dollars are invested annually in producing and maintaining spatial data. Sound spatial decision making often requires integration of spatial datasets. An organization may need access to external spatial data, and data sharing is essential for efficient and effective decision making. Proper functioning of spatial data infrastructures requires a positive attitude towards data sharing. Therefore, understanding the mechanisms behind spatial data sharing is crucial.

Understanding spatial data sharing is much more complicated than simply determining how data created by one organization or individual can be used by other organizations or individuals. Although interactions among strangers on the Web suggest certain models for sharing, in many traditional government and business contexts the sharing of spatial data requires existing relationships. The ability of different individuals and organizations to cooperate determines what spatial data is available.

This paper presents a conceptual model for spatial data sharing and its social and cultural aspects. A model is always an abstraction of reality, and no one model applies equally well to all situations. Quiun (1988) indicates that overemphasizing one model will only lead to failure. Scott (1987, 1992) recommends integration of valuable insights from different theories. The proposed model is based on three theories—TPB, culture theory, and Hofstede's cultural dimensions—which provide valuable insights into SDS.

The model makes a clear distinction between individual and organizational SDS behaviors. The individual and organizational submodels are linked through 5 cultural dimensions and 6 motivational factors. In the model the relations between all the factors are presented in the form of 23 hypotheses. These hypotheses describe expected relations between sociocultural factors and spatial data sharing. The formulation of the relations is based on evidence from the literature and our own reasoning. Some of the formulated hypotheses are clear and well supported by literature, while for others the relations are not so obvious. For instance, the positive effect of trust on spatial data sharing has been documented by many authors; the influence of cultural factors on SDS, however, might not always be as clear as stated in the hypotheses. The hypotheses may need to be reworded, qualified, and retested. Are the proposed relations really there?

A questionnaire designed to test the hypotheses has been administered in Egypt and in the Netherlands. The primary results provide support for most of the proposed hypotheses (Omran et al., submitted for publication), emphasizing the influence of cultural differences and motivational factors on the individual and organizational SDS. The model is valid in Egypt and the Netherlands, where it explains 79 percent and 77 percent, respectively, of the variation in SDS behavior at the individual level. However, at the organizational level, the model explains 39 percent of the variation in SDS behavior in Egypt and 70 percent in the Netherlands. In Egypt, hierarchy and fatalism were the dominant organizational patterns. In the Netherlands, the dominant patterns were hierarchy, individualism, and egalitarianism.

SDS is essential for spatial data infrastructures. Future research should investigate actual SDS behavior and test the validity of the proposed model.

ACKNOWLEDGMENTS

This work is part of El-Sayed Ewis Omran's PhD dissertation and was funded by the Egyptian Ministry of Higher Education and Research (project 51499961/610). Dr. Erik de Man is acknowledged for comments on a draft of this article.

REFERENCES

Adobor, H. 2005. Trust as sensemaking: The microdynamics of trust in interfirm alliances. *Journal of Business Research* 58 (3): 330–37.

Adobor, H. 2006. Optimal trust? Uncertainty as a determinant and limit to trust in interfirm alliances. *Leadership & Organization Development Journal* 27 (7): 537–53.

Ajzen, I. 1985. From intentions to actions: A theory of planned behavior. In *Action-control: From cognition to behavior,* eds. J. Kuhl and J. Beckman, 11–39. Heidelberg, Germany: Springer.

Ajzen, I. 1988. *Attitudes, personality, and behavior.* Milton Keynes, UK: Open University Press.

Ajzen, I. 1991. The theory of planned behavior. *Organizational Behavior and Human Decision Processes* 50: 179–211.

Ajzen, I. 2002. Perceived behavioral control, self-efficacy, locus of control, and the theory of planned behavior. *Journal of Applied Social Psychology* 32: 665–83.

Azad, B., and L. Wiggins. 1995. Dynamics of inter-organizational geographic data sharing: A conceptual framework for research. In *Sharing geographic information,* eds. H. Onsrud and G. Rushton, 22–43. New Brunswick, NJ: Center for Urban Policy Research.

Bandura, A. 1982. Self-efficacy mechanism in human agency. *American Psychology* 37: 122–47.

Bandura, A. 1986. *Social foundations of thought and action: A social cognitive theory.* Englewood Cliffs, NJ: Prentice Hall.

Bandura, A. 1991. Social cognitive theory of self-regulation. *Organization Behavior and Human Decision Processes* 50.

Bensaou, M., and N. Venkatraman. 1995. Configurations of inter-organizational relationships: A comparison between U.S. and Japanese automakers. *Management Science* 41 (9): 1471–92.

Bhattacharya, R., T. Devinney, and M. Pillutla. 1998. A formal model of trust based on outcomes. *Academy of Management Review* 23 (3): 459–72.

Bond, R., and P. Smith. 1996. Cross-cultural social and organizational psychology. *Annual Review of Psychology* 47: 205–35.

Bradley, S. P., and R. L. Nolan, eds. 1998. *Sense and respond: Capturing value in the network Era.* Boston: Harvard Business School Press.

Buchanan, D. A. 1993. The organizational politics of technological change. In *Human factors in geographical information systems,* eds. D. Medyckyj-Scott and H. M. Hearnshaw, 211–22. London: Belhaven Press.

Calkins, H., and R. Weatherbe. 1995. *A case study approach to the study of institutions sharing spatial data.* Washington, DC: Urban and regional information systems association (URISA) proceedings.

Calkins, J., E. Epstein, J. Estes, H. J. Onsrud, J. Pinto, G. Rushton, and L. Wiggins. 1991. Sharing of geographic information: Research issues and a call for participation. I-9 Specialist Meeting Report. National Center for Geographic Information and Analysis, Santa Barbara, CA.

Chiasson, M. W., and C. Y. Lovato. 2001. Factors influencing the formation of a user's perceptions and use of a DSS software innovation. *Database for Advances in Information Systems* 32 (3): 16–35.

Coulter, K. S., and R. A. Coulter. 2002. Determinants of trust in a service provider: The moderating role of length of relationship. *Journal of Services Marketing* 16 (1): 35–50.

Craig, W. J. 1995. Why we can't share data: Institutional inerta. In *Sharing geographic information,* eds. H. Onsrud and G. Rushton, 107–18. New Brunswick, NJ: Center for Urban Policy Research.

Deshpande, R., and G. Zaltman. 1987. A comparison of factors affecting use of marketing information in consumer and industrial firms. *Journal of Marketing Research* 21:114–18.

Doney, P. M., and J. P. Cannon. 1997. An examination of the nature of trust in buyer-seller relationships. *Journal of Marketing* 61 (1): 35–51.

Douglas, M. 1970. *Natural symbols: Explorations in cosmology.* London: Routledge.

Douglas, M. 1978. *Cultural bias.* London: Royal Anthropological Institute.

Dueker, K., and R. Vrana. 1995. Systems integration: A reason and a means for data sharing. In *Sharing geographic information,* eds. H. Onsrud and G. Rushton, 149–71. New Brunswick, NJ: Center for Urban Policy Research.

Eason, K. D. 1993. Planning for Change: Introducing a geographical information system. In *Human factors in geographical information systems,* eds. D. Medyckyj-Scott and H. M. Hearnshaw, 199–210. London: Belhaven Press.

Ford, D., C. Catherine, and M. Darren. 2003. Information systems research and Hofstede's culture's consequences: An uneasy and incomplete partnership. *IEEE transactions on engineering management* 50 (1): 8–25.

Galbraith, J., and D. Nathanson. 1978. *Strategic implementation: The role of structure and process.* Dallas: Business Publications.

Grandori, A. 1997. An organizational assessment of interfirm coordination modes. *Organization Studies* 18 (6): 897–925.

Gross, J., and S. Rayner. 1985. *Measuring culture: A paradigm for the analysis of social organization.* New York: Columbia University Press.

Hofstede, G. 1980. *Culture's consequences.* Beverly Hills: Sage Publications.

Hofstede, G. 1991. *Cultures and organizations: Software of the mind.* Berkshire, UK: McGraw-Hill.

Hofstede, G. 2001. *Culture's consequences: Comparing values, behaviors, institutions and organizations across nations,* 2nd ed. Thousand Oaks: Sage Publications.

Hofstede, G., and M. Bond. 1988. The Confucius connection: From cultural roots to economic growth. *Organizational Dynamics* 16:4–21.

Hofstede, G., and G. J. Hofstede. 2005. *Cultures and organizations: Software of the Mind,* 2nd ed. New York: McGraw-Hill.

Karahanna, E., D. W. Straub, and N. L. Chervany. 1999. Information technology adoption across time: A cross-sectional comparison of pre-adoption and post-adoption beliefs. *MIS Quarterly* 23 (2): 183–213.

Kevany, M. J. 1995. A proposed structure for observing data sharing. In *Sharing geographic information,* eds. H. Onsrud and G. Rushton, 76–100. New Brunswick, NJ: Center for Urban Policy Research.

Kramer, R. M., and T. R. Tyler. 1996. *Trust in organizations: Frontiers of theory and research.* London: Sage.

Mathieson, K. 1991. Predicting user intentions: Comparing the technology acceptance model with the theory of planned behavior. *Information Systems Research* 2 (3): 173–91.

Mayer, R. C., J. H. Davis, and F. D. Schoorman. 1995. An integrative model of organizational trust. *Academy of Management Review* 20 (3): 709–34.

McCann, J., and R. Galbraith. 1981. Interdepartmental relations. In *Handbook of organizational design,* eds. P. C. Nystrom and W. H. Starbuck, 60–84. New York: Oxford University Press.

McKnight, D. H., and N. L. Chervany. 2002. What trust means in e-commerce customer relationships: An interdisciplinary conceptual typology. *International Journal of Electronic Commerce* 6 (2): 35–72.

Medyckyj-Scott, D., and H. M. Hearnshaw, eds. 1993. *Human factors in geographical information systems.* London: Belhaven Press.

Meredith, P. 1995. Distributed GIS: If its time is now, why is it resisted? In *Sharing geographic information,* eds. H. Onsrud and G. Rushton, 7–21. New Brunswick, NJ: Center for Urban Policy Research.

Miles, R. W., and C. C. Snow. 1992. Causes of failure in network organizations. *California Management Review* 34 (4): 53–72.

Moenart, R., and E. Souder. 1990. An analysis of the use of extra functional information by R&D and marketing personnel: Review and model. *Journal of Product Innovation Management* 7:91–107.

Molnar, J. 1978. Comparative organizational properties and interorganizational interdependence. *Sociology and Social Research* 63:24–48.

Nakata, C., and K. Sivakumar. 2001. Instituting the marketing concept in a multinational setting: The role of national culture. *Journal of the Academy of Marketing Science* 29 (3): 255–75.

Nedovic-Budic, Z., J. Pinto, and L. Warnecke. 2004. GIS database development and exchange: Interaction mechanisms and motivations. *URISA journal* 16 (1): 15–29.

Obermeyer, N. J. 1995. Reducing inter-organizational conflict to facilitate sharing geographic information. In *Sharing geographic information,* eds. H. Onsrud and G. Rushton, 138–48. New Brunswick, NJ: Center for Urban Policy Research.

Oliver, C. 1990. Determinants of inter-organizational relationships: Integration and future direction. *Academy of Management Review* 15 (2): 241–65.

Omran, El-Sayed E., A. Bregt, and J. Crompvoets. Spatial data sharing: Test and validation of cross-cultural model. Submitted for publication.

Omran, El-Sayed E., J. Crompvoets, and A. Bregt. 2006. Benefits and bottlenecks for SDI development in Egypt. *GIS Development*, 2, no. 1 (January/February): 32–35.

Park, S. H., and G. Ungson. 1997. The effect of partner nationality, organizational dissimilarity, and economic motivation on the dissolution of joint ventures. *Academy of Management Journal* 39:279–307.

Pavlou, P., and L. Chai. 2002. What drives electronic commerce across cultures? A cross-cultural empirical investigation of the theory of planned behavior. *Journal of Electronic Commerce Research* 3 (4): 240–53.

Pavlou, P. A. 2002. What drives electronic commerce? A theory of planned behavior perspective. *Best Paper Proceedings of the Academy of Management Conference*, Denver, CO. August 9–14.

Pfeffer, J., and G. Salancik. 1978. *The external control of organizations: A resource dependence perspective.* New York: Harper and Row.

Pinto, J. K., and H. J. Onsrud. 1995. Sharing geographic information across organizational boundaries: A research framework. In *Sharing geographic information*, eds. H. Onsrud and G. Rushton, 44–64. New Brunswick, NJ: Center for Urban Policy Research.

Porter, M. E. 1987. From competitive advantage to corporate strategy. *Harvard business review* 65:43–59.

Provan, K. 1982. Interorganizational linkages and influence over decision making. *Academy of Management Journal* 25:443–51.

Quiun, E. R. 1988. *Beyond rational management.* San Francisco: Jossey-Bass.

Rajabifard, A., and I. Williamson. 2004. Regional SDI development: A fundamental framework. *Journal of Geospatial Today* 2 (5).

Rayner, S. 1988. The rules that keep us equal: Complexity and the costs of egalitarian organization. In *Rules, decisions and inequality in egalitarian societies*, eds. J. Flanagan and S. Rayner, 20–42. Aldershot: Averbury.

Ruekert, R., and O. Walker. 1987. Interactions between marketing and R&D departments in implementing different business strategies. *Strategic Management Journal* 8:233–48.

Scott, R. W. 1987. *Organizations: Rational, natural, and open systems.* Englewood Cliffs, NJ: Prentice Hall.

Scott, R. W. 1992. *Organizations: Rational, natural and open systems*, 3rd ed. New Jersey: Prentice Hall.

Smith, P., eds. 1973. *Groups within organizations.* London, NY: Harper and Row.

Straub, D., M. Keil, and W. Brenner. 1997. Testing the technology acceptance model across cultures: A three country study. *Information management* 33:1–11.

Tayeb, M. H. 1988. *Organizations and national culture. A comparative analysis.* London, NP: Sage Publications.

Taylor, S., and P. A. Todd. 1995. Understanding information technology usage: A test of competing models. *Information Systems Research* 6 (3): 144–76.

Thompson, M., R. Ellis, and A. Widavsky. 1990. *Cultural theory.* Boulder, Colorado: Westview Press.

Tjosvold, D. 1988. Cooperative and competitive dynamics within and between organizational units. *Human Relations* 41:425–36.

Tulloch, D., and F. Harvey. 2006. When data sharing becomes institutionalized: Best practices in local government GI relationships. *URISA journal.* (In press.)

Warnecke, L., J. Beattie, K. Cheryl, W. Lyday, and S. French. 1998. *Geographic information technology in cities and counties: A nationwide assessment.* Washington, DC: American Forests.

Wehn de Montalvo, U. 2001. Crossing organizational boundaries: Prerequisities for spatial data sharing in South Africa. D. Phil. thesis, SPRU-Science and Technology Policy Research, University of Sussex, Brighton.

Wehn de Montalvo, U. 2003a. In search of rigorous models for policy-oriented research: A behavioral approach to spatial data sharing. *URISA Journal* 15 (APA I).

Wehn de Montalvo, U. 2003b. *Mapping the determinants of spatial data sharing.* Aldershot, England: Ashgate Publishing Ltd.

Weick, K. E., K. Sutcliffe, and D. Obstfeld. 1999. Organizing for high reliability. In *Research in organization behavior,* eds. R. Sutton and B. Staw, 81–123. Stamford, Connecticut: JAI Press.

INSPIRE: An Innovative Approach to the Development of Spatial Data Infrastructures in Europe

MAX CRAGLIA AND ALESSANDRO ANNONI

EUROPEAN COMMISSION, DG JOINT RESEARCH CENTRE, INSTITUTE FOR ENVIRONMENT AND SUSTAINABILITY, SPATIAL DATA INFRASTRUCTURES UNIT, ISPRA, ITALY

ABSTRACT

First-generation spatial data infrastructures (SDIs) were product oriented and focused on databases. Second-generation SDIs are more process oriented and emphasize partnerships and stakeholder involvement. Nevertheless, most spatial data infrastructures are still led by public-sector organizations, with limited involvement by the private sector or society at large and the involvement of user groups often sporadic and poorly organized. INSPIRE, a spatial data infrastructure for Europe, is innovative in two respects: (1) it is based on existing resources at the national and subnational levels, and (2) it engages user communities and geographic information stakeholders by organizing them in spatial data interest communities. This approach poses challenges at both technical and organizational levels but also offers important opportunities for sustainability.

INTRODUCTION

Development of spatial data infrastructures (SDIs) across the world has shifted from the first generation, which was product oriented and focused on the development or completion of databases, to the second generation, which is more process oriented and emphasizes partnerships and stakeholder involvement. With respect to coordination activities, which are crucial to the development and management of an SDI, the first generation was largely led by the national mapping agencies while the second generation has seen an increasing role of organizational models which are often independent of the mapping agencies and seek to be more representative of the stakeholder communities.

Despite the shift from a product-centered to a process-centered approach, most spatial data infrastructures are still led by public-sector organizations, with limited involvement by the private sector or society at large and often sporadic and poorly organized involvement of user groups. The spatial data infrastructure for Europe INSPIRE, proposed by the European Commission, is innovative in two respects: (1) it is based on existing resources at the national and subnational levels, and (2) it engages user communities and geographic information stakeholders by organizing them in spatial data interest communities. This approach poses challenges at both technical and organizational levels but also offers important opportunities for sustainability.

The paper describes the nature of SDI initiatives and the shift from first- to second-generation SDIs; evaluates the findings of a recent survey of SDIs in 32 European countries; examines the development of INSPIRE with a focus on spatial data interest communities; and discusses opportunities and challenges.

THE CHANGING NATURE OF SDIs

SDI developments have been well documented by Masser (1999, 2005), Williamson et al. (2003), Craglia et al. (2003), Vandenbroucke (2005), and Crompvoets and Bregt (2003). While there are many definitions of SDIs, a useful framework is the one put forward by Rajabifard et al. (2003) (figure 1), which places particular emphasis on the dynamic relationship between data, people, and technology-policy standards. The authors argue that the relationship between these categories is dynamic because communities require access to different sets of data mediated by the ever-changing technology. The interactions among these components in turn put new demands on rights, restrictions, and responsibilities enshrined in policy, as there is a constant need for interpreting and responding to political and technological changes and new user needs, which may have been unforeseen at the initial stage of development. While information systems that respond to clearly defined tasks and user groups within organizations have well-documented challenges (Jirotka and Goguen 1994; Dittrich et al. 2002), additional challenges exist for SDIs because their Internet-based nature makes identifying user communities and responding to their needs more difficult. This is also why coordination is so critical. Without effective coordination, different components—reference data, metadata, clearinghouses—may be in place without a cohesive whole.

International experiences in 2002 and 2003 (Craglia et al. 2003, p. 240) indicated that coordination is one of the most important aspects in the development of an SDI. Coordination involves:

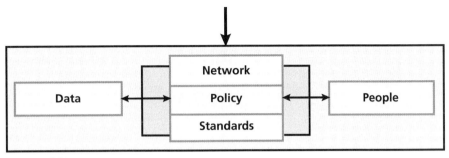

Figure 1. SDI components.
Reprinted from Rajabifard et al. 2003 with permission of Taylor and Francis Books.

- Leadership
- Mediating interagency conflicts
- Sustaining political support
- Selling the benefits to multiple audiences
- Providing technical guidance and enforcement of common standards
- Raising awareness and disseminating the results

In addition, coordination can also play a very useful role in identifying gaps or inconsistencies in the legal and organizational framework, and suggesting remedial action to the government. This is particularly important as the legal framework within which SDIs operate is strongly affected by many other policy areas, such as Public Sector Information legislation, Freedom of Information, international conventions (e.g., Aarhus), competition law, and so on. Moreover, all these areas of policy may have some variation not only at national but also across subnational levels.

The more dynamic the social, political, and technological environment in which the SDI is embedded and the more distributed the framework upon which it is built, the greater the need for coordination.

A complementary perspective for analyzing the importance of coordination comes from a review on the diffusion of SDIs and their evolving nature in responding to social and technological changes. Masser (2005) uses the classical diffusion model originally put forward by Rogers, who defines diffusion as "the process by which an innovation is communicated through channels over time among the members of the social system" (Rogers 1995). This model takes the form of a bell-shaped curve, with innovators and early adopters at the beginning of the process, followed by an early majority, late majority, and laggards at the tail end. Masser argues that the 11 SDIs he reviewed in 1998 represent the group of innovators and early adopters: Canada and the United States in North America; Qatar, Indonesia, Japan, Australia, and Korea in Asia; and the Netherlands, Portugal, and United Kingdom in Europe. In spite of their differences in size, wealth, scope, and organization, this group was defined as representing the first generation of SDIs, characterized as having a specifically national focus, an emphasis on the development of databases, and often (but not always) leadership provided by national mapping agencies.

The transition towards the second generation of SDIs, or the early majority in the Rogers model, is placed by Rajabifard et al. (2003) around the year 2000, when a consolidation of the SDI community (the social system in Rogers' definition) took place with the establishment of a series of international conferences, the publication of shared experiences in the SDI Cookbook, and the strengthening of channels of communication that enable diffusion. The second generation of SDIs is characterized by an increasing recognition of geographic information stakeholders within society. Hence, the emphasis moves from the development of products toward a process involving partnerships, agreements, and a broader set of applications. Instead of being led by data producers, second-generation SDIs use organizational models designed to be representative of the different stakeholders. It could be argued that the sharing of experiences and consolidation of technologies and standards such as ISO and OGC have also freed newcomers from having to worry about technology and enabled them to pay more attention to the institutional and organizational arrangements. Table 1 summarizes the key features of the two generations of SDIs.

Comprehensive studies of SDIs in 32 European countries were conducted by the University of Leuven in 2003, 2004, and 2005 with funding from the European Commission. In his overview of the 2005 results, Vandenbroucke identified the following key features:

1. Increasing regional and local contributions to national SDIs

2. Greater involvement of stakeholders other than the main data producers

3. Increasing adoption of international standards and specifications (ISO, OGC) and availability of Web-based services and portals

Characteristic	First generation	Second generation
Nature	Explicitly national	Explicitly national in a hierarchical context and therefore more flexible for cross-jurisdictional collaboration
Goal	Integration of existing data	Establishing linkages between people and data
Expected outcome	Linkage into a seamless database	Knowledge infrastructures, interoperable data and resources
Participants	Mainly data providers	Cross-sectoral: data providers, integrators, users
Funding	Mainly no specific or separate budget	Mostly included in national mapping programs or having separate budget
Coordinating agency	Mainly national mapping organizations	More independent organizational committees/ partnership groups
Public awareness	Low at the beginning, then growing gradually	High
Capacity building	Very low	Communities are more prepared to engage in ongoing activities
Number of SDI initiatives	Very low	High
Model	Predominantly product based	Increasingly process-based or product-process hybrid
Relationship with other SDI levels and international initiatives	Low	High
Value measurements	Productivity, savings	Holistic sociocultural value versus the expense of not having an NSDI

Table 1. Key features of the two generations of SDIs.
Reprinted from Rajabifard et al. 2003 with permission of Taylor and Francis Books.

Vandenbroucke also notes that "the large majority of countries do not yet have an integrated approach in which the tasks for building and maintaining the NSDI are well defined and divided amongst the different stakeholders" (page 12) and that "one of the conclusions that can be drawn . . . is that clear mandates for building (parts) of the components of the NSDI are often lacking or that some mandates are rather fuzzy in relation to the NSDI" (page 1).

Vandenbroucke classifies the countries surveyed into two groups. In the first group, a national data provider (national mapping and/or cadastral agency) is the officially mandated or *de facto* leading organization for the establishment of the NSDI, with a further subdivision based on whether users are involved. In the second group, NSDI initiatives are led by a council of ministries or administrative departments, by a (nongovernmental) GI association, or by a partnership of data users; this group is further subdivided on the basis of the presence of a formal mandate for SDI coordination (*ibid.*, page 15).

This classification mirrors to a large degree the two generations identified by Masser (2005) and Rajabifard et al. (2003). We can reclassify the 32 European countries in Vandenbroucke's study into 3 groups of similar sizes: first generation (data producer led, users not involved), second generation (user led), and intermediate (users are involved but do not lead the process).

While the generational view is helpful for understanding the evolution of SDIs and the new challenges, Rogers himself recognized the pro-innovation bias of his model, in other words, the tendency to assume a linear transition between one stage and the next. This is clearly not the case for infrastructures such as SDIs, which are embedded in social and political processes. For example, the Portuguese infrastructure launched on the Internet in 1995 experienced very limited progress for almost five years due to budgetary constraints and a reorganization of the coordinating structures (Juliao 2005). Similarly, the United Kingdom SDI (classified as first generation by Masser [1999]) has suffered setbacks and is now fragmenting, with independent strategies for Scotland, Wales, and Northern Ireland and with England struggling to define its own strategy due to lack of political leadership (Masser 2005). Even in the United States, which was seen by many as a leading example, particularly because of the high-level political commitment in President Clinton's Executive Order, progress has been difficult at times. Engaging state and local jurisdictions as well as the private sector in the development of a truly national SDI has been particularly problematic (Urban Logic 2000; National Research Council 2001). For example, Harvey and Tullock (2003) report that almost half of the local organizations contacted did not know about the national SDI and did not rely on any standard for their geospatial activities. The authors conclude that a data-centric approach is unlikely to succeed and that much more emphasis is needed on establishing and supporting social networks.

The above review of the literature indicates that the changing nature of SDIs requires greater involvement of stakeholders and user communities, not just from the public sector but also from the private sector and society at large. More effort is therefore needed in building and maintaining social networks, understanding needs, evaluating effects on society, and delivering results to heterogeneous user groups with often conflicting objectives. Addressing coordination and social contextualization challenges in the multicultural and multilingual context of Europe is particularly difficult and requires a fresh approach as discussed below.

INSPIRE is a directive establishing the legal framework for setting up and operating an Infrastructure for Spatial Information in Europe based on spatial information infrastructures in European Union member states. The purpose of INSPIRE is to support the formulation, implementation, monitoring, and evaluation of European Community environmental policies. The component elements of INSPIRE include:

- Metadata
- Key spatial data themes and services
- Network services and technologies
- Agreements on sharing and access
- Coordination and monitoring mechanisms
- Process and procedures

The general background of INSPIRE was described by Annoni and Craglia (2005) at GSDI-8. Despite significant progress, major barriers remain:

- Inconsistent data collection: spatial data are often missing or incomplete, or the same data are collected by different organizations
- Inadequate documentation: description of available spatial data is often incomplete
- Incompatible datasets: datasets often cannot be combined with other datasets
- Incompatible geographic information initiatives: the infrastructures for accessing spatial data often function in isolation only
- Barriers to data sharing: cultural, institutional, financial, and legal barriers prevent or delay the sharing of spatial data

From the outset of this initiative it was recognized that, to overcome some of the barriers highlighted above, it would be necessary to develop a legislative framework requiring member states to coordinate their activities and agree on a minimum set of common standards and processes. This required wide support of member states for the objectives of INSPIRE. Therefore, a very collaborative process was put in place to formulate the INSPIRE proposal. This process involved the establishment of an expert group composed of official representatives from all member states and working groups composed of experts in environmental policy and geographic information to formulate proposals and forge consensus. The groups agreed to base INSPIRE on the following key principles:

- Spatial data should be collected once and maintained at the level where this can be done most effectively
- It must be possible to seamlessly combine spatial data from different sources across the EU and share it between many users and applications
- It must be possible for spatial data collected at one level of government to be shared between all levels of government
- Spatial data needed for good governance should be available without restrictions on extensive use
- It should be easy to discover which spatial data is available, to evaluate its fitness for purpose, and to know which conditions apply for its use

Ten words and phrases used by the Commission that are not used in the Council draft	Ten words and phrases not used by the Commission but used in the Council draft
Accessibility Commercial activities Common licensing Competition Decision 1692/96/EC25[a] Distortion Focuses[AG1] Harmonized specifications Requisite Rights of use	Apply charges Click licences Corresponding fees Cost-benefit Excessive costs Limit sharing Payment Precondition Reciprocal Viability
[a]European Transport Networks	

Table 2. Comparison between the Council and Commission versions of INSPIRE.
Courtesy of Christopher Edward Henry Corbin.

Following three years of intensive consultation among the member states and their experts, a public consultation, and assessment of likely impacts (see http://www.ec-gis.org/inspire), the European Commission adopted the INSPIRE proposal in July 2004 (CEC 2004). The European Parliament expressed its favourable opinion on the Commission's proposal in June 2005 and introduced clarifying amendments. The European Council in January 2006 introduced limitations to the data-sharing arrangements put forward by the Commission. An analysis of the original proposal and the one adopted by the Council (Corbin 2006) (table 2) clearly showed the difference in emphasis between the two, demonstrating once more that SDIs are strongly embedded in a political process, in which public-sector organizations are funded in different ways in different member states.

A compromise agreement between the Council and European Parliament was reached in November 2006, and the directive was approved in February 2007. The Joint Research Centre of the European Commission and Eurostat have been coordinating the drafting of the implementation rules envisaged by the directive.[1]

Implementation rules are needed for each of the key components of the infrastructure: metadata, data specifications and harmonization, network services, data and service sharing, and monitoring and reporting. Given the political context of the proposal, the drafting of implementation rules requires not only a high level of technical competence but above all the participation and engagement of all the key geographic information stakeholders in Europe. To organize this process, two mechanisms have been put in place. The first is engaging national and subnational European organizations that already have a formal legal mandate for the coordination, production, or use of geographic and environmental information. The second is facilitating the self-organization of stakeholders, including both data providers and users of spatial data, in spatial data interest communities (SDICs) by region, societal sector, and thematic issue. SDICs should naturally form strategic partnerships—public-public, public-private, and private-private—to align the demand for spatial data and services with the necessary investments.

The central role played by SDICs in the development of implementation rules consists of:

- Identifying and describing user requirements (understood as being in line with environmental policy needs, as opposed to "maximum" requirements beyond the scope INSPIRE and beyond realistically available resources)
- Providing expertise to INSPIRE drafting teams
- Reviewing proposed implementation rules
- Developing, operating, and evaluating pilot implementation projects
- Developing initiatives for guidance, awareness raising, and training

In addition, legally mandated organizations (LMOs) play a central role in reviewing and testing proposed implementation rules and in assessing potential costs and benefits.

An open call was announced on March 11, 2005 for the registration of interest by SDICs and LMOs, which were also asked to put forward experts and reference materials for preparing implementation rules. By April 29, 2005, the INSPIRE Web site recorded the following registrations:

- Spatial data interest communities: 133
- Legally mandated organizations: 82
- Proposed experts: 180
- Referenced materials: 90
- Identified projects: 91

The majority of LMOs are national in character and dominated by producers of reference data (figure 2). SDICs, on the other hand, characterize themselves primarily as research organizations and GIS coordinating bodies, with each SDIC bundling together many organizations representing different viewpoints and interests.

An example of a regional SDIC is GDI NRW, which is a nonprofit initiative of the state of North Rhine–Westphalia, Germany, in which representatives from the business, government, and science communities work together in a public–private partnership as geoinformation providers, enablers, brokers, and users. GDI NRW has more than 100 members, including various state authorities in North Rhine–Westphalia (land surveying office, geological survey, ministry of the environment, etc.), about 20 local authorities, several research institutions, and multiple companies.

SDICs organized by thematic issue include the European Soil Bureau Network (ESBN), the European environment information and observation network (EIONET), and the European Meteorological Infrastructure (EMI).

ESBN was created in 1996 as a network of national soil science institutions to collect, harmonize, organize, and distribute soil information for Europe. Soil information is used to address leaching of agrochemicals, deposition of heavy metals, disposal of waste, degradation of soil structure, risk of erosion, immobilization of radionuclides, water levels at catchments, assessment of suitability and sustainability for traditional and alternative crops, and estimation of soil stability.

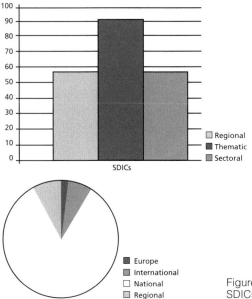

Figure 2. Key features of registered SDICs and LMOs.

EIONET was established in 1990 and aims to provide timely and accurate data, information, and expertise for assessing the state of the environment in Europe and the pressures acting upon it. EIONET connects the European Environment Agency (EEA), a number of European Topic Centres, and a network of about 900 experts from over 300 national and regional environmental agencies in 37 countries.

EMI is an operational infrastructure established by the European National Meteorological Services to deliver information services to decision makers, customers, and users throughout Europe. EMI is part of the World Meteorological Organization's telecommunication system for the European region.

The above examples show not only the thematic breadth of the SDICs but also the extent to which INSPIRE needs to build on existing infrastructures and interact effectively with their architectures, technologies, standards, protocols, and organizational frameworks.

Whereas the communities described above already existed, the INSPIRE open call also led some organizations to create new groups of interest or to join existing ones. As a consequence, user groups not previously considered by data producers in strategic decisions now have greater influence in defining priorities and needs.

With the growing awareness of INSPIRE, the numbers of SDICs and LMOs increased to 160 and 98, respectively, as of March 2006 and the numbers of registered projects and reference materials almost doubled.

Of significant interest is the high number of experts (180) proposed by the SDICs and the LMOs. Considering that the experts are not paid by the European Commission but are supported instead by the organizations and communities that nominated them, these figures indicate strong support of INSPIRE.

From the large pool of experts available, 70 were selected with the goal of balancing the perspectives of data producers, users, and solution providers from the private sector. With the advice of member state representatives, teams for drafting implementation rules for the following 5 areas were established: metadata, data specifications, network services, data and service sharing, and monitoring and reporting. From the time they began their work in October 2005, the teams have been setting an example of how European Union legislation can be developed with stakeholder contributions.

Several aspects are particularly important in understanding the work of the drafting teams. First, each expert represents a community of interest and therefore has the responsibility to bring to the table the expertise, expectations, and concerns of this community. Second, each drafting team has to reach out to all the thematic communities that are addressed by INSPIRE. This is no small undertaking, as the proposed directive covers more than 30 different data themes which are the responsibility of multiple national, regional, and local agencies across 27 countries. In comparison, the U.S. NSDI defined only 7 framework themes—geodetic control, orthoimagery, elevation, transportation, hydrography, governmental units, and cadastral information—for each of which a designated federal agency takes the lead in data collection and management. The European Union has no supranational institutions in charge of data collection.

The drafting teams have a difficult task in collecting and summarizing reference material, seeking common denominators and reference models, and developing recommendations which satisfy user requirements without imposing undue burden on organizations responsible for data collection and management. Seeking compromise between different requirements and perspectives is crucial to the work of each drafting team. Last but not least, the drafting teams have ownership of their work. They make recommendations and submit them for review to all the registered SDICs and LMOs and the representatives of the member states. Only after all reviewer comments have been taken into account does the European Commission submit a proposed implementation rule to public consultation.

The complexity of this participatory approach is certainly innovative not only for SDIs but for European Union public policy in general. The resulting consensus-based policy and a strong network of stakeholders will be instrumental in the implementation of this distributed SDI.

ANALYSIS AND CONCLUSIONS

First-generation SDIs were product oriented. Process-based initiatives emphasizing partnerships, social networks, and multisectoral collaboration are more common today.

INSPIRE has engaged hundreds of stakeholder organizations across Europe in the drafting of the legislative framework. Involving all interested parties from the very beginning and giving them a role in shaping the infrastructure is in line with best practices and the literature on participatory approaches (Arnstein 1969; Thomas 1996), which emphasize the advantages of real empowerment over mere tokenism. Moreover, a network of stakeholders representing different regions and thematic areas will contribute to INSPIRE's sustainability. The wide-ranging project also faces complex challenges. As with other European legislation, INSPIRE development is a long process, spanning some 15 years from inception to full implementation. Sustaining the momentum, mediating the different interests, coordinating the activities, managing the expectations, and delivering meaningful value to all the stakeholders is a very complex undertaking, particularly in the constantly changing political and technological environment.

In addition to internal organizational challenges (SDICs and LMOs), INSPIRE must also develop effective organizational and technical relationships with existing national and subnational SDIs; other important European initiatives such as e-government, thematic information networks, and infrastructures such as those of the International Hydrographic Organization and the World Meteorological Organization; and global initiatives such as GSDI and GEOSS (Global Earth Observation System of Systems) (http://earthobservations.org/). This will require coordination, identification of synergies, harmonization of data and practices, and interoperability of services.

Last but not least, the drafting of INSPIRE implementation rules needs to balance changing technologies, practices, and requirements across the different geographic and thematic layers with the need to encode agreements into legal text, backed up where necessary by European (CEN) or international (ISO) standards and industrial specifications. The tension between the need to accommodate change and retain flexibility and the need to "freeze" practices and agreements into standards and legislation characterizes all information infrastructures (Hanseth and Lyytinen 2005), but the complexity of the participatory process and the multiplicity of actors, languages, and cultures make these conflicting pressures particularly important for INSPIRE to balance effectively.

In conclusion, we have emphasized the importance of building a modern spatial data infrastructure through a combination of bottom-up participatory approaches across multiple stakeholder communities and careful coordination backed up by a legal framework. Creating a broad social network with empowered stakeholders and building on existing infrastructures, professional practices, and agreements are central features of INSPIRE. This approach entails multiple challenges, which we are striving to address together with our partners.

ENDNOTE

1. A directive is a piece of legislation defining general principles and objectives and allowing member states to determine their own means of reaching these objectives through national legislation. Implementation rules, in contrast, specify technical details mandated by the Commission for all member states to ensure the coherent implementation of the directive.

REFERENCES

Annoni, A., and M. Craglia. 2005. Towards a Directive establishing an infrastructure for spatial information in Europe (INSPIRE). In *Proceedings of GSDI-8 from Pharaohs to Geoinformatics: The role of SDIs in an information society,* Cairo, April 16–25. http://gsdidocs.org/gsdiconf/GSDI-8/papers/ts_47/ts47_01_annoni_graglia.pdf.

Arnstein, S. R. 1969. A ladder of citizen participation. *Journal of American Institute of Planners* 35 (4): 216–24.

Commission of the European Communities. 2004. Proposal for a Directive of the European Parliament and of the Council establishing an infrastructure for spatial information in the Community (INSPIRE). COM (2004) 516 final. Brussels, Commission of the European Communities.

Corbin, C. 2006. INSPIRE Word-Phrase Analysis v. 1. Document posted to the european-gi-policy@jrc.it list on March 2, 2006.

Craglia, M. et al., eds. 2003. *GI in the Wider Europe*. http://www.ec-gis.org/ginie/doc/ginie_book.pdf.

Crompvoets, J., and A. Bregt. 2003. World status of national spatial data clearinghouses. *URISA Journal* 15:43–50.

Dittrich, Y., C. Floyd, and R. Klischewski, eds. 2002. *Social thinking, software practice*. London: MIT.

ESA. 2005. Global Earth Observation System of Systems (GEOSS) 10-Year Implementation Plan. http://earthobservations.org.

Harvey, F., and D. Tulloch. 2003. Building the NSDI at the base: Establishing best sharing and coordination practices among local governments. http://www.tc.umn.edu/~fharvey/research/BestPrac7-03.pdf.

Hanseth, O., and K. Lyytinen. 2005. Theorizing about the design of information infrastructures: Design kernel theories and principles. Paper presented at the 1st workshop on cross-learning research workshop on cross-learning between spatial data infrastructures (SDI) and information infrastructures (II), ITC, Enschede, March 30–31, 2005.

Jirotka, M., and J. A. Goguen, eds. 1994. *Requirements engineering: Social and technical issues*. London: Academic Press.

Juliao, R. P. 2005. Rebuilding an SDI: The Portuguese Experience. Presented at 11th EC-GIS Workshop, Alghero, June 2-July 1. http://www.ec-gis.org/Workshops/11ec-gis/presentations/14juliao.pdf.

Masser, I. 2005. *GIS Worlds: Creating spatial data infrastructures*. Redlands, CA: ESRI Press.

Masser, I. 1999. All shapes and sizes: The first generation of national spatial data infrastructures. *International Journal of Geographical Information Science* 13:67–84.

National Research Council. 2001. National spatial data infrastructure programs: Rethinking the focus. Washington, DC: National Academy Press.

Rajabifard, A., M.-E. F. Feeney, and I. Williamson. 2003a. Spatial data infrastructures: Concept, nature, and SDI hierarchy. In *Developing spatial data infrastructures: from concept to reality*, eds. I. Williamson, A, Rajabifard, and M.-E. F. Feeney, 17–40. London: Taylor & Francis.

Rajabifard, A., M-E. F. Feeney, I. Williamson, and I. Masser. 2003. National SDI-initiatives. In *Developing spatial data infrastructures: from concept to reality*, eds. I. Williamson, A. Rajabifard, and M.-E. F. Feeney, 95–109. London: Taylor & Francis.

Rogers, E. 1995. *Diffusion of Innovations*, 4th ed. New York: Free Press.

Thomas, H. 1996. Public Participation in Planning. In *British planning policy in transition: Planning in the 1990s*, ed. M. Tewdwr-Jones, 168–88. London: UCL Press.

Urban logic. 2000. Financing the NSDI: *National spatial data infrastructure.* www.fgdc.gov.

Vandenbroucke, D. 2005. *Spatial data infrastructures in Europe: State of play Spring 2005.* Report by the Spatial Applications Division, K.U. Leuven Research and Development. http://www.ec-gis.org/inspire/reports/stateofplay2005/rpact05v42.pdf.

Williamson, I., A. Rajabifard, and M.-E. F. Feeney, eds. 2003. *Developing spatial data infrastructures: From concept to reality.* Boca Raton: CRC Press.

Providing Spatial Data Infrastructure Services in a Cross-Border Scenario: The SDIGER Project

F. J. ZARAZAGA-SORIA,[1] JAVIER NOGUERAS-ISO,[1] M. Á. LATRE,[1] A. RODRIGUEZ,[2] E. LÓPEZ,[2] P. VIVAS,[3] AND P. R. MURO-MEDRANO[1]

COMPUTER SCIENCE AND SYSTEMS ENGINEERING DEPARTMENT, UNIVERSITY OF ZARAGOZA, ZARAGOZA,[1] INSTITUTO GEOGRÁFICO NACIONAL,[2] AND CENTRO NACIONAL DE INFORMACIÓN GEOGRÁFICA,[3] MADRID, SPAIN

ABSTRACT

SDIGER is a Eurostat-funded pilot project of the Infrastructure for Spatial Information in Europe (INSPIRE) designed to test proposed solutions, estimate the costs, and identify obstacles. The project is focused on developing an SDI to support access to geographic information resources for the Water Framework Directive in an interadministration, cross-border scenario. The project area consists of the Adour-Garonne and Ebro river basins in France and Spain. Problems identified by the project are discussed.

INTRODUCTION

SDIGER (Spatial Data Infrastructure for Adour-Garonne and Ebro River basins) is a cross-border pilot project of the Infrastructure for Spatial Information in Europe (INSPIRE) that supports the Water Framework Directive (CEC 2004). Funded by Eurostat, the project aims at testing solutions for sharing spatial data and services based on the INSPIRE position papers of 2002, estimating the costs of implementing interoperability-based solutions, and identifying obstacles which might be encountered during the implementation of INSPIRE.

The "call for tender" for this project required the cross-border application to be focused on an environmental subject. The SDIGER project includes the development of a spatial data infrastructure (SDI) to support access to geographic information resources for the Water Framework Directive (WFD) (OJ 2000) within an interadministration and cross-border scenario. The project area covers two river basins in France and Spain: Adour-Garonne River basin district, managed by the Adour-Garonne Water Agency (L'Agence de l'Eau Adour-Garonne), and the Ebro River basin district, managed by the Ebro River Basin Authority (Confederación Hidrográfica del Ebro).

This project has been developed by a consortium consisting of IGN France International (Institut Géographique National France International), the National Geographic Institute of France (Institut Géographique National), the National Centre for Geographic Information of Spain (Centro Nacional de Información Geográfica), and the University of Zaragoza (together with experts from the University Jaume I). Additionally, this consortium counts on the help of the National Geographic Institute of Spain (Instituto Geográfico Nacional), the Adour-Garonne Water Agency, the Ebro River Basin Authority, the Regional Direction of the Ministry of Environment for the Midi-Pyrenees region, and the French GIS-ECOBAG association.

The above entities (most of them public institutions) are the main providers of topographic and hydrographic data in the project area. SDIGER is a two-year project launched by Eurostat directed at problems that may arise in the large-scale implementation of INSPIRE. The activities are presented below, and all of them, except for the last one, correspond to the first year of the project.

- **Definition of a cross-border scenario.** Two uses have been defined for the project. Both uses are focused on the environmental domain and take into consideration the problems of at least two adjacent countries and at least two different languages.
- **Metadata-related activities.** Three metadata profiles with full technical documentation and user guides have been developed: geographical data mining, assessing and using geographical data, and Water Framework Directive. An open-source metadata management tool must be developed for the above profiles.
- **Multilingual access portal.** A multilingual portal providing access to the geographic information and services available from organizations participating in or collaborating with SDIGER has been created. The services accessed through the portal have been configured according to the standards (Open Geospatial Consortium and ISO 19100) described in the INSPIRE Architecture and Standards Position Paper (CEC 2002).

- **Multilingual aspects of the application.** French and Spanish are the official languages of the two countries directly involved in the project. An English version of the geoportal is also available. Multilingual thesauri and gazetteers are used to facilitate the creation of metadata and the development of ergonomic search interfaces for data and service catalogs (Nogueras-Iso et al. 2004).
- **Common data model.** A common data model has been created in UML for mapping data from local repositories into the agreed common model.
- **Configuration of the servers.** It is necessary to provide access to data and services covered by the application according to the ISO TC211 and Open Geospatial Consortium (OGC) standards.
- **Internet application.** The applications proposed in the application scenario have been implemented using the elements established in the previous activities.
- **Study report for implementation of INSPIRE at the European level.** The report identifies problems, solutions, and the costs of using configurations commensurate with the European scale of INSPIRE.
- **Maintenance for the second year.** Network services and applications must be maintained during the second year of the project.

More details of these activities can be found elsewhere (Latre et al. 2005) or in the documents published on the SDIGER Web site (http://www.sdiger.net). Problems stemming from the cross-border nature of the project are discussed below.

CROSS-BORDER SCENARIO

The area covered (figure 1) by SDIGER is particularly interesting because, although most of the Adour and Garonne river basins are in France and the Ebro River basin is in Spain, some stream and river headwaters are located across the border. For instance, the source of Garonne River is in Spain and is managed by the Ebro River Basin Authority, while the headwaters of Irati River, an Ebro River tributary, are in France and are managed by the Adour-Garonne Water Agency. Cross-border information is, thus, of great importance for each of the basin authorities in order to ensure that the Water Framework Directive requirements are fulfilled in each river basin district. Additionally, this cross-border area includes several protected areas included within Natura 2000, the network of protected areas in the European Union.

Within this scenario, two applications have been developed on the basis of INSPIRE principles: WFD Reporting and Water Abstraction Request. The first one is devoted to generating WFD-mandated reports for the member states. It introduces a new approach to data and information collection and reporting. Data required by articles 3 and 5 of the WFD are used for implementing the reporting mechanisms in an INSPIRE-compliant way; in other words, the required data and information are directly accessible from an SDI. The second one is oriented toward improving the administrative process for obtaining a water abstraction authorization.[1] In France and Spain, the use of both surface and ground waters for private purposes requires an authorization. The administrative process for a water abstraction request requires users to provide an application form specifying the characteristics of the water abstraction point, water use, and water discharge point. The objective of this application is to provide

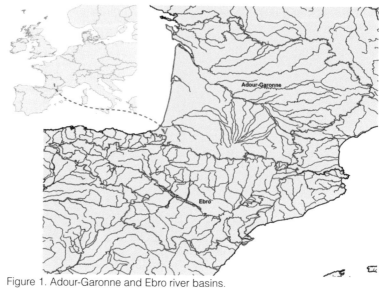

Figure 1. Adour-Garonne and Ebro river basins.
SDIGER project map layers used with permission.

users with some guidance, data, and documentation needed to follow the administrative process of a water abstraction request. Apart from solving a real need, this application demonstrates that the WFD data required by the European Commission can facilitate decision making in other scenarios. SDIGER 2005 provides a full description of these use cases and the datasets involved.

The proposed architecture for the implementation of the proposed applications is displayed in figure 2. The main purpose of this architecture was to demonstrate that the SDIGER project could be easily implemented with the services provided by existing SDI initiatives in the participant member states, which have created geoportals with attractive map viewers and search services for data holdings based on metadata. However, INSPIRE must go further and prove that SDIs solve real problems in an easier way than developing stand-alone applications from scratch. Thus, the main principle for the design of SDIGER has been the reuse of services, models, and data offered by public administrations at the local, national, or European level with responsibilities in the reference data field (e.g., national mapping agencies) or in the thematic data field (e.g., environment ministries or river basin authorities).

PROBLEMS DETECTED

Several problems related to the cross-border scenario have been identified. Critical issues are related to the different political and administrative institutions of the two countries, reflected in the organizational processes and data models.

Another important source of problems is the multilingual heterogeneity of the study area. A cross-border scenario does not necessarily involve a multilingual requirement because multiple languages can be found within a country. For instance, several regions in Spain have more than one official language. The European Union (EU) currently has 20 official languages. One of the requirements of the SDIGER

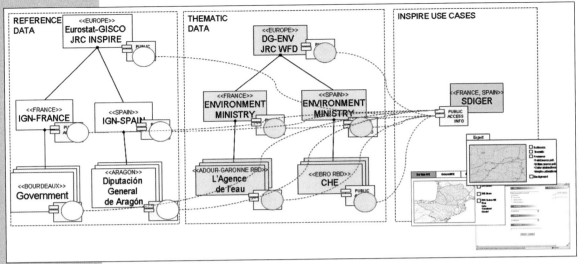

Figure 2. Architecture of SDIGER.

system is to support multilingualism (Spanish, French, and English) in all services and functionality and simulate provision of these services to the whole EU.

Problems detected in administrative processes organization, data model harmonization, multilingual adaptation, and project management are discussed below.

Administrative processes organization. The different legislation and regulations that exist in the two countries prevented the design of a seamless and harmonized solution. For instance, the procedures required in Spain and France for water abstraction requests are not the same. Although harmonization usually means interfacing the greatest common divisor, a set of parameters was needed in order to provide a useful and realistic application. If the system is extended (i.e., additional river basin districts are integrated) the number of parameters may increase.

This administrative heterogeneity also makes defining useful applications in a cross-border scenario very difficult. For instance, SDIGER provided an opportunity to facilitate the administrative processes of water agencies. The complex and heterogeneous administrative processes in both river basin districts were thoroughly analyzed. This analysis led to some unexpected results, indicating that the application may need to be redesigned:

- **Access restrictions.** Access to information about water abstraction points in some areas may be restricted. However, in adjacent areas, this information is available for public viewing. This led to a customization of the Web application that, depending on the area, considers or omits this factor in the administrative processes.

- **Computer security issues.** Because of security restrictions (e.g., firewalls), Web Map Service proxies were installed to allow access to services provided by public institutions that have not made their Web Map Service available via the Internet.

Harmonization of data models. The participation of different administrative regions in a single application usually requires data model harmonization. The following four harmonization processes have been identified for SDIGER.

Common models for geographic data. Ideally, data from one side of the border should be compatible with data from the other side. However, the actual situation is very different:

- Conceptual schema languages for describing local models differ.

- Spanish, French, and English describe features differently.

- Most of data models used by water agencies in Europe are based on the recommendations provided by the Water Framework Directive GIS guidelines. In those cases the definitions of common models and the conversion of data to the common models are relatively straightforward. However, when they don't follow these recommendations, the models of different water agencies are so divergent that the common model can be agreed upon only at a very high level. This limits harmonization processes. Where data has not been created in accordance with a predefined model (such as pressure data for the Water Framework Directive), differences among datasets can limit harmonization efforts even more.

- In general, servers used for providing data access, usually offering OGC-compliant Web Feature Service (Vretanos 2005), may return data with some structural differences (e.g., GML data with different structures). This necessitates the creation of a software layer that puts requests in the required format for the servers.

- Problems with the availability or completeness of information on the servers may result in significant delays in task completion. For instance, initial GML files from France did not have relevant information for the application and did not provide information about centroids, the geographic locations that should be shown for each water body. In other cases, the lack of some information had disabled the use of some input parameters. For instance, as the units of measure for French data are unknown, it makes no sense to use the field "size."

- Spatial reference systems (SRS) differ. Multiple Web Map Services can interoperate only if they share at least one SRS as a common denominator. Lack of support of specific or common projection systems may prevent the visualization of more than one WMS at a time. The map viewer client integrated within the SDIGER geoportal acts as a front service that deals with geometry and image transformation matters. However, if noncommon SRS are provided by Web Map Services, at very small scales the transformations performed result in a rough overlay of layers originally using different SRS. Establishing common SRS for Web Feature Services (WFS) is similarly challenging.

Common models for metadata. During many years there has been a wide range of metadata standards (regional, national, and international) proposed. ISO 19115 became an international metadata standard for geographic information in 2003 (ISO 2003). However, this provides only a conceptual model for metadata and does not provide an encoding model. The implementation model, ISO technical specification 19139 (ISO 2006), became a standard at the end of 2006 (when the SDIGER project ended). In addition, most of the countries and international organizations have proposed different application profiles. In a cross-border scenario, different metadata application profiles (usually different subsets of ISO 19115 with different semantics and using different controlled vocabularies and thesauri) need to be harmonized in accordance with ISO 19139.

Common models for gazetteer data. There is no internationally agreed-upon model for the exchange of data used for gazetteers. Therefore, geographic name data at local repositories must be transformed into an agreed-upon common model. Differences in feature typologies need to be reconciled. For instance, the French National Geographic Institute provides 74 feature types, the Spanish National Geographic Institute provides 50 feature types, and the Ebro River Basin Authority provides 20 feature types, for a combined total of 144 different types with some overlaps. After harmonization, the current gazetteer common data model contains 129 unique feature types. Another way to align the different types would be to use a common thesaurus. The following 2 thesauri for gazetteer typology can be considered:

1. NGA (GNS 2006; NGA 2006) is a plain list of two levels (8 classes and 600 types of features) for military use. The advantage is that it has been widely used for the GeoNames server; people have usually taken this as input for different applications. The problem is that it does not have a hierarchy.

2. The ADL Feature Type Thesaurus (Hill et al. 1999; Hill and Zheng 1999) offers the advantages of having a good hierarchy and having been used by the ADL Gazetteer, which integrates data of different types and from different sources. The disadvantage may be that it does not interoperate with other services and applications.

In both cases and whenever possible, the alignment of types should be verified by matching feature instances, in other words, comparing mappings based on different sources that share toponyms (some of them also available in different languages) and geographic locations.

Common models for thesaurus data. Several multilingual thesauri have been integrated in SDIGER with two main aims: (1) to facilitate the creation of metadata and gazetteer data with appropriate content and (2) to facilitate searches performed with the catalog client and the gazetteer client. A common format was needed for the exchange of these thesauri and their upload into a thesaurus service. The agreed-upon common model for thesaurus data has been the Simple Knowledge Organization System (SKOS) core format. The SKOS project forms part of W3C Semantic Web Activity and has proposed a model to represent lexical ontologies in RDF (Resource Description Framework) (Miles and Brikley 2005). SKOS core is an RDF vocabulary for expressing the basic structure and content of concept schemes such as thesauri, classification schemes, subject heading lists, taxonomies, terminologies, glossaries, and other types of controlled vocabulary (Alistair et al. 2005). This format was selected because it is becoming the de facto standard for the interchange of thesauri, with many thesauri covering different areas being generated in this format.

Although it seems clear that the adoption of SKOS seems appropriate (justified by its wide use), additional work is required for the adaptation of thesauri not directly available in this format. Moreover, it may be necessary to develop a special thesaurus where there is no existing solution. For instance, it has been necessary to build a thesaurus for names of administrative units in Europe. This thesaurus has been created to ease the creation of toponyms and the search interface used for the gazetteer. This thesaurus was based on two main sources. The names of Spanish municipalities were obtained from the National Statistics Institute (Instituto Nacional de Estadística [INE]), with names of provinces and autonomous communities compiled manually from the information given by the Spanish Public Administration. The names of French regions, departments, and communes came from the National Institute for Statistics and Economic Studies (Institut National de la Statistique et des Études Économiques [INSEE]). Both INE and INSEE data were formatted as tabular text. A specific program was needed to transform this into the SKOS core format.

Multilingual adaptation. Multilingual adaptation presents an important problem for configuration management. The software must support a multiple-value variable for each of the different languages in the system and additional languages in the future.

Software. The software must support multilingual capabilities. The internationalization of desktop and Web applications is error prone and increases the complexity of development, requiring the use of an existing infrastructure for internationalization. The software has to take into account the restriction imposed by multilingual support. It is nearly impossible or very expensive to provide multilingual support without a native capability for internationalization, making it necessary to manage the use of differently encoded character sets. Some languages (e.g., Polish) require a character set encoding (UTF-8[2]) different from the one commonly used for other European languages (e.g., ISO8859 for English, French, and Spanish).

Another problem related to multilingual adaptation is software version management. One of the main problems in the creation and evolution of software systems concerns configuration management. However, multilingual support issues are not usually considered part of the configuration management process. The translation capability is traditionally provided by people other than those developing software. It is necessary to identify clearly the new or modified functionality in order to be able to coordinate the translation processes. Tools for distinguishing items already translated from ones that have just been incorporated into the system can be very valuable.

Web portal contents. The SDIGER Web portal provides access to data and services of institutions that are partners or collaborators of the SDIGER consortium. The portal has been structured in four main sections:

- The **general information** section describes project objectives, partners, results, and so on, and lists useful links.
- The **generic services** section offers access to the geodata catalog search application, gazetteer application, and geographic information visualization application.
- The **application** section provides the tools for WFD Reporting and Water Abstraction Requests.
- The **private area** is a restricted section (with login and password) providing access to deliverables and internal documents.

The portal offers these functionalities in three languages: Spanish, French, and English. Graphic user interface (GUI) components (labels, buttons, value lists, etc.) must also be displayed in the language specified by the user. Java and XML technologies (including XSLT) have been used to dynamically internationalize the software components, load Web pages stored as XML documents, and apply appropriate style sheets for displaying the required portal style with the appropriate language for text labels.

This model can be extended to other portals, taking into account the difficulties of developing a Web portal infrastructure with capabilities for internationalization and the difficulties of translating the contents into different languages. The latter problem is very similar to the management of the multilingual evolution of the software. The translation is usually done by people outside the development team. It is necessary to identify clearly the new or modified contents of the portal in order to be able to coordinate the translation processes. Also, suggestions for additional tools to identify the elements translated and the ones that still require translation are welcome.

Multilingual support of the data offered by Web applications presents several problems. For instance, catalog or gazetteer services should support cross-language information retrieval. There are a lot of geographic information resources that are cataloged using only one language, but users that make their queries in one language may be interested in candidate resources that have been described in another language. The user is more interested in the resource (map, image, or multimedia) than in the metadata describing it. Thus, catalogs must provide users with mechanisms facilitating multilingual search without forcing cataloging organizations to describe their resources in all possible languages. For that purpose, the following multilingual resources are used in the SDIGER project:

- Multilingual thesauri like GEMET (EEA 2001), UNESCO (UNESCO 1995), EUROVOC (EC 2006), and AGROVOC (FAO 2006) are used for creating metadata and developing ergonomic search interfaces for data and service catalogs (Nogueras-Iso et al. 2004).

- In addition to using multilingual thesauri to facilitate cross-language information retrieval, it is also important to help the user understand metadata records written in a language different from the user query language. It would be desirable to translate the records online by a machine translation service. SYSTRANLinks from SystranSoft (http://www.systransoft.com) has been selected as the machine translation service for the project.

Most standards for service specifications do not take into account the problems of internationalization. The SDIGER Web application provides a map viewer allowing data layers to be viewed at different scale levels (European, national, and regional). This map viewer is a client application compliant with the OGC Web Map Service specification (Beaujardiere 2004). The main difficulties have been in the internationalization of legends in Web Map viewers because the Web Map Service standard does not support the management of names of layers in multiple languages.

Documents. A software project not only consists of software and portal contents but also involves all the documents that must be produced and published: user manuals, general guidelines, tutorials, and other technical information. In most cases, all these documents should be generated in the different languages of the project. To be able to create these documents and especially to be able to manage their evolution in coordination with the rest of the project, it is necessary to provide a structured and organized process based on a set of procedures and tools adopted from outside or developed within the project. The main objective of these procedures and tools is to make the configuration management system easier for documents and for software and data contents.

General aspects. Several noteworthy aspects of multilingualism must be addressed. Firstly, all translation work must be accomplished by people with enough technical knowledge. Otherwise, the results would have poor quality and could be incomprehensible. Secondly, multilingual support increases the work without offering new functionality. Every time a new language must be incorporated in the project, all the software has to be translated into this new language. For SDIGER, the following tasks must be accomplished:

1. Internationalization of the metadata editing tool. This includes the internationalization of the GUI as well as translation of user manuals and profile guidelines.

2. Translation of the contents of the geoportal. This includes translation of textual contents, internationalization of the map viewer GUI, internationalization of the catalog client GUI (both general and thematic), internationalization of the gazetteer client GUI, and internationalization of the Web application GUI.

Project management. The general management of an international and inter-administrative project includes project planning and coordination of activities, preparation of meetings, creation of a technical framework for the collaborative work of different institutions (e.g., mailing lists, FTP sites, etc.), and activities concerned with the diffusion of information about the project (e.g., participation in workshops and attendance of meetings on related projects).

The participation of different organizations in the development of a project is always a factor that increases the complexity of project coordination. This is especially problematic for SDIGER due to its transinstitutional and transnational nature. The management of the project must take into account the different multilingual and multicultural contexts. Moreover, all project participants work in a foreign language: English.

Additional logistical problems were related to the dependence of scheduling and meeting participation on the various commitments of the participants and their travel arrangements.

CONCLUSIONS

SDIGER has explored the feasibility of a cross-border SDI, identifying potential problems for the implementation of INSPIRE at the European Union level. Several problems were related to the cross-border nature of the project. At the technical level, the initial planning of SDIGER as an SDI built on existing SDIs and standardized services has remained a dream. Apart from the expected lack of stability in nascent public SDI initiatives, the immature status of some service specifications and harmonizing initiatives has led to the proposal of new specifications, their approval, and setting up of new services specifically for this project. In addition to technical complexity, problems stemming from differences in the political and administrative institutions of the two countries were identified (administrative processes, organization, and data models).

The coordination of multiple partners proved to be another challenge for this transnational and transinstitutional project. For instance, the reconciliation of different data access policies has been especially problematic. Whereas some institutions provide unimpeded data access, others are not allowed to provide free public access to their data, not even for display. This forced the development of special services verifying client permissions and restricting access to data on specific geographic areas.

Last but not least is the multilingual heterogeneity. This is a special issue to take into account for SDI initiatives in countries like Spain, where there are several regions with more than one official language.

Despite the obstacles, the project has demonstrated that SDIs provide important benefits. An SDI avoids (or reduces) the bureaucracy between organizations; in other words, it reduces the communication channels and gives access to up-to-date data. The project has raised the awareness of the potential of SDIs, not only within the institutions involved in the project but also by providing feedback to other projects and initiatives. SDIGER results have been used as input for the WFD GIS working group, the INSPIRE drafting teams, and CEN/TC 287 WG 5 work on metadata. SDIGER also will likely encourage the development of local and national projects with similar functionality or similar architecture. The project can be extended to other European river basin districts with minimal work. The authorities can profit from the documented know-how, established architecture, agreed-upon common models, and a set of complementary tools and procedures.

Finally, two main issues have been identified as upcoming challenges and future actions. First, the further analysis of maintenance reports produced during the second year and similar reports in equivalent projects could help to answer important questions about the technical configuration (number and technical characteristics of servers) of an SDI. SDIGER maintenance statistics can serve as baselines for estimating the maximum number of simultaneous users allowed, bandwidth, need for backup servers, and up-time. Second, for border area scenarios, geometric and semantic harmonization should be further investigated. The main problem in geometric harmonization is how to accomplish edge matching (the comparison and graphic adjustment of features to obtain agreement along the edges of adjoining map sheets) between datasets produced by different organizations. Edge matching involves both geometric edge matching, semantic edge matching, and tabular data edge matching. It is a prominent research topic among national mapping agencies, and it would be interesting to work further on some open issues: study the influence of edge matching errors on visual analysis, check whether it is acceptable to ignore these errors (as is often done), and investigate whether there is a smaller scale (e.g., 1:400,000) at which higher-scale datasets (e.g., 1:200,000) can be visualized without edge matching error distortions. A fair amount of work remains to be done on facilitating edge matching agreements between national mapping agencies.

ENDNOTES

1. The European Environment Agency (EEA) multilingual environmental glossary (http://glossary.eea.europa.eu/EEAGlossary) defines water abstraction as follows: "Water removed from any sources, either permanently or temporarily. Mine water and drainage are included. Similar to water withdrawal."

2. UTF8 stands for unicode transformation format. It is one of the three possible encodings of UNICODE.

ACKNOWLEDGMENTS

This work was funded by the European Commission through the Statistical Office of the European Communities (Eurostat), contract 2004 742 00004, for the supply of informatics services in the Community Statistical Programme. Additionally, this work was partially supported by the Spanish Ministry of Education and Science through project TIC2003-09365-C02-01 (National Plan for Scientific Research, Development and Technology Innovation).

REFERENCES

Alistair, M., B. Matthews, and M. Wilson. 2005. SKOS Core: Simple knowledge organization for the WEB. Proceedings of the International Conference on Dublin Core and Metadata Applications, Madrid, Spain, September 5–13.

Beaujardiere, J., ed. 2004. Web Map Service, v. 1.3. Open Geospatial Consortium Inc, OGC 04–024.

Commission of the European Communities (CEC). 2004. Proposal for a Directive of the European Parliament and of the Council establishing an infrastructure for spatial information in the Community (INSPIRE). COM (2004) 516 final, 2004/0175 (COD).

Commission of the European Communities (CEC). 2002. INSPIRE Architecture and Standards Position Paper, EUR 20518 EN. Available at http://inspire.jrc.it/reports/position_papers/inspire_ast_pp_v4_3_en.pdf.

European Communities (EC). 2006. EUROVOC Thesaurus. http://eurovoc.europa.eu.

European Environment Agency (EEA). 2004. General Multilingual Environmental Thesaurus (GEMET), 2004 version. http://www.eionet.europa.eu/gemet.

Food and Agriculture Organization (FAO) of the United Nations. 2006. AGROVOC Thesaurus. http://www.fao.org/agrovoc.

Geonet Names Server (GNS). 2006. http://earth-info.nga.mil/gns/html.

Hill, L., J. Frew, and Q. Zheng. 1999. Geographic names: The implementation of a gazetteer in a georeferenced digital library. *D-Lib Magazine* (January). http://www.dlib.org/dlib/january99/hill/01hill.html.

Hill, L. L., and Q. Zheng. 1999. Indirect geospatial referencing through place names in the digital library: Alexandria digital library experience with developing and implementing gazetteers. *Knowledge: Creation, organization and use.* Proceedings of the 62nd Annual Meeting of the American Society for Information Science, Washington, DC. October 1999.

International Organization for Standardization (ISO). 2003. Geographic information: Metadata. ISO 19115:2003.

International Organization for Standardization (ISO). 2006. DTS 19139 geographic information: Metadata XML schema implementation. ISO/TC 211 Doc. Nr 2049.

Latre, M. A., F. J. Zarazaga-Soria, J. Nogueras-Iso, R. Béjar, and P. R. Muro-Medrano. 2005. SDIGER: A cross-border inter-administration SDI to support WFD information access for Adour-Garonne and Ebro River Basins. Proceedings of the 11th EC GI & GIS Workshop, ESDI Setting the Framework, Alghero (Sardinia), Italia.

Miles, A., and D. Brikley, eds. 2005. SKOS Core Vocabulary Specification W3C Editor's Working Draft. http://www.w3.org/TR/2005/WD-swbp-skos-core-spec-20050510.

NGA GeoNet Designations. 2006. http://libraries.mit.edu/gis/data/findingaids/geonet-names/designations.html.

Nogueras-Iso, J., F. J. Zarazaga-Soria, J. Lacasta, R. Tolosana, and P. R. Muro-Medrano. 2004. Improving multilingual catalog search services by means of multilingual thesaurus disambiguation. Proceedings of the 10th European Commission GI and GIS Workshop, ESDI: The State of the Art, Warsaw, Poland.

Official Journal (OJ) of the European Union. 2000. The EU Water Framework Directive: Integrated river basin management for Europe, L 327:0001–0073.

SDIGER. 2005. Application scenario of the SDIGER project, version 2.0. http://sdiger.unizar.es/public_docs/ApplicationScenario_v2.0.pdf.

United Nations Educational, Scientific and Cultural Organization (UNESCO). 1995. UNESCO thesaurus: A structured list of descriptors for indexing and retrieving literature in the fields of education, science, social and human science, culture, communication and information. Paris: UNESCO Publishing.

Vretanos, P., ed. 2005. Web feature service implementation specification, v. 1.1.0. Open Geospatial Consortium Inc, OGC 04–094.

The Role of Spatial Data Infrastructures in Establishing an Enabling Platform for Decision Making in Australia

IAN WILLIAMSON, ABBAS RAJABIFARD, AND ANDREW BINNS

CENTRE FOR SDIs AND LAND ADMINISTRATION, DEPARTMENT OF GEOMATICS, UNIVERSITY OF MELBOURNE, VICTORIA, AUSTRALIA

ABSTRACT

In order to deliver a greater range of services and information to users across jurisdictions, the concept of spatial data infrastructures (SDIs) is beginning to progress towards the development of an enabling platform, helping to link services across national, state, and local jurisdictions, organisations, and disciplines. This cross-jurisdictional approach aims to provide users with the ability to access and utilise precise information in real time about both the built and the natural environments within the sphere of decision making—something that is beyond the ability of single organisations to deliver. The article describes the changing role that spatial data infrastructures are playing in the development of such an enabling platform in Australia, which is a federation of states. The changes include the growing demand for subnational-government and private-sector involvement in SDI development. Technical and institutional challenges are discussed.

INTRODUCTION

The role that spatial data infrastructure (SDI) initiatives are playing within society is changing. SDIs were initially conceived as a mechanism to facilitate access and sharing of spatial data for use within a GIS environment. This was achieved through the use of a distributed network of data custodians and stakeholders in the spatial information community. Users, however, now require the ability to gain access to precise spatial information in real time about real-world objects, in order to support more effective cross-jurisdictional and interagency decision making in priority areas such as emergency management, disaster relief, natural resource management, and water rights. The ability to gain access to information and services has moved well beyond the domain of single organisations, and SDIs now require an enabling platform to support the chaining of services across participating organisations.

The ability to generate solutions to cross-jurisdictional issues has become a national priority for countries such as Australia, and the development of effective decision-making tools is a major area of business for the spatial information industry. Much of the technology needed to create these solutions already exists; however, the solutions also depend on an institutional and cultural willingness to share outside of one's immediate work group. This creates the need for jurisdictional governance and interagency collaborative arrangements to bring together both information and users.

This article outlines the role of spatial data infrastructures in creating more effective decision-making processes to deal with cross-jurisdictional issues through the creation of an enabling platform that links services and information across jurisdictions and organisations. The creation of an enabling platform will be more than just the representation of feature-based structures of the world and will also include the administrative and institutional aspects of such features, enabling both technical and institutional considerations to be incorporated into decision making. This would support a knowledge base to access information derived from a model of integrated datasets from different perspectives such as the natural and built environments and support the creation of a virtual jurisdiction.

**CHANGING ROLE
OF SDIs**

SDIs aim to facilitate and coordinate the sharing of spatial data between stakeholders, based on a dynamic and multihierarchical concept that encompasses the policies, organisational mandates, data, technologies, standards, delivery mechanisms, and financial and human resources necessary to ensure that those working at the appropriate (global, regional, national, local) scale are not impeded in meeting their objectives (GSDI 1997). This in turn supports decision making at different scales for multiple purposes and increases benefits to society arising from the availability of spatial data. "The benefits will accrue through the reduction of duplication of effort in collecting and maintaining of spatial data as well as through the increased use of this potentially valuable information" (MSC 1993).

First-generation SDIs, developed starting in the mid-1980s, were designed to promote economic development, to stimulate better government, and to foster environmental sustainability (Masser 1998). Countries like the United States and Australia relied on developing data access relationships, which became the precursor to the development of national SDI initiatives. These countries designed and developed SDIs based on their specific national characteristics, requirements, and priorities, paving the way for the documentation of experiences through status reports on SDI initiatives (such as Onsrud 1998). From this documentation, most countries developed a product-based approach to SDI development driven largely by national governments (Rajabifard et al. 2003).

In 2000, the second generation of SDI development began to appear, with some of the leading countries changing and updating the SDI conceptual model. This was brought about by the creation of a much more user-oriented SDI perspective which was more effective in maximising the added value of a nation's spatial-information assets and more cost-effective as a data dissemination mechanism (Masser 2005). The second generation, as witnessed in Australia and the United States, focused much more on facilitating the management of information assets instead of accessing databases and using a process-based approach. The first- and second-generation models are shown in figure 1 (Rajabifard et al. 2006a).

For first-generation SDIs, data was the focus, and initial development was driven by top-down national governments. The second generation is driven by the needs of users, with the focus on the use of data and data applications as opposed to the data itself, with one result being that subnational governments and the private sector have greater influence. This influence can be seen in the Federal Geographic Data Committee (FGDC) Future Directions Project, which states that "the continued development of the NSDI requires that the private sector, academia, the utility industries as well as state, tribal and local governments play a major role" (FGDC 2004).

SDI development over the past 15 years has seen 3 main players emerge— national governments, subnational governments, and the private sector—with the role of each being quite different. As shown in figure 2, initial SDI development was the domain of national governments, which played both strategic and operational roles in mapping and collecting small-scale data about nations. As policy development came from the national level, subnational governments and the private sector had no clear role.

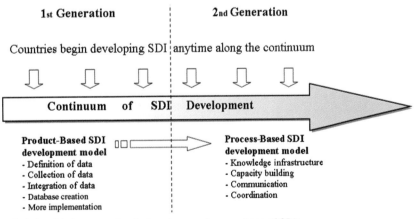

Figure 1. Relationship between the first and second generations of SDIs.
Reprinted from Rajabifard et al. 2006 with permission of the International Journal of GIS.

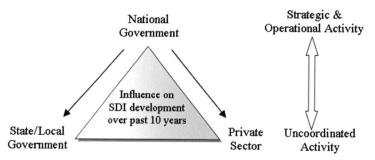

Figure 2. Roles of national governments, subnational governments, and the private sector in SDI development over the past decade (Rajabifard et al. 2006a). Reprinted from Rajabifard et al. 2005 and 2006 with permission of the International Journal of GIS.

The roles are now changing, however, as can be seen in Australia, with the national focus moving from being both strategic and operational to primarily strategic. This is especially so for countries that are federations of states. National datasets are generally small-scale, which lessens the need for updating, maintenance, and infrastructure development. The operational responsibility for SDIs is moving to subnational governments, where large-scale data is being used for everyday decision making in emergency management, natural resource management, and policy development. This data is highly detailed and dynamic, requiring systems for updating and maintenance. In countries such as Australia this does not mean that national government agencies are not involved in operational activities—just that subnational governments are now playing a larger role.

In the United States, local governments and the private sector are devoting considerable resources to obtaining spatial data they need to serve citizens and clients (National Research Council 2001). The operational role of the Australian private sector is increasing, as it leads the drive for greater access to large-scale "people-relevant" data (property and socioeconomic data). Subnational governments are also moving forward in creating policies and initiatives that aid in the development of SDIs and utilise the expertise and cooperation of the private sector. These two sectors are now responsible for building infrastructures in a collaborative manner, with national government providing the overall framework in which such infrastructures can operate. In Australia, communication now flows from these three players (figure 3) rather than only from a top-down national government (figure 2) (Rajabifard et al. 2006a). This overcomes problems inherent in purely top-down or bottom-up approaches. As described by van Loenen (2006), in a top-down approach decision makers believe in the potential of an SDI, without actual commitment. A bottom-up approach has the opposite problem: the bottom acknowledges some successful experiences but lacks support from the top for broad-scale implementation.

This changing approach to SDI development has also been the driving force behind governments moving forward in creating policies and initiatives that open up more information to the public, and this change needs to continue. According to Radwan et al. (2005), to address today's information needs, the role of the traditional SDI needs to be adjusted continuously. Citizens and organisations need infrastructures on which they can rely for provision of services. This goes

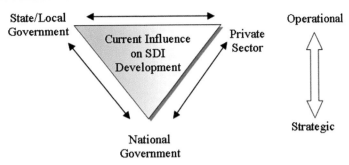

Figure 3. Current roles of national governments, subnational governments, and the private sector in SDI development.
Reprinted from Rajabifard et al. 2005 and 2006 with permission of the International Journal of GIS.

beyond current first- and second-generation SDIs, which were designed primarily for data discovery and retrieval.

ENABLING PLATFORM

Current SDI models have not met all user needs, offering mainly access to and retrieval of spatial data. Hence the concept of an SDI needs to be expanded so that it allows for more than just access to spatial information. SDIs need to be able to facilitate the sharing of data, business goals, strategies, processes, operations, and value-added products and services in order to provide an enabling platform supporting government and business activities. Such an enabling platform will serve the broader objectives of creating services and functions that can be utilised at higher levels in the information chain (Rajabifard et al. 2005).

The rapid advancement and development of information and communications technologies (ICT) and changes in business opportunities make today's spatial information market very dynamic. This has forced spatial data stakeholders (data producers, value adders, and data users) to change the way they deliver and use services to facilitate better decision making. The result is that meeting user needs with a variety of spatial data and information services across various jurisdictions, in large volumes and in near real time, goes beyond the capacity of single organisations or government agencies. Therefore, many jurisdictions are investing in mechanisms that encourage their stakeholders (both government and private sector) to work together in a more collaborative way. This has also led to an increased involvement of subnational governments and the private sector in SDI development, as described above.

A key element in the increased involvement of the private sector is the requirement that SDIs meet business needs, with focus moving from data to services, similarly to service-oriented spatial infrastructures as defined by Todd (2005). This will aid in meeting the long-term objective of creating a virtual jurisdiction or environment—a Virtual Australia.

The enabling-platform concept broadens current SDI practice in order to better support the vision of a virtual jurisdiction. According to Radwan et al. (2003), in a virtual jurisdiction, individual (small as well as large) organisations or partners work as a collaborative network to deliver specialised products or services on the basis of common standards and business understanding. The virtual jurisdiction is structured and managed in such a way that it is seen by third parties as one single enterprise. According to Rajabifard et al. (2006b) an enabling platform would serve as a knowledge base and a major resource for discovery and communication of complete, correct, and current information about the environment and related spatial information applications.

In Australia, the Cooperative Research Centre for Spatial Information (CRC-SI) is also investing in activities that contribute to the creation of a Virtual Australia, defined as a "virtual [digital] model containing and representing all non-trivial objects and their contextual environment—from blue sky to bedrock—in [real world] Australia" (CRC-SI 2005). CRC-SI brings together over 40 small to medium spatial information companies as well as federal and state governments to create spatial information applications that are affordable, useful, and readily available to all (CRC-SI 2006). The ability to create a Virtual Australia is limited if SDIs are utilised only in their traditional sense as underlying infrastructures. If applications, services, business models, and functions are created only for specific users, as illustrated in figure 4, uses that span applications and services may be difficult to facilitate.

An enabling platform aims to support a virtual jurisdiction, forming the underpinning structure. It is more than simply an access mechanism, although this feature of an SDI is prominent in any move to an enabling platform. It enables the linking of data, services, products, and real-world objects through the creation of appropriate governance and legal support, built on an open-source, distributed technical infrastructure. Applications and tools such as those being developed by the CRC-SI need to be linked together if the Virtual Australia concept is to become a reality.

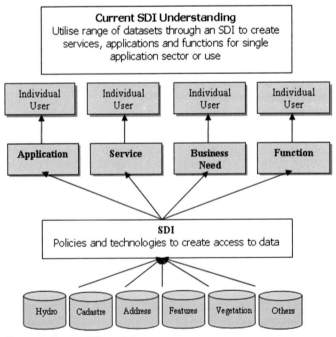

Figure 4. Current SDI model.

As pointed out by van Loenen (2006), both technical and nontechnical characteristics of a dataset are important, and the same is true for an enabling platform. An enabling platform for Australia is defined as technical, governance, and legal structures linking data, services, products, and real-world objects underpinning a virtual jurisdiction. The platform itself is the next step in the SDI process. It will facilitate interoperability of functional entities within a heterogeneous environment through the use of both technical characteristics and appropriate access policies. This creates multiple uses for individual applications and services, as shown in figure 5.

The information and services an enabling platform would make available should be combined in such a way that all data can be analysed and acted on together within a single environment, subject to appropriate security, privacy, and commercial considerations. These considerations will need to be carefully investigated in order to break down institutional barriers, as SDIs have been doing.

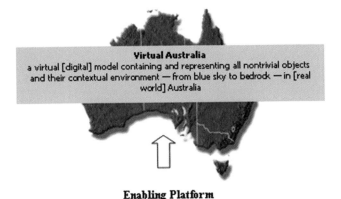

Virtual Australia
a virtual [digital] model containing and representing all nontrivial objects and their contextual environment — from blue sky to bedrock — in [real world] Australia

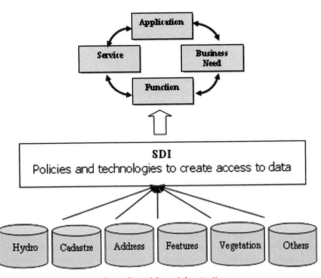

Enabling Platform

Facilitating interoperability and interworking of functional entities within a heterogeneous environment. This will link data, services, products and real world objects through the creation of appropriate technical, governance and legal support that underpins a virtual jurisdiction.

Figure 5. Enabling platform for a Virtual Australia.

This will ensure that information from a range of jurisdictions and applications such as land administration, environmental management, statistics, social development, land use patterns, and so forth, is accessible and usable both within each application sector and between all sectors.

Overall, development of an enabling platform aims to build a common rail gauge and reduce barriers to information and services for government agencies, industry, and the public. If barriers are minimised, organisations will be able to pursue their core business objectives with greater efficiency and effectiveness. In particular, reduced costs will encourage industry to invest in capacity building for generating and delivering a wider range of spatial information products and services to a broader market, both in Australia and internationally.

AUSTRALIAN EXPERIENCE

Current arrangements for accessing and sharing spatial data in Australia generally meet only very basic user requirements (Rajabifard et al. 2006b). To address this concern, some national and subnational agencies have taken progressive approaches. National leadership is being provided by ANZLIC (Australian and New Zealand Spatial Information Council), and various whole-of-government spatial information initiatives are being developed, especially at the subnational level.

The major aim of these initiatives is to give cross-jurisdictional agencies access to spatial information in a more efficient and effective manner. Instead of people having to deal with several different agencies to obtain information, one online system will eventually provide access to key information controlled by various agencies. Replacing a "silo" mentality (agencies keeping their expertise, data, and knowledge to themselves) with a more open approach would reduce duplication and inefficiencies, improve service delivery, and yield strategic and commercial benefits to government (Rajabifard et al. 2006b). It would also help develop systems and policies to spatially enable government and broader society (Wallace et al. 2006).

Duplication of effort and expense in creating and sharing spatial data still occurs at all levels of administration in Australia. Existing national, state, and local spatial data initiatives and policies for access and delivery of data and information need to be integrated. This will enhance the capabilities of government, the private sector, and the general community to engage in systems-based, integrated, and holistic decision making about the future of Australia.

Technical issues. Services and data are delivered in Australia by online systems using an interoperability architecture based on distributed, custodial data management and open standards. Authorised custodians operate distributed Web services providing uniform and consistent managed access (Staling et al. 2004). The aim of this architecture is to allow initiatives to grow in an open environment that gives agencies the ability to operate in an integrated manner.

An enabling platform should be built on a network of state systems (currently based on SDI technology), the installation and operation of which shall be the responsibility of each Australian state. The costs associated with building a new platform would be very high, and if the platform was not compatible with current spatial information initiatives, it would not be utilised.

Harmonization of data standards and specifications through the adoption of common data definitions, formats, models, and exchange formats will be crucial to the success of an enabling platform. This will ensure unimpeded flow of data and information between the various levels of government and the private sector. An enabling platform must also be able to monitor change and create realistic simulations of the evolving world. This is possible if information and applications for both the built (legal, land rights, etc.) and the natural (environmental, natural resource, etc.) environments are available in an integrated form. Generally these two forms of data have been developed to serve different purposes and are usually managed separately. An enabling platform must provide the technical ability to integrate not only these forms of data but applications and services that utilise these different datasets. This will enable more holistic decision making.

Data storage also needs to be investigated. New database management software and technology, along with virtual libraries, GRID computing technologies, and super servers, are changing the way in which data (especially spatial data) is stored, managed, and used. This in turn will have an impact on the development of an enabling platform.

Institutional issues. An enabling platform would need to be implemented progressively, with data and services populating and linking off a basic technical platform. This will allow the funding, technology, and institutional arrangements needed to create such a platform to be implemented as they are finalised. Technically, existing state-based spatial information and SDI initiatives have the potential to contribute to the development of an enabling platform. However, sociotechnical issues (economic, educational, cultural, institutional, legal, political, and organisational) currently impede the sharing and accessing of spatial information just as technical issues can (Groot and McLaughlin 2000; Crompvoets 2006). A lot of work needs to be done on such sociotechnical issues, and both top-down and bottom-up approaches based on current spatial information initiatives are needed.

A common policy framework for good information management (e.g., metadata custodianship and data sharing agreements) is also needed. Several Australian states have implemented such policies, which provide a platform from which the spatial information industry can expand and enable organizations from different jurisdictions to bring their spatial information, tools, and services together into interoperable formats.

In summary, enabling-platform development has to take into account the following institutional issues (Rajabifard et al. 2006b):

- SDI goals are changing from data access to service delivery, and development should be driven by user needs
- An open-source, interoperable enabling environment supporting government and business activities is needed
- Lack of spatial awareness and education: existing work practices do not include a strong culture of using maps and spatial information
- Lack of cross-jurisdictional relationships
- Proprietary restrictions of some agencies limit data availability

- Institutional and cultural barriers
- Governance models and organisational structures

The business needs of the stakeholders and the broader spatial information community will also need to be investigated. This will allow the private sector to serve as a catalyst in the creation of an enabling platform.

FUTURE DIRECTIONS

In order to meet today's information needs, the traditional role of SDIs is changing. Service-oriented infrastructures that go beyond the first and second SDI generations are needed. Spatial information managers need to deliver a virtual world which facilitates decision making at a community level within a national context.

SDI initiatives driven by subnational governments differ from the top-down national SDIs. The new bottom-up subnational approach reflects the different aspirations of various stakeholders. The challenge is to ensure some measure of standardisation and uniformity while recognising the diversity and heterogeneity of all stakeholders. The use of open standards and an interoperable enabling platform will allow functions and services that meet business needs to be brought together at a subnational level, reducing duplication of effort and furthering the development of a spatially enabled society.

Institutional practices need to focus on making existing and future technology more effective. Very few jurisdictions have developed a framework for establishing a spatial infrastructure that comprehensively addresses operational, organisational, and legal issues. It is these processes that will enable the infrastructure to be readily useable and available to all stakeholders.

CONCLUSIONS

An enabling platform aims to facilitate interoperability of functional entities within a heterogeneous environment, enabling the linking and sharing of cross-jurisdictional services and functions that meet business needs. The inclusion of the private sector is important. The development of effective institutional practices and the creation of linkages between government agencies, SDI initiatives (both national and subnational), and the private sector will enable the delivery of a range of products and services from different agencies through an enabling platform.

Design strategies for an enabling platform need to link current subnational SDIs with new functions identified by research, end users, and service providers and to overcome any resistance to change that current systems may have developed. An enabling platform is largely dependent on collaboration between all parties and requires effective SDIs to support efficient access, retrieval, and delivery of spatial information.

ACKNOWLEDGMENTS

We acknowledge the support of the members of the Centre for SDIs and Land Administration at the Department of Geomatics, University of Melbourne; the Australian Research Council (ARC); Geoscience Australia; and the Department of Sustainability and Environment, Victoria. However, the views expressed in this paper are those of the authors.

REFERENCES

CRC-SI. 2005. Know, think, communicate: Key elements for Virtual Australia: A discussion paper, eds. B. Thomson, and T. O. Chan. Cooperative Research Centre for Spatial Information, Melbourne. http://www.crcsi.com.au.

CRC-SI. 2006. Vision and statement of purpose of operation. Cooperative Research Centre for Spatial Information, Melbourne. http://www.crcsi.com.au.

Crompvoets, J. 2006. National spatial data clearinghouses: Worldwide development and impact. PhD thesis. Wageningen University, Netherlands.

FGDC. 2004. NSDI future directions initiative: Towards a national geospatial strategy and implementation plan. Submitted by the NSDI future directions planning team commissioned by the FGDC. http://www.fgdc.gov/policyand planning/future-directions/reports/FD_Final_Report .pdf.

Groot, R., and J. McLaughlan, eds. 2000. *Geospatial data infrastructure: Concepts, cases and good practice.* Oxford: Oxford University Press.

GSDI. 1997. Global Spatial Data Infrastructure conference findings and resolutions. Chapel Hill, North Carolina. http://www.gsdi.org/docs1997/97_gsdi97r.html.

Mapping Sciences Committee. 1993. Towards a coordinated spatial data infrastructure for the nation. Washington, DC: National Research Council, National Academy Press.

Masser, I. 1998. The first generation of national geographic information strategies. In Proceedings of Selected Conference Papers of the Third Global Spatial Data Infrastructure Conference, Canberra.

Masser, I. 2005. *GIS worlds: Creating spatial data infrastructures.* Redlands, CA: ESRI Press.

National Research Council. 2001. National spatial data infrastructure programs: Rethinking the focus. Washington, DC: Mapping Science Committee, National Research Council, National Academy Press.

Onsrud, H. 1998. Survey of national and regional spatial data infrastructure activities around the globe. In Selected Conference Papers of the 3rd Global Spatial Data Infrastructure Conference, Canberra, ANZLIG.

Radwan, M., A. Alvarez, R. Onchaga, and J. Morales. 2003. Designing an integrated enterprise model to support partnerships in the geo-information industry. MapAsia.

Radwan, M., R. Onchaga, and J. Morales. 2005. The design requirements for service-oriented spatial information infrastructures. Global Spatial Data Infrastructure 8 and FIG Working Week Conference, Cairo, Egypt.

Rajabifard, A., A. Binns, I. Masser, and I. Williamson. 2006a. The role of sub-national government and the private sector in future Spatial Data Infrastructures. *International Journal of GIS* 20 (7): 727–41.

Rajabifard, A., A. Binns, and I. Williamson. 2006b. Virtual Australia: An enabling platform to improve opportunities in the spatial information industry. *Journal of Spatial Science* Special Edition (in press 2006).

Rajabifard, A., A. Binns, and I. Williamson. 2005. Creating an enabling platform for the delivery of spatial information. Proceedings of SSC 2005 Spatial Intelligence, Innovation and Praxis: The national biennial Conference of the Spatial Sciences Institute, Melbourne.

Rajabifard, A., M.-E. F. Feeney, I. Williamson, and I. Masser. 2003. National spatial data infrastructures. In *Developing spatial data infrastructures: From concept to reality*, eds. I. Williamson, A. Rajabifard, and M.-E. F. Feeney. London: Taylor & Francis.

Starling, R., M. Wilson, and R. Mason. 2004. Notional architecture for a geo-enabled enterprise portal platform for the AusIndustry spatial interoperability demonstration project, vol. 1, version 1.0.

Todd, P. 2005. Impact of uncertainty on governance strategies for integrated coastal zone management. Presentation at the Marine Cadastre Workshop 2005, Department of Lands, Sydney, Australia. http://www.geom.unimelb.edu.au/maritime/workshop2005.htm.

van Loenen, B. 2006. Developing geographic information infrastructures: The role of information policies. PhD thesis. The Netherlands: Delft University Press.

Wallace, J., I. Williamson, A. Rajabifard, and R. Bennett. 2006. Spatial information opportunities for Government. *Journal of Spatial Science* 51 (1): 79–100.

National Spatial Data Clearinghouses, 2000 to 2005

JOEP CROMPVOETS AND ARNOLD BREGT

WAGENINGEN UNIVERSITY, CENTRE FOR GEO-INFORMATION, WAGENINGEN, THE NETHERLANDS

ABSTRACT

One of the key features of a national spatial data infrastructure is a national clearinghouse for spatial data, which can be regarded as a network facilitating access to spatial data and related services. Between April 2000 and 2005, a longitudinal Web survey was undertaken to assess all national clearinghouses throughout the world and to identify critical factors for coordinators and policy makers. By April 2005, 83 countries had established national clearinghouses. However, low suitability and declining trends in use, content, and management were found. The reasons for these troubling trends could be the dissatisfaction of the GI community with the functional capability of clearinghouses and the piecemeal funding of the majority of these facilities. The main critical factors for success were identified as public awareness, Web services, user-friendly interfaces, metadata standard ISO19115, and continuous funding.

INTRODUCTION

Over the last two decades, many governments and the private sector have invested tens of billions of Euros in the development of geographic information, largely to serve specific communities (forestry, agriculture, urban/rural planning, land records management, military, security service, health care, development aid, emergency services, retail, etc.), within local, national, international, and even global frameworks (Groot and McLaughlin 2000). The focus is increasingly shifting towards a framework for integrating geographic information by means of spatial data infrastructures (SDIs). SDIs facilitate access to spatial data and services, improving on the existing complex, multistakeholder decision-making process (Feeney 2003). Moreover, SDIs facilitate (and coordinate) the exchange and sharing of spatial data (at the local, national, and international levels) between stakeholders within the geoinformation (GI) community. This community includes mapping agencies, universities, governmental and nongovernmental organisations, and public and private institutions.

Over the last few years many countries have spent considerable resources on developing national spatial data infrastructures in order to manage and utilise spatial data assets more efficiently, reduce the costs of data production, and eliminate duplication of data acquisition efforts (Groot and McLaughlin 2000; Bernard et al. 2005; Williamson et al. 2003; Masser 2005). National SDIs have, according to Masser (2005), three common characteristics: (1) they are explicitly national in nature, (2) refer to either geographic information or spatial data, and (3) imply the existence of some form of a coordinating mechanism for policy formulation and implementation.

National SDIs are also facilitative in nature (making data accessing and sharing easier) and are set up with long-term development in mind. A key feature of a national SDI is the national spatial data clearinghouse (Crompvoets and Bregt 2003), which can be defined as an electronic facility for searching, viewing, transferring, ordering, advertising, and/or disseminating spatial data from numerous sources via the Internet. Such a facility usually consists of a number of servers which contain information (metadata) about available digital data (Crompvoets 2006). It provides complementary services and improves the exchange and sharing of spatial data between suppliers and users.

The concept of a clearinghouse originated in the financial world. With respect to financial transactions between banks, the clearinghouse keeps the data on mutual debts. At the end of each day, banks are informed about the final amounts to be transferred between banks. Every day there is a "clearing" between them (Bogearts et al. 1997). The first clearinghouse was the London Banker's Clearinghouse, which was established in 1773. The New York Clearinghouse Association described its clearinghouse role in 1853 as simplifying the chaotic exchange between New York City banks (Clearinghouse Payments Company 2005). This clearinghouse regards itself as the place where payments meet, mix, and move expeditiously to their final destination. In 1994, the U.S. Federal Geographic Data Committee (FGDC) established the National Geospatial Data Clearinghouse, aimed at facilitating efficient access to the overwhelming quantity of spatial data (from federal agencies) and coordinating its exchange, with the objective of minimizing duplication (in the collection of expensive spatial data) and assisting partnerships where common needs exist (Rhind 1999; FGDC 2000; Crompvoets et al. 2004).

A national clearinghouse for spatial data can be considered the access network (or window) of a national SDI, focusing on the facilitation of spatial data discovery, access, and related services. It is not a national repository where datasets are simply stored. It can be seen as a one-stop shop for all national spatial data, sourced from governmental agencies and/or industrial bodies (Crompvoets et al. 2004). National clearinghouse implementation can vary enormously. The way in which a national clearinghouse is set up depends on technological, legal, economic, institutional, and cultural factors within the country. A national clearinghouse differs from local, state, international, and global clearinghouses in that it's embedded in a national SDI. As of April 2005, 83 national clearinghouses were established on the Internet. Examples include MIDAS (MetaInformacni Databazovy System) in the Czech Republic, Geodata-info.dk in Denmark, India NSDI portal, Spatial Data Catalogue in Malawi, Russian GIS Resources, Geocat.ch in Switzerland, and the Clearinghouse Nacional de Datos Geograficos in Uruguay.

National clearinghouses are evolving worldwide in tandem with national SDIs. A body of literature has been compiled on national experiences (e.g., Spatial Applications Division of Catholic University of Leuven 2003 and papers of Global Spatial Data Infrastructure Association from 2002 to 2005). The majority of this literature focuses on the technical aspects of clearinghouses and does not take into account the evolutionary nature of these facilities. A longitudinal perspective is needed. A detailed study of all national clearinghouses worldwide could identify the critical factors behind the success or failure of a clearinghouse. Knowledge thus obtained could be applied in future implementation strategies. However, simply consolidating the best practices of a few well-operating national clearinghouses (Australia, Canada, and United States) gives no guarantee of sustainability for other national clearinghouses. Such best practices cannot necessarily be applied equally in other countries, due to societal differences (Crompvoets and Kossen 2002; Delgado et al. 2005).

The article focuses on the worldwide development of national spatial data clearinghouses between 2000 and 2005, identifies critical factors for current and future implementations, and offers suggestions for overcoming obstacles. National clearinghouses are a key feature of national SDIs, and they can help improve the availability and accessibility of spatial data and services.

WEB SURVEY

A longitudinal Web survey of national clearinghouses was conducted systematically and periodically starting in 2000 (April and December 2000, 2001, and 2002 and April 2005). The procedure consisted of taking an inventory of all national clearinghouses on the Internet and measuring several characteristics.

The inventory was compiled by extensive browsing of the Internet, reading related literature, and contacting experts and webmasters. Clearinghouses were evaluated on history, use, content, and management. The following 11 characteristics were measured:

A. Year of first implementation on the Web

B. Number of data suppliers

C. Monthly number of visitors

D. Level of data and metadata accessibility

E. Number of datasets

F. Most recently produced dataset

G. Availability of view (Web mapping) services

H. Frequency of Web updates

I. Web address change

J. Funding continuity

K. Metadata standard

Almost all the above information was collected from the clearinghouse Web pages. The history of the clearinghouses is described by characteristic A; use is described by B and C; content is described by D, E, F, and G; and management is described by H, I, J, and K. Each characteristic is explained below.

Next, the suitability of each national clearinghouse was calculated on an index scale using the Web survey results of December 2002 and April 2005. This suitability index provides a measure of the overall quality of a national clearinghouse. (For more information on this index, see Crompvoets 2006.)

MAIN RESULTS

History. History was assessed as the year of first implementation on the Web (characteristic A). Since 1994, the number of countries with national clearinghouses has been steadily increasing (figure 1). As of April 2005, 83 countries had a functional clearinghouse on the Web and 25 countries had initiated implementation projects (table 1). National clearinghouses are being implemented throughout the world (figure 2), but numbers vary considerably between regions. More than 60 percent of the countries in Europe and the Americas have established national clearinghouses, whereas less than 20 percent of African countries have such clearinghouses. The many project initiatives in Africa are promising, but 85 countries throughout the world have not initiated any plan for such a national facility, likely due to a low standard of living. Countries with a high standard of living are more likely to establish national clearinghouses than countries whose citizens are primarily concerned with the daily problems of survival (Crompvoets 2006).

National clearinghouses vary significantly in use, content, and management. The results of the survey are presented as averages and medians, since the distribution of some variables is highly skewed. Because the median is less sensitive to extreme values, it is more informative than the average.

Use. A clearinghouse serves as a central facility for data suppliers to disseminate their products. Characteristic B, number of data suppliers, is decreasing (figure 3), possibly due to national clearinghouses losing their popularity for supplying spatial data or newly established clearinghouses having fewer suppliers.

Characteristic C, monthly number of visitors, does not identify users by behaviour or type. The slightly decreasing numbers (figure 4) could mean that clearinghouses are losing popularity.

Status	Number of countries				
	Total	Africa	Asia-Pacific	Europe	Americas
Implemented[a]	83	11	21	27	24
Planned	25	12	5	5	3
No initiative	85	28	38	11	8
a Between 1994 and April 2005					

Table 1. National clearinghouses by continent.

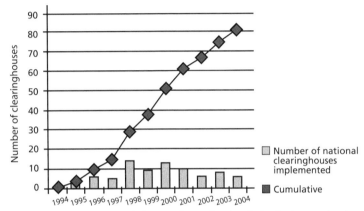

Figure 1. Yearly and cumulative numbers of national clearinghouses implemented.

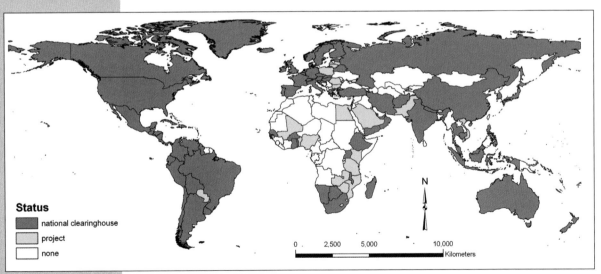

Figure 2. Distribution of national clearinghouses as of April 2005.
Courtesy of ESRI Data & Maps 2005.

Country	Score		Difference	Country	Score		Difference
	Dec. 2002	April 2005			Dec. 2002	April 2005	
Argentina	43	19	−24	Ireland	46	31	−15
Austria	46	29	−17	Italy	10	44	+34
Australia	83	76	−7	Japan	43	71	+28
Barbados	31	22	−9	Luxembourg	12	7	−5
Belgium	28	24	−4	Malaysia	61	74	+13
Bolivia	34	26	−8	Mexico	46	44	−2
Brazil	37	60	+23	Netherlands	42	37	−5
Brunei	33	21	−12	New Zealand	63	54	−9
Canada	94	91	−3	Nicaragua	37	31	−6
Chile	56	49	−7	Norway	45	60	+15
Czech Republic	51	42	−9	Panama	9	4	−5
China	44	47	+3	Peru	31	12	−19
Colombia	46	48	+2	Philippines	36	26	−10
Costa Rica	33	30	−3	Portugal	48	48	0
Croatia	37	21	−16	Qatar	27	52	+25
Denmark	57	62	+5	Russia	12	51	+39
Dominican Republic	33	24	−9	Senegal	24	21	−3
Dominica	49	31	−18	Singapore	44	54	+10
Ecuador	13	7	−6	Slovakia	22	42	+20
El Salvador	38	36	−2	Slovenia	45	61	+16
Estonia	41	56	+15	South Africa	64	62	−2
Ethiopia	35	35	0	South Korea	17	61	+44
Finland	72	62	−10	Spain	34	57	+23
France	29	18	−11	Sweden	33	42	+9
Germany	62	74	+12	Switzerland	36	68	+32
Ghana	35	21	−14	Trinidad and Tobago	31	22	−9
Greece	8	5	−3	Turkey	14	29	+15
Guatemala	27	35	+8	Uganda	18	13	−5
Guyana	18	18	0	United Kingdom	70	66	−4
Honduras	35	26	−9	United Arab Emirates	17	12	−5
Hungary	35	37	+2	Uruguay	50	59	+9
Iceland	37	34	−3	United States	97	97	0
Indonesia	62	72	+10	Venezuela	41	43	+2
Iran	28	17	−11				
				Average	39	41	+2
				Median	37	37	0

Table 2. Suitability indices of 67 national clearinghouses established before December 2002.

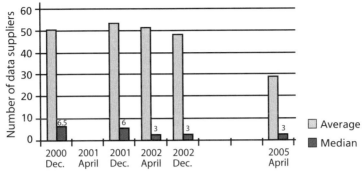

Figure 3. Numbers of data suppliers.

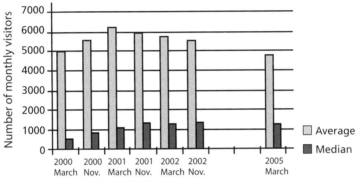

Figure 4. Numbers of monthly visitors.

Content. Characteristic D, level of (meta)data accessibility, refers to three classes: nonstandardised metadata, standardised metadata, and data with standardised metadata (direct access). The number of clearinghouses with standardised metadata has remained stable, and the number of clearinghouses providing direct access to data using what the Open Geospatial Consortium (OGC) calls a Web Feature Server has slightly increased (figure 5).

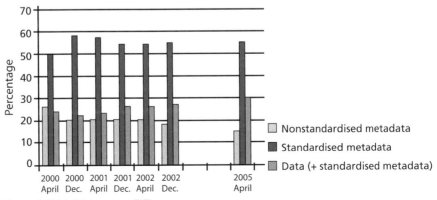

Figure 5. (Meta)data accessibility.

Characteristic E, number of datasets, refers to metadata records, each of which describes one dataset. The significant decrease (figure 6) could mean that the quality of clearinghouse contents has decreased and could be related to the decreasing numbers of data suppliers and monthly visitors.

The total number of datasets for national clearinghouses worldwide increased slightly to 225,000 (figure 7).

Characteristic F, most recently produced dataset, was defined as the number of months between the last dataset produced and the date of measurement. The increased numbers (figure 8) suggest that national clearinghouses are becoming less up-to-date and provide another indication that the quality of clearinghouse contents has decreased. Weaker content could also have a negative impact on clearinghouse use.

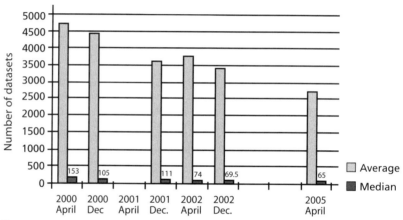

Figure 6. Numbers of datasets.

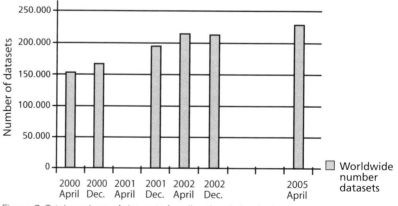

Figure 7. Total numbers of datasets for all national clearinghouses.

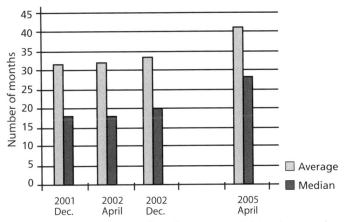

Figure 8. Numbers of months since the most recent dataset was produced.

National clearinghouses provide standardised view services for documented data (INSPIRE Architecture and Standards working group 2002). Viewing geographic information over the Internet is also referred to as Web mapping (characteristic G). The technology behind view services is aimed at displaying spatial information quickly and easily and requiring only basic map-reading skills. The number of clearinghouses providing view services has increased significantly. This could increase use (Crompvoets et al. 2004).

Management. Characteristic H, frequency of Web updates, was defined as the number days between the last Web update and the date of measurement. Low numbers of days correspond to high frequency of Web updates, which can be a sign of a well-managed clearinghouse. This characteristic refers not to updates of the data but to updates of the clearinghouse Web site. The decreased frequency of updates (figure 9) could indicate that clearinghouse managers are losing interest in updating their sites regularly.

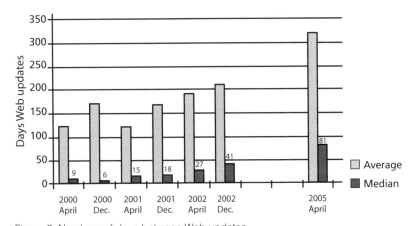

Figure 9. Numbers of days between Web updates.

Characteristic I, Web address change, refers to change in the URL address of the clearinghouse between two measurements (for the April 2005 measurement, this refers to change since December 2004). It was measured for six periods. The stable percentage of changes in Web addresses (figure 10) could indicate that clearinghouse managers are still struggling to create the right technological environment. Moreover, 8 percent of the national clearinghouses operating in December 2002 were not functioning in April 2005, which is another indication of poor management.

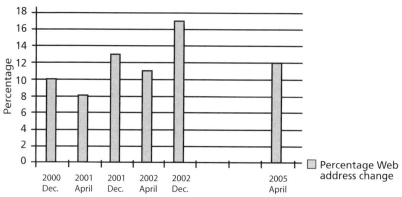

Figure 10. Changes in Web addresses.

Funds are used for designing and establishing discovery, transfer, and access services; preparing, validating, and publishing metadata; and creating the legal and institutional environment. Characteristic J, funding continuity, was measured twice. The increasingly piecemeal funding of national clearinghouses (figure 11) has a negative impact (Crompvoets et al. 2004). It appears that funds are not frequently used for maintenance purposes.

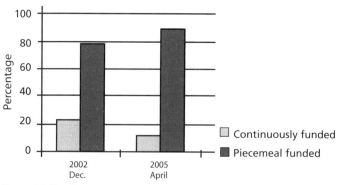

Figure 11. Funding.

In order to ensure interoperability amongst the datasets and to develop common access mechanisms, standards are essential. Metadata contains information about the content, quality, source, and lineage of spatial data. A number of organisations have developed metadata standards. Characteristic K, metadata standard, was measured five times. Standard ISO19115 is used more commonly than CEN (Comité Européen de Normalisation), FGDC, ANZLIC, national, and other standards (figure 12). The widespread use of ISO19115 could provide a strong foundation for establishing interoperability between datasets and creating a culture of sharing within the geographic information community.

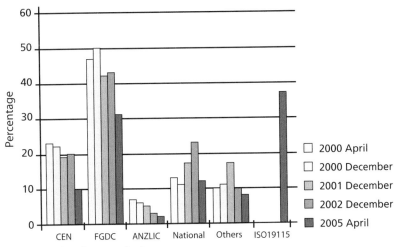

Figure 12. Metadata standards.

Clearinghouse suitability index. By using the method described by Crompvoets (2006), clearinghouse suitability indices were calculated for December 2002 and April 2005 (tables 2 and 3). The national clearinghouses of the United States, Canada, Australia, Germany, Malaysia, Indonesia, and Japan rated the highest in April 2005. The range of scores was very large, reflecting significant differences among countries. Between December 2002 and April 2005, 38 scores decreased, 25 scores increased, and 4 scores remained unchanged. This confirms the slightly negative developments identified above. Nevertheless, average suitability slightly increased.

Country	Score[a]	Country	Score[a]
Afghanistan	44	Madagascar	53
Belarus	17	Malawi	26
Botswana	26	Namibia	42
Burkina Faso	10	Nepal	19
Cambodia	8	Oman	17
Cuba	48	Thailand	39
India	48	Togo	12
Israel	18	Yemen	18
		Average	28
		Median	22.5
a April 2005.			

Table 3. Suitability indices of 16 national clearinghouses established since December 2002.

EVALUATION OF RESULTS

The number of national spatial data clearinghouses will likely continue to increase. On the basis of their proliferation, national clearinghouses can be considered a worldwide success.

The increase in the number of clearinghouses contrasts with the slight declines in use, content, and management. The trend observed between 2000 and 2002 (Crompvoets et al. 2004) continued in 2005. The decline in use mainly reflects decreasing numbers of suppliers and visitors. Another troubling trend is the diminishing quantity and quality of content, as evidenced by the decreased numbers and less frequent updating of datasets. The lower frequency of Web updates, the high percentage of changed Web addresses, and the lower prevalence of continuous funding reflect poor management by clearinghouse coordinators and policy makers.

The declines may reflect the dissatisfaction of the GI community with clearinghouses that do not fit current expectations (Crompvoets et al. 2004). Functional capabilities should perhaps focus on users and applications rather than on data, in line with the objectives of second-generation spatial data infrastructures (Rajabifard et al. 2003). Another reason for the declines might be the piecemeal funding of the majority of clearinghouses, which makes establishing a long-term framework difficult.

On the positive side, more view (Web mapping) services are available, the number of data download facilities has increased slightly, and metadata standard ISO19115 has been adopted widely. These developments indicate that clearinghouses may be becoming more service-oriented. Introduction of Web services is the main technological hallmark of second-generation spatial data infrastructures.

The low suitability scores could be due to the fact that the functional capabilities of clearinghouses do not fit the expectations of the GI community (e.g., more Web services and user-friendly interfaces) (Crompvoets et al. 2004). Average suitability slightly increased between December 2002 and April 2005, mainly because of technological improvements (e.g., introduction of Web services) that counterbalanced the slight declines in management, use, and content. The suitability scores of most national clearinghouses in Southeast Asia increased, whereas those of clearinghouses in the Americas decreased, possibly because coordinators of older clearinghouses are less motivated to maintain their facilities.

The following factors are critical to the success of national clearinghouses: creating awareness of functional capabilities, introducing Web services, building user-friendly interfaces, applying ISO19115, funding continuously, motivating data suppliers to participate, and updating regularly. National clearinghouses are expected to provide more Web services in the near future. These services could widen the use of spatial data (Maguire and Longley 2005; Beaumont et al. 2005; Foust et al. 2005). We recommend that case studies be conducted to explore the trends identified in this article.

ACKNOWLEDGMENTS

We gratefully acknowledge financial support from the Dutch innovation programme Space for Geo-Information.

REFERENCES

Beaumont, P., P. A. Longley, and D. J. Maguire. 2005. Geographic information portals: A UK perspective. *Computers, Environment and Urban Systems* 29:49–69.

Bernard, L., J. Fitzke, and R. M. Wagner, eds. 2005. *Geodaten-infrastruktur, grundlagen und anwendungen.* Heidelberg: Herbert Wichmann Verlag (in German).

Bogaerts, T. J. M., H. J. G. L. Aalders, and J. Gazdzicki. 1997. Components of geo-information infrastructure. Proceedings ELIS'97, Prague.

Crompvoets, J., and H. Kossen. 2002. The impact of culture on national spatial data clearinghouses, 9.1-9.3. Proceedings of GISDECO-conference, governance and the use of GIS in developing countries, Enschede, the Netherlands.

Crompvoets, J., and A. Bregt. 2003. World status of national spatial data clearinghouses. *URISA Journal* 15:43–50.

Crompvoets, J., A. Bregt, A. Rajabifard, and I. Williamson. 2004. Assessing the worldwide developments of national spatial data clearinghouses. *International Journal of Geographical Information Science* 18 (7), 665–689.

Crompvoets, J. 2006. National spatial data clearinghouses, worldwide development and impact. PhD thesis, Wageningen University, the Netherlands.

Delgado Fernández, T., K. Lance, M. Buck, and H. J. Onsrud. 2005. Assessing an SDI readiness index. Proceedings from pharaohs to geoinformatics, FIG Working Week 2005 and 8th International Conference on Global Spatial Data Infrastructure, Egypt, Cairo.

Feeney, M.-E. F. 2003. SDIs and decision support. In *Developing Spatial Data Infrastructures: From Concept to Reality,* eds. I. Williamson, A. Rajabifard, and M.-E. F. Feeney, 195–210. London: Taylor & Francis.

FGDC. 2000. Questions and answers about clearinghouses. Washington, DC: Federal Geographic Data Committee.

Foust, J., W. S. M. Tang, and J. Selwood. 2005. Evolving infrastructure: Growth and evaluation of spatial portals. Proceedings from pharaohs to geoinformatics, FIG Working Week 2005 and 8th International Conference on Global Spatial Data Infrastructure, Egypt, Cairo.

Groot, R., and J. McLaughlin, eds. 2000. *Geospatial data infrastructure: Concepts, cases and good practice.* Oxford: Oxford University Press.

INSPIRE Architecture and Standards working group. 2002. INSPIRE architecture and standards position paper, ed. P. Smits. Infrastructure for spatial information in Europe. Commission of the European Communities.

Maguire, D. J., and P. A. Longley. 2005. The emergence of geoportals and their role in spatial data infrastructures. *Computers, Environment and Urban Systems* 29:3–14.

Masser, I. 2005. *GIS Worlds: Creating spatial data infrastructures.* Redlands, CA: ESRI Press.

Rajabifard, A., M.-E. F. Feeney, I. Williamson, and I. Masser. 2003. National SDI-initiatives. In *Developing spatial data infrastructures: From concept to reality,* eds. I. Williamson, A. Rajabifard, and M.-E. F. Feeney, 95–109. London: Taylor & Francis.

Rhind, D. 1999. National and internal geospatial data policies. In *Geographical Information Systems: Principles, Techniques, Management and Applications,* eds. P. Longley, M. Goodchild, D. Maguire, and D. Rhind, 767–87. New York: John Wiley & Sons.

The Clearinghouse Payments Company. 2003. New York Clearinghouse: Historical perspective. In *Developing spatial data infrastructures: From concept to reality,* eds. I. Williamson, A. Rajabifard, and M.-E. F. Feeney. London: Taylor & Francis.

Proposal for a Spatial Data Infrastructure Standards Suite: SDI 1.0

DOUG NEBERT,[1] CARL REED,[2] AND ROLAND M. WAGNER[3]

FEDERAL GEOGRAPHIC DATA COMMITTEE, RESTON, VIRGINIA, UNITED STATES,[1] OPEN GEOSPATIAL CONSORTIUM, FT. COLLINS, COLORADO, UNITED STATES,[2] AND CON TERRA GMBH, MUNSTER, GERMANY[3]

ABSTRACT

The successful implementation of Internet-based spatial data infrastructures (SDI) requires the adoption of a compatible suite of standards to enable interoperability. The proliferation of new standards and new versions of old standards raises issues of dependency and compatibility that may impede the implementation of SDI architectures. This article proposes a suite of geospatial standards—SDI standards suite version 1.0—to facilitate the description and acquisition of compatible technology for SDIs worldwide. The application of a common set of standards for SDIs may reduce life cycle costs, enhance interoperability, decrease implementation risk, and improve services, particularly in the developing world. However, to date, no consistent guidelines exist to define the SDI standards baseline.

For over 20 years, SDI activities have been progressing at the local, regional, and national levels. Some efforts have been extremely successful, while others have been more limited in scope. Few of the current SDI implementations can seamlessly interoperate. Spatial data infrastructures are the realization of technical and human efforts to coordinate and provide geospatial information and services for multiple purposes. The SDI Cookbook (Nebert 2004) introduces SDIs as follows:

> The term "Spatial Data Infrastructure" (SDI) is often used to denote the relevant base collection of technologies, policies and institutional arrangements that facilitate the availability of and access to spatial data. The SDI provides a basis for spatial data discovery, evaluation, and application for users and providers within all levels of government, the commercial sector, the non-profit sector, academia, and by citizens in general.

Today's innovative SDI implementations provide data and service discovery and access and the ability to invoke and execute a growing number of geospatial analysis services.

An SDI may be defined in broad social terms as a framework for collaboration. The technical framework of an SDI enables interoperability for the access and exchange of geospatial resources. The problem is that most current SDI activities operate as independent application "silos" with little (or no) interoperability between them. Each initiative develops its own best practices with regard to what standards are used, what version of a given standard is used, and so forth. The result is that, while any given SDI may be interoperable within its own community, it might not interoperate with its Internet neighbors. This limits our ability to implement a virtual global SDI (GSDI).

After a year of concentrated effort, the U.S. National Geospatial Intelligence Agency (NGA) recently announced approval of an OGC spatial data infrastructure 1.0 (SDI 1.0) specification baseline (NGA 2005). The problem still facing NGA is that they now have to carefully consider and specify the actual baseline of standards (including versions) and interdependencies. Further, there is currently no general reference architecture for defining a framework in which these standards can work together to enable a standards-based SDI.

At the same time, dozens of SDI initiatives are implementing a variety of international standards for data and service discovery, data access, visualization, and analysis. The use of different versions of these standards limits interoperability between systems and initiatives. Further, different SDI initiatives are using different content models for key data themes, such as land cover and land ownership. Guidance on best practices and approaches to solving these interoperability issues is critical to our ability to define and implement a GSDI.

This article focuses on the identification of compatible, mature geospatial standards that will allow maximum technical interoperability based on general evaluation criteria—SDI 1.0. It also discusses a candidate suite of standards for future SDI deployments or enhancements. SDI 1.0 is intended for local, regional, national, and multinational communities interested in providing and accessing geospatial data over the Internet. Transnational SDIs, also known as geospatial data infrastructures (GDI), are loosely defined environments in which participants interact to develop and share geospatial data and services to the common benefit of a nation or, in the case of Europe, a continent.

A coordinated SDI standards suite is intended to manage the complexity of available standards and version change and to encourage global compatibility of solutions.

Complexity. Nearly 100 standards can be identified that directly or indirectly support the deployment of geospatial solutions, including various standards in information technology. The selection of an appropriate technical architecture can be daunting, and independent selection of standards may lead to incompatibilities between adjacent SDI implementations. The definition of a relatively small suite of standards allows a shorthand reference for nominal capabilities in an SDI environment, with provision for identifying optional supplemental standards.

Evolution cycles. Standards will change version numbers on a regular basis, but they are rarely coordinated with changes in other standards. The identification of a specific set of standards (and their version numbers) that are known to work well together is of great benefit to implementers and adopters. Adapting to frequent changes in standards is expensive and prone to issues of incompatibility. Minimizing the number and frequency of version changes is a goal of this proposal. An SDI standards suite version number, that is, 1.0, endeavors to package such compatible versions and in the future will need to be incremented itself to incorporate revisions.

Global compatibility. Through the identification of a common set of standards for SDI usage, the development of software that supports an SDI in one part of the world can be readily deployed for another SDI. This broadens the market reach of solution providers and reduces the cost of software development through targeted support of specific standard versions.

STANDARDS CONSIDERED

Geospatial standards are primarily developed by the International Organization for Standardization (ISO) Technical Committee 211 (TC 211) and the Open Geospatial Consortium (OGC) and are often dependent on other industry standards, such as those of the World Wide Web Consortium (W3C) and OASIS, which develop e-business standards. International standards for country codes and coordinate reference systems existed prior to the 1990s, but the more detailed standardization efforts began in earnest in 1994 with the formation of TC 211 and the OGC.

The geospatial standards development process has advanced over the past 12 years largely in the context of the World Wide Web and its emerging standards and infrastructure. Including underlying Internet standards, well over 75 standards may be relevant to the geospatial domain. Versions of these standards exist in various states—development, endorsement, implementation, or deprecation—

so deployment of all standards in a coordinated manner is not practical. Further, there is no assurance that they will function well together.

The identification of a common set of standards is already a practice in national SDI/GDI contexts. The Canadian Geospatial Data Infrastructure (CGDI) recognizes and promotes the use of a selected suite of standards through its Technology Advisory Panel. The U.S. National Spatial Data Infrastructure (NSDI) supports selected standards through its Geospatial One-Stop portal, geodata.gov. In both national contexts, such standards allow for the federation of a large number of provider-operated services and for data to be discovered, visualized, and accessed by Web browsers and software applications.

CRITERIA FOR INCLUSION

Given the proliferation of standards, it is useful to define a core suite for compatible implementation. Inclusion of a standard in SDI 1.0 is based on the following criteria:

Evidence of implementation. The implementation of given standards takes time and understanding before adoption and deployment take place; the latest version of a standard may be too complex or new to be widely adopted and deployed in software. Basing multiple software products and applications on a given version of a standard greatly increases the opportunities to build on existing capabilities. The SDI standards suite can also be thought of as an enabling set of standards for defining a service-oriented architecture as a reference pattern for implementing an SDI. This approach reduces costs and risks and increases value by leveraging existing services and content.

Commercial and noncommercial software solutions and documentation (publications, how-to guides, and workbooks) are useful metrics in identifying mature standards. For example, the OGC lists providers that have implemented OGC standards. Also, several OGC members have developed tools that search the Web looking for publicly available OGC Web Service-enabled servers. Based on this search capability, there are over 1,000 operational instances of the OpenGIS WMS Implementation Specification (Refractions 2006) (these numbers do not include instances of OGC standards that are hidden behind a firewall).

Dependencies. Standards rarely are monolithic or stand-alone and frequently have implicit and explicit dependencies on other standards. Hierarchies of standards, such as the Open Systems Interoperability (OSI) stack, describe vertical hardware, operating system, protocol, and application relationships. There are horizontal and containment relationships or dependencies as well. The latest version of a standard is not necessarily the one that will work with a selected set of other standards. Successful application of standards must clearly define the type, context, and version of related standards and their usage. Dependencies on other standards that are not mature or as widely adopted may cause problems with interoperability. Minimizing the number of dependencies can facilitate migration to newer versions of standards, considering that related standards may evolve on an independent schedule.

Stability and conformance. The implementation of technical standards to ensure interoperability requires that the standards have some means of being assessed or tested for conformance or compliance. The availability of tests—testing service, assessment methodology, model assertions, or testing software—promotes the adoption of interoperable solutions. An example of a compliance testing environment is the OGC Compliance and Interoperability Testing and Evaluation (CITE) capability for testing WMS and WFS compliance (http://www.opengeospatial.org/resources/?page=testing).

Core or supplemental status. Whereas several geospatial standards appear to be common to local, regional, and national SDIs, a number of other standards may be optional. The core standards should be viewed as the most widely implemented standards that provide baseline functionality in an SDI. Supplemental standards may not be required for SDI implementation; however, they identify optional but well-known capabilities.

Reference matrix. Table 1 lists the standards used by four major SDI projects. The first group is formal standards referenced by SDI initiatives, whereas the last five represent prototype implementations not yet approved as final standards. The CGDI selections represent both endorsed and recommended standards. The U.S. NSDI selections represent the standards required by the Geospatial One-Stop portal in its interaction with community data and services. The NRW (North Rhein–Westphalia [Germany]) selections represent a combination of standards in use by the local GDI project as well as a cross-border project operated in a partnership with the Netherlands. The Catalonian selections represent current technology implemented in the first phase of the SDI. Not intended to be an exhaustive exploration of adopted standards, the table illustrates commonalities and differences between national and regional SDI environments.

Standard	Canada CGDI	U.S. NSDI	GDI NRW	Catalonia
Formal				
OGC Web Map Service	●	●	●	●
OGC Web Feature Service	●	●	●	●
OGC Filter Encoding	●	●	●	
OGC Style Layer Descriptor	●		●	
OGC Geography Markup Language	●	●	●	●
OGC Web Map Context	●	●		●
OGC Catalogue Service 2.0 Z39.50 protocol binding	●	●		
FGDC Content Standard for Digital Geospatial Metadata	●	●		
OGC Web Coverage Service		●		●
OGC Catalogue Service 2.0 HTTP protocol binding (CS-W)	●	●	(●)	
Tentative				
OGC Web Coordinate Transformation Service				
OGC Gazetteer Profile of WFS	●	●		
OGC Web Pricing and Ordering Service			●	
ISO Metadata DTS 19139			(●)	
OGC Web Processing Service	●			

Table 1. Standards used in SDIs.

Table 2 lists fundamental standards on which the geospatial standards may be dependent. Not all of these standards are required for implementation of SDI 1.0 standards, but they may be required or expected to be present in a community's operating environment.

Information content standards. The following standards apply to information content.

ISO IS19115/DTS 19139 metadata standard. Metadata standard ISO 19115:2003 contains an abstract model represented in UML depicting the content and relationships of descriptions of geographic data and services. The 19115 standard does not provide guidance on the encoding or exchange of metadata but serves as a guide for what information should be documented for data and services. ISO Draft Technical Specification 19139 was scheduled for release in late 2006. The primary content of interest in this specification is a set of XML schema documents that can be used in the validation and structuring of compliant ISO metadata, derived from the 19115 metadata model.

Although a number of software packages and systems claim to support 19115 metadata, the delayed availability of the official encoding in DTS 19139 means that there will be few compliant implementations until the new schemas are adopted and implemented. Due to the lack of implementation practice, 19115/19139 should be considered supplemental to the SDI 1.0 standards suite.

FGDC Content Standard for Digital Geospatial Metadata. The U.S. Federal Geographic Data Committee (FGDC) approved version 1.0 of the Content Standard for Digital Geospatial Metadata (CSDGM) in 1994 and version 2.0 in 1998. The standard includes only an abstract model of content, relationships, obligation, and repeatability of properties that describe geospatial data. The FGDC has published schemas (XML document type declaration and XML schema documents) on its Web site to facilitate the validation and processing of metadata according to the standard. A metadata parser (mp) program is available from the U.S. Geological Survey for stand-alone and Web usage to validate metadata according to this standard. Extensions for biological and remote sensing data are available as well.

CSDGM is the most widely deployed metadata standard in the world. As of March 2006, over 8 million metadata records existed on the Internet in searchable collections that support the CSDGM. Metadata collections supporting this

W3C Recommendation: eXtensible Markup Language (XML), version 1.1
W3C Recommendation: XML Schema, version 1.0
W3C Recommendation: Hypertext Transport Protocol (HTTP), version 1.1
W3C Recommendation: Simple Object Access Protocol (SOAP), version 1.2
W3C Note: Web Services Description Language (WSDL), version 1.1
Oil and Gas Producer (OGP, formerly EPSG) Geodetic Parameter Dataset, version 6.9 (2006)
Geographic Tagged Image File Format (GeoTIFF), version 1.0
JPEG-2000 (ISO/IEC 15444-1:2004)
Information retrieval (Z39.50)-application service definition and protocol specification (ISO 23950:1998)

Table 2. Foundations for SDI standards.

standard have been developed for 32 countries in at least four languages: English, French, Spanish, and Portuguese.

Since its acceptance as an international standard in 2003, ISO 19115 has been slowly replacing the CSDGM internationally, with validation through ISO DTS 19139 XML encoding. By 2008 the ISO metadata standard will likely supersede CSDGM in a future SDI standards suite version. Until there is wide adoption and deployment of 19115/19139, CSDGM remains a primary vehicle for the description of geospatial data used in SDIs. It is recommended for inclusion in the SDI 1.0 standards suite; the ISO 19115/19139 standards are recommended as supplemental SDI 1.0 standards.

OGC Geography Markup Language. The OGC Geography Markup Language (GML), also an ISO draft international standard (19136), provides a means of encoding geographic features and their properties using XML. GML is the expected packaging for features requested from an OGC Web Feature Server (WFS). Data encoded in GML versions 2 and 3 can be validated using XML schemas published with the standard and maintained in a schema repository by OGC.

The OGC community uses three major versions of GML that are significantly different and are not backwardly compatible. Version 2.1.2 is currently the most widely deployed, as it is frequently paired with implementations of WFS, although the WFS standard does not preclude the service of GML 3.1.1-encoded data. GML version 3 provides new capabilities, including support of profiles that constrain the expression of feature data to more predictable combinations, for example, features with simple geometry. It also supports simpler expressions of point, line, and area geometries. GML 3.1.1 is being used to express basic data themes, known as framework data in the United States, and for similar data modeling efforts in Australia.

Given its prevalence, GML version 2.1 is recommended for the core SDI 1.0 standards suite. GML 3.1.1, with its new capabilities, is recommended as a supplement to the SDI 1.0 standards suite.

OGC Filter Encoding specification. The OGC Filter Encoding (FE) specification is used to express a query, or filter, using a predicate language or terms and operators that are stored in XML elements. FE used in the request messaging sent to WFS and in the query sent to the OGC Catalogue Service CS-W. OGC hosts reference XML schema documents that can be used to validate queries structured according to the standard. FE version 1.1 was approved in 2004 and is recommended for inclusion in the SDI 1.0 standards suite.

OGC Styled Layer Descriptor. The OGC Styled Layer Descriptor (SLD) standard defines the structure of an XML file that applies rendering or symbolization rules to features. An SLD can be invoked as an argument to a Web Map Service (WMS) to present a requested map according to submitted style rules. SLD support is an optional feature of WMS and as such should be considered supplemental to the SDI 1.0 standards suite.

OGC Web Map Context. According to the OGC adopted-specifications page, "The . . . Context specification states how a specific grouping of one or more maps from one or more map servers can be described in a portable, platform-independent format for storage in a repository or for transmission between clients. This description is known as a 'Web Map Context Document' [WMC] or simply a 'Context.'" Version 1.1 of WMC is coordinated with WMS version 1.1.1. Like the SLD, WMC version 1.1 support is an optional feature of WMS and as such should be considered supplemental to the SDI 1.0 standards suite.

Service and interface standards. The following standards apply to access to geospatial information and build upon the information content standards above.

OGC Catalogue Service specification. The Catalogue Service specification provides both an abstract model and protocol-specific solutions for the discovery of geospatial resources. Catalogues contain some form of metadata (searchable descriptive information) and a query interface (for returning the metadata properties to the requestor). Often embedded in these metadata are links to actual data or services that allow the catalogue to act as a referral service to other information resources.

Three protocol bindings are described in Catalogue Service version 2.0: CORBA, Z39.50, and HTTP (the latter is also known as Catalogue Services for the Web [CS-W]). The HTTP binding requires the declaration of an additional application profile to define the specifics of interaction within a community. Two major application profiles are formally proposed: one for a general registry information model (ebRIM) and the other for data and service objects based on the semantics and structures of ISO 19115/19119/19139 metadata. Schemas for metadata responses are published with the draft profile documents and can support limited validation testing; however, a compliance test suite has not yet been formally developed or endorsed. A third ad hoc profile of CS-W has been drafted to query and present FGDC CSDGM metadata. Given the early stages of adoption and uncertain interoperability of these CS-W profiles, the CS-W protocol binding is recognized as an emerging candidate for a future SDI standards suite and is recommended as a supplement to the SDI 1.0 standards suite.

Of the three protocols, Z39.50 (also adopted as ISO 23950) has been implemented most widely, with over four hundred registered servers from seven vendors supporting the geospatial query and response rules. Although no official conformance suite exists for the protocol, Z39.50 server compliance is tested by the FGDC using online query tools and a validation suite executed within the Geospatial One-Stop portal. OGC Catalogue Service Z39.50 protocol binding is recommended for the SDI 1.0 standards suite.

The following issues impede adoption of Z39.50 in a suite of Web service standards: Z39.50 is TCP/IP-based and is therefore not a conventional Web service, it requires the use of a unique TCP/IP communication port that is not commonly configured for public access, and it requires operating a different service and software than those used by other Web protocols. Given these issues and increased implementation testing of CS-W, it is likely that CS-W and its application profiles will supersede Z39.50 as the preferred standard in the future.

OGC Web Map Service. By far the most popular and widely implemented of the geospatial standards, the OGC Web Map Service (WMS versions 1.1.1 and 1.3; ISO 19128) supports the request and display of maps derived from data accessed by the service. Maps, delivered as graphical images (GIF, JPEG, TIFF, etc.), may be requested from one or more WMSs overlaid in browsers or client applications. Features "behind" the map can also be queried, and their properties can be returned to a requesting client. As discussed above, SLD and WMC files are used optionally to interact with the rendering or recall of maps, respectively.

Schemas for validating the "capabilities" of an XML file returned from a WMS service exist, and compliance testing is available through the OGC for assessing WMS performance on all key functionalities.

WMS version 1.1.1 is the most widely deployed (ISO 19128, however, is harmonized with WMS version 1.3 but is not yet widely deployed) and is recommended for inclusion in the SDI 1.0 standards suite.

OGC Web Feature Service. According to the OGC adopted-specifications page, "the OGC Web Feature Service allows a client to retrieve and update geospatial data encoded in Geography Markup Language (GML) . . . from multiple Web Feature Services. The . . . interfaces must be defined in XML . . . GML must be used to express features within the interface . . . the predicate or filter language will be defined in XML and be derived from CQL [Common Query Language] as defined in the OpenGIS Catalogue Interface Implementation Specification." The WFS provides an abstraction of the underlying data store, expressed in GML, as defined through GML application schemas referenced by the service.

The most common implementation of WFS is version 1.0, which typically returns features encoded using GML 2.1. A growing number of services based on WFS 1.1 will also return a more recent version of GML-3.1.1-a capability that is advertised through the getCapabilities response. It is recommended that WFS version 1.0 be included in the SDI 1.0 standards suite, with required support for GML 2.1 and optional support for GML 3.1.1 response encoding.

OGC Web Coverage Service. The OGC Web Coverage Service (WCS) ". . . extends the Web Map Server (WMS) interface to allow access to geospatial 'coverages' that represent values or properties of geographic locations, rather than WMS generated maps (pictures)," according to the OGC adopted-specifications page. WCS can return different representations of continuous data surfaces (coverages) for any location: grids, triangulated irregular networks (TINs), point sets. Most commonly, however, the form of coverage most often returned is a grid in a declared coordinate reference system and common format such as GeoTIFF. WCS version 1.0 has been available since 2003 and is recommended for inclusion in the SDI 1.0 standards suite for the exchange of raster or grid data (not rendered imagery).

Candidate SDI 1.0 standards. Table 3 lists the standards for SDI 1.0 and for future versions.

Establishing an SDI standards baseline serves many market purposes. Some fundamental relationships that best illustrate the need for a well-defined and well-managed SDI 1.0, and some of the market and policy forces dictating the requirements for an SDI standards suite, are discussed below. Analogies to other information technology communities that faced a similar set of issues and market drivers are also drawn.

Evolution of the SDI standards suite. The coordination of the release cycles of the various standards is currently limited. This lack of coordination can impede

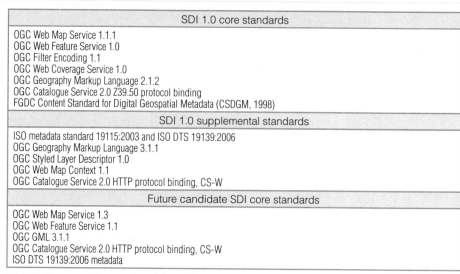

SDI 1.0 core standards
OGC Web Map Service 1.1.1 OGC Web Feature Service 1.0 OGC Filter Encoding 1.1 OGC Web Coverage Service 1.0 OGC Geography Markup Language 2.1.2 OGC Catalogue Service 2.0 Z39.50 protocol binding FGDC Content Standard for Digital Geospatial Metadata (CSDGM, 1998)
SDI 1.0 supplemental standards
ISO metadata standard 19115:2003 and ISO DTS 19139:2006 OGC Geography Markup Language 3.1.1 OGC Styled Layer Descriptor 1.0 OGC Web Map Context 1.1 OGC Catalogue Service 2.0 HTTP protocol binding, CS-W
Future candidate SDI core standards
OGC Web Map Service 1.3 OGC Web Feature Service 1.1 OGC GML 3.1.1 OGC Catalogue Service 2.0 HTTP protocol binding, CS-W ISO DTS 19139:2006 metadata

Table 3. Core, supplemental, and future SDI standards.

maintenance of operational capabilities as new versions of standards are made available. The problem is compounded when interdependent standards are not revised and released in a coordinated manner. This situation is not much different from issues related to software product development and release cycles. Therefore, one major reason for having a well-defined and agreed-to SDI standards suite is to support software (and standards) life cycle management.

Any release of a new version of the SDI standards suite needs to be predictable and coordinated. Backward compatibility is a key requirement for preserving customer investments in the overall technology. Exceptions to backward compatibility may be tolerated by users if the relationship of new functions creates enough value to fully compensate customer investments in change management. These considerations also apply to new versions of SDI applications.

The period between standard releases is also relevant to software life cycle management. Factors that need to be considered include relationship investments, the added value of the new suite, return on investment, and the ability to enhance SDI applications in a timely manner. A new SDI suite needs to add enough value to justify upgrading an SDI application or portal. Based on previous studies, we know that the initial investment in using SDI 1.0 will add value and reduce life cycle management costs (NASA 2005). Over time, more and more users understand the value and potential of using a standards-based approach. The value may be expressed as money but could also be measured by other indicators. Only if the first generation has generated enough value and passed the return on investment (ROI) decision point will new investments for implementing the next generation of the SDI suite be available. The expression "generation of value" applies to SDIs and SDI applications.

Although the longevity of a given version of an SDI standards suite cannot be determined yet, the concept of generations is helpful for identifying which functions should be included in the first release and which could be moved to a future release (candidates).

SDI zones. User interface requirements, pricing, processing functions, security, and rights management requirements can vary broadly from region to region. These variations are due to local customer requirements, government policies, legal systems, and so forth. A monolithic approach may not work for an SDI that crosses jurisdictional boundaries. This market force is very similar to what the telecommunications industry has to deal with on a daily basis. In telecommunications, standards-based infrastructures effectively deal with different policy, tax, legal, and pricing requirements by accommodating regional variations.

SDIs face similar regional variations. We call them "SDI zones." Rather than forcing every jurisdiction to use the same SDI implementation, architecture, and policy framework, we suggest creating SDI zones to meet local requirements while maintaining interoperability between zones. The advantages of distinct but connected zones are lower dependencies and reduced risks of bottlenecks. If an SDI zone or a connection to it is not operational, other zones are not affected directly. The current SDI 1.0 design should take zones into account and offer a zone-to-zone connection mechanism.

Different zones may implement different versions of the SDI standards suite, or they may implement the same version at different times. Therefore, the connection mechanism also needs to connect zones with different versions.

Zone compatibility. Applications often address a specific user need and thereby create great value for that specific purpose. A classic example in the PC world is graphics programs for professionals. On the one hand, these programs create enormous value for the professional; on the other hand, the number of potential (professional) users and their total investment potential are relative small.

Compatibility between SDI zones is a key requirement for selling an application to a larger market. Conversely, the number of (addressable) early adopters for a specific application has a direct impact on the willingness of developers to invest.

SDIs versus SDI applications. Although the interface between an infrastructure and an application is often expressed technically, it can be defined at a specific point in the value chain on the basis of organizational and economic criteria. Electricity, a classic example, is produced in a power plant, distributed over a network, and then used in an application, for example, a radio or a heater. The infrastructure interface is located at some well-defined point behind an electric meter. Therefore, the responsibility of the infrastructure ends at this interface point, where the downstream power supply is measured for upstream money compensation.

Using this analogy, an SDI is a transport mechanism for spatial data and services. Therefore, a defined gateway is needed to act as the organizational interface between the SDI operator and the SDI application customer.

The distinction between operating systems (OS) and applications in computers provides another analogy. This distinction is defined by a set of application programming interfaces (APIs), that connect programs and allow software applications to be written. The suppliers of operating systems often are not the suppliers of applications. In operating systems, a specific function can be augmented and made reusable if many applications use it. Classic examples are video and audio drivers. These functions were once considered application specific, but because they were used by many applications, a standardized API was eventually included in the operating system.

Another well-known example is the SUN Java Developer Kit (JDK). Although it consists of a large number of functions and interface specifications, the package is released with a single number, for example, JDK 1.4. Application developers and operators can simply define the requirements and state, "JDK 1.4 is required." The Java Community Process (JCP) is used for new projects, demonstrating the value of collaboration between institutions.

In the future, the concept of SDI should include SDI applications as well as the SDI interfaces. OGC Geospatial Decision Support (GeoDSS), for example, could easily be implemented using an SDI 1.0 standards suite coupled with the OASIS Business Process Execution Language (BPEL) standard. The GeoDSS application would load data from different repositories, perform an analysis, load more data, perform another analysis, and so forth. However, the requirements for service chaining are outside the scope of SDI 1.0. A future generation of SDIs could include additional functions.

Governance. SDI 1.0 requires an international consensus process to properly define, document, and manage the standards framework in order to ensure that the needs of the many constituents are properly represented. A structured and open process will facilitate dialogue, approval of the SDI 1.0 framework and future revisions, and effective life cycle management.

GSDI, INSPIRE, ANZLIC, CGDI, GDI NRW, the U.S. NSDI, and various e-government initiatives can provide an excellent forum for refining SDI 1.0, having identified best practices for developing standards-based SDIs and having played active roles in the OGC.

The OGC Architecture Working Group could take responsibility for documenting and reviewing the standards baseline for SDI 1.0. The formal vetting of the SDI 1.0 framework would occur in the OGC Architecture Board (OAB), a new key committee responsible for enforcing consistency and ensuring proper life cycle management of the OGC standards baseline. Life cycle management rigor will ensure that SDI 1.0 is coordinated effectively and that revisions are carefully considered and documented.

CONCLUSIONS

SDIs are becoming a major resource for access to geospatial data and services. Partnerships between the public and private sectors are paying off in higher returns on investment. Perhaps even more importantly, SDIs are contributing to sound decision making, enhanced e-government applications, and better services and are poised to take the next step in their evolution: SDI networks. SDI networks are necessary for emergency preparedness and response, counterterrorism, monitoring of and response to pandemics, and environmental protection. In order for these transnational applications to be effective, SDIs (or SDI zones) must interoperate. The interoperability can be achieved only through consistent and structured implementation of interface and encoding standards. This article proposes a suite of standards for all SDIs.

We recommend that the concept of an SDI standards suite be considered by the global SDI community as an important work item. OGC members agree that SDIs need a well-defined and well-managed suite of standards. We therefore propose that the OGC take formal responsibility for providing life cycle management and documentation of the suite and that the global SDI community take responsibility for defining the actual standards.

REFERENCES

Federal Geographic Data Committee. http://www.fgdc.gov.

Java Community Process. http://jcp.org.

National Geospatial Intelligence Agency. 2005. NGA Announces Requirement for OGC and Complementary Standards. http://www.nga.mil/NGASiteContent/StaticFiles/OCR/nga0518.pdf.

National Aeronautical and Space Agency (NASA): Geospatial Interoperability Office. 2005. Geospatial Interoperability Return on Investment Study. Study performed by Booz Allen Hamilton.

Nebert, Douglas. 2004. Developing Spatial Data Infrastructures: The SDI Cookbook. GSDI. www.gsdi.org/docs2004/Cookbook/cookbookV2.0.pdf.

Moore, Gordon E. 1965. Gordon E. Moore's Law. *Electronics Magazine* (April 19, 1965). http://en.wikipedia.org/wiki/Moore's_law.

Refractions. 2006. Personal correspondence with Paul Ramsey.

A Metaphor-Based Sociotechnical Perspective on Spatial Data Infrastructure Implementations: Some Lessons from India

SATISH K. PURI,[1] SUNDEEP SAHAY,[1] AND YOLA GEORGIADOU[2]

DEPARTMENT OF INFORMATICS, UNIVERSITY OF OSLO, NORWAY,[1] AND INTERNATIONAL INSTITUTE FOR GEO-INFORMATION SCIENCE AND EARTH OBSERVATION (ITC), ENSCHEDE, THE NETHERLANDS[2]

ABSTRACT

Some of the more popular metaphors that have been used to frame and understand spatial data infrastructures (SDIs) are those of the superhighway and the marketplace. These metaphors do not bring out the sociotechnical nature of SDI design and implementation. On the contrary, they emphasize top-down approaches, centralized control, and the view of information as a tradable commodity, while marginalizing the role of communities in shaping infrastructure development. These metaphors also tend to ignore the historically contingent nature of SDI development and the existing ground realities in developing countries like India, for which the rainbow metaphor provides a more appropriate and fuller perspective. The relevance and effectiveness of the rainbow metaphor are demonstrated by a case study in India.

INTRODUCTION

Metaphors have been used for analyzing social phenomena in various fields including organization studies, law, information systems, and more recently the information infrastructure[1] (II) domain. We draw upon some of these experiences in examining the phenomenon of spatial data infrastructure (SDI) implementation. We consider some of the dominant metaphors being used to guide SDI research and practice: information superhighway, marketplace, and rainbow. We believe such an analysis will contribute to developing a stronger sociotechnical perspective on the dynamics of SDI implementation. We discuss how different metaphors can shape implementation processes in different ways and empirically apply the rainbow metaphor to the national spatial data infrastructure (NSDI) of India.

Infrastructures, whether physical (roads, rail, ports, airports, telecommunications, electricity, etc.) or virtual (digital libraries, health care infrastructures, information infrastructures, SDIs, etc.), have important differences, but also similarities. In this article, we argue that the key metaphors guiding SDI implementation to date have focused on the D in SDI, as evidenced by the dominance of the information superhighway metaphor (Mosco 1998). We see an urgent need to shift the focus from the D to the I in SDI so as to give the issues of usage proper weight. Focusing on the I in SDI can also facilitate exploration of new metaphors that transcend some of the constraints of the superhighway metaphor. This refocusing can provide richer sociotechnical insights into the dynamics of SDI implementation.

Below, we discuss the power of metaphors in social analysis and draw upon the rainbow metaphor to analyze the Indian NSDI.

POWER OF METAPHORS

A metaphor is a linguistic mechanism (a figure of speech) used to aid the understanding of one phenomenon in terms of another (Drummond and Hodgson 2003), "a way of thinking and a way of seeing that pervade how we understand our world generally" (Morgan 1986, p. 12). For example, the description of a heart as a "pump" and the corrective treatment of clogged arteries that supply blood to this pump as "bypass" surgery helps nonmedical people visualize the procedure and relate it to their everyday experience. Giddens (1991) and Castells (1996) have developed the metaphors of the "runaway world" and "network society," respectively, to analyze different dynamics around modernity and globalization: our world's out-of-control nature in the case of the former and interconnectedness in the case of the latter.

Metaphors provide an analytical bridge between positivist epistemology and everyday subjective experiences in the process of knowledge development (Ortony 1979, p. 1). Metaphors do not merely embellish language but play a fundamentally constitutive role in our comprehension schemata (Sfard 1998). Checkland (1989) suggests that facts and logic alone do not present the full import of human situations; myths and meanings by which humans make sense of this world are equally important. Lakoff and Johnson (2003, p. 3) also argue that metaphors are pervasive in everyday life, not just in language but in thought and action. Similarly, Ortony (1979, p. 3) emphasizes the communicative potential of metaphors in language, politics, poetry, psychology, cognitive linguistics, philosophy, religion, and architecture (Lakoff and Johnson 2003). Metaphors may often hide as much as they reveal (*ibid.*). For example, the description of a man as a lion emphasizes his physical prowess but may hide his tender and affable disposition. Hirschheim and Newman (1991, p. 37), drawing from the information systems (IS) domain, also caution us that metaphors, although pervasive and sometimes helpful, may also mislead and can be "really dangerous fantasies." For example, Drummond and Hodgson (2003) use a "chimpanzees' tea party" metaphor to explain how efforts to exercise increased control in IT-related project management may actually be counterproductive and lead to chaos. This metaphor tends to ignore the predominant role of institutional politics and power in shaping the eventual success or failure of such projects (Beck 2002).

The power of metaphors is being recognized in various domains, including law. In her study on the use of metaphors in jurisprudence, Gore (2003, p. 406) argues that "no matter how apt or misguided the analogy may prove to be, the spoils of legal victory often go to the proponent of the most persuasive metaphor." Judges, courts, and juries regularly resort to analogies and metaphors of existing technologies to gain some level of comfort and decide to what extent established legal principles cover the new technology (*ibid.*). According to Blavin and Cohen (2002), courts and commentators employ metaphors as heuristics to generate hypotheses about the application of law to novel, unexplored domains, such as the Internet. These researchers argue that when "courts encounter new technologies not yet anticipated by law, their reliance on analogical reasoning plays a profoundly important role in the application of legal rules. By failing to adopt appropriate metaphors in regulating new technologies, courts risk creating bad law" (*ibid.*, p. 268). IS and II research has also in recent years started to explore the power of metaphors, as discussed below.

METAPHORS FOR INFORMATION INFRASTRUCTURES

The use of metaphors as a heuristic device for thinking about national information infrastructures and SDIs has received considerable attention in recent times (e.g., Sawhney 1992, 1996; NRC 1994; Rohrer 1995; White 1996; Clement and Shade 1998). We analyze the use of three of these metaphors: superhighway, marketplace, and rainbow.

The information superhighway metaphor derives its power from its apparent ability to fill the "investment-benefit" conceptual gap of IIs. It evokes an easy-to-grasp, clean, obedient, and mechanistic model of an infrastructure, with emphasis on predictability, procedures, and efficiency of delivery (OECD 2000; Streeter 2003). The metaphor implies a key role for central government in promoting II development, a top-down "construction" with intimate government direction, the assumption of harmonious collaboration between various governmental agencies (NRC 1994), and an emphasis on technical access guided by the needs of commerce and government (White 1996). For example, the U.S. General Accounting Office (GAO) (1995) refers to "a meta-network that will seamlessly link thousands of broadband digital networks, [. . .] allow a two-way flow of information, with users being able to receive and transmit large volumes of digital information, and enable equal access for service and network providers" (p. 11). The GAO report projects the investment required for the information superhighway to be several billion dollars (p. 2), with the implicit assumption that the financial venture will eventually pay off similarly to the transportation infrastructure of the past.

The information marketplace metaphor evokes an image of a market where buyers meet sellers across the boundaries of space and time to engage in commercial transactions (Mosco 2003). The metaphor goes beyond the mechanical access to information to encompass the utility of information, traded by e-literate people in a laissez-faire, market-driven environment, with the role of governmental regulation largely ambiguous (NRC 1994). Because of the metaphor's emphasis on the utility of information, ethical questions (such as which uses are encouraged, allowed, discouraged, or forbidden) become prominent (*ibid.*).

The marketplace metaphor also casts a shadow on the civic life aspect of a marketplace, better captured in the "electronic agora" and "marketplace of ideas" variations (NRC 1994). It also treats information primarily as a commodity and downplays the fact that geographic information in particular possesses the classic characteristics of a public good[2] (Georgiadou and Groot 2002). Onsrud (2004) suggests that the digital library is a far more appropriate metaphor for exploring possible future directions for SDIs and for providing incentives to public data collectors to document their spatial datasets.

The marketplace rhetoric was also evident in the report of the Task Force on Financial Mechanisms (TFFM 2004) commissioned by the Secretary General of the United Nations to study the issue of financing the information society and bridging the global digital divide. The TFFM report (2004, pp. 73–74) acknowledges that rural areas and poor communities around the world pose a challenge to the marketplace approach for the information society. However the report's discourse adheres to the prevalence of the market and glorifies the 50 percent of the world's population that lacks access to information as a vast, untapped opportunity for market expansion. Various civil society groups have criticized the market-driven, private-investment, private-ownership model of the information and communication technology (ICT) infrastructure promoted in this key report. For example, Peyer et al. (2005, p. 12) note that:

> [. . .] apart from the ongoing success of some decades-old publicly owned networks, this ignores more recent waves of experience. India is making significant strides in rural and village access by putting in place publicly-owned networks. In Stockholm, Amsterdam and many U.S. urban and rural municipalities, publicly-owned broadband networks and services are offering services, either retail or wholesale, designed to maximize overall benefits to the city . . . Cooperative ownership is a growing option in many rural areas. Some are there because the private sector has failed; others are simply providing a service by the best means possible, and sometimes competing with private networks.

The marketplace metaphor emphasizes private investment and ownership, thereby undermining the vast and largely untapped potential of harnessing local community resources to support network development. Besides lowering costs and improving services, participatory development of such networks also enhances community empowerment (Peyer et al. 2005, p. 19). The marketplace metaphor tends to ignore history and how the current marketplace is shaped by legacy systems, both technical and institutional.

Sawhney (2003, p. 25) makes a similar point with respect to information infrastructures, arguing that "a new technology does not strike roots and grow on a virgin ground. Instead, it encounters a terrain marked by old technologies. The new technology's growth then is shaped not only by its own potentialities but also the opportunities and restraints created by the systems based on old technologies."

Table 1 lists the key II features emphasized or overlooked by the superhighway metaphor and the marketplace metaphor. We posit that the rainbow metaphor, discussed in reference to the Indian NSDI below, may help address some of the concerns with respect to the neglect of history (by the information infrastructure metaphor) and the assumptions of universal access and connectivity (by the marketplace metaphor).

Metaphor	II implementation features	
	Emphasized	Overlooked
Information superhighway	Interconnectedness (like that of a road network), top-down design, government control	History (in trying to build from scratch)
Marketplace	Importance of user (market) needs in shaping demand	Public good, history, role of communities in shaping development, inequities in access and connectivity

Table 1. Key features of the information superhighway and marketplace metaphors.

INDIAN NSDI

The NSDI of India was inspired in part by the superhighway and marketplace metaphors. We base this statement on secondary data (such as conference presentations and minutes of meetings held at various stages of implementation), informal meetings with officials, and focused interviews with members of the national task force in 2005.

Development of the Indian NSDI was initiated in 2000 jointly by the Department of Science and Technology (DST) and the Indian Space Research Organization (ISRO) through the establishment of a national task force to prepare an action plan under the aegis of DST. The influence of the information highway metaphor is evident in the following statement of the secretary of DST (DST 2001, p. 5, Foreword by Secretary, DST):

> There is a widespread consensus, internationally, that spatial data sets need to be integrated to create what is called a geo-spatial data infrastructure. Such infrastructures have been likened to information highways, linking a variety of databases and providing for the flow of information from local to national levels and eventually to the global community.

The NSDI initiative can be viewed as largely technology driven (Winner 1989). The influence of the marketplace metaphor can be seen in the following statement of the secretary of DST (DST 2001, p. 5, foreword by Secretary, DST):

> In the emerging market-place, geographic or geo-spatial information occupies a pre-eminent position. In fact, the use of high quality, reliable, geo-spatial information is crucial for every sphere of socio-economic activity—disaster management, forestry, urban planning, land management, agriculture, infrastructure development, business demographics etc.

A comprehensive review conducted in 2003 at Agra (http://www.nsdiindia.org/publication/) described the NSDI as "another Taj in the making," to metaphorically reflect the rather grandiose, top-down nature of the vision (similar to an information superhighway).[3]

In an interview with one of the private-sector members of the national task force, we were told that:

> The scientific institutions are still contesting who should be in charge. Each is saying that "they are the SDI" rather than discussing how they can facilitate the establishment of an effective SDI.

The CEO of a large GIS consultancy service, whom we interviewed in February 2005, was critical of the approach adopted by the government scientists and technocrats in DST and ISRO for involving the private sector in the NSDI. She said, "Those in charge of NSDI have formulated no clear-cut policy or tested a business model for data sharing amongst the NSDI stakeholders. The fact is that those not used to sharing are now in charge of the entire process, with no viable strategy in sight." A senior executive in another large private-sector organization expressed his disenchantment as follows:

> NSDI was conceptualized and is being implemented by the government, for the government, within the bureaucratic framework of the government. . . We would not participate in NSDI unless it is established outside the pale of the government, and functions as an enlightened, independent body.

The above brief description of the Indian NSDI highlights the information superhighway and marketplace metaphors emphasized in the rhetoric of the implementation planning and their negation in actual practice. For example, the information superhighway metaphor is espoused by the policy makers in emphasizing universal connectivity, while historically, the government's control over topographic maps makes this promise difficult to implement in practice. Similarly, the marketplace metaphor emphasizes that the users will have their own needs and demands which will drive the process of supply. However, in practice, the users have been almost totally neglected in the designing of NSDI, which makes it difficult for them to understand or shape the supply dynamics. Also, in a state-controlled domain where the private sector has literally had no role to play until recently and where the use of maps is not historically evident (Sahay and Walsham 1997), the assumptions of a marketplace approach contradict the historical realities on the ground.

Below we examine how the use of the rainbow metaphor may be more relevant in the Indian case, given the historical and contextual realities.

RAINBOW METAPHOR

Gurstein (2004) has criticized the overwhelming emphasis of the superhighway metaphor on access. This emphasis undermines the importance of analyzing how, by whom, and for what this access might be used. The information superhighway metaphor fails to clarify what access to the information infrastructure encompasses and does not account for the intricate relationships between the social and technical architectures of the information infrastructure (Clement and

Shade 1997, p. 34). Some aspects of the importance of people and their transactions may be better captured by the rainbow metaphor.

The rainbow metaphor for access to information infrastructure was proposed by Clement and Shade (1997) with the intention to strengthen public policy perspectives in the Canadian II debate. Clement and Shade are at odds with the corporate view of cyberspace as a shopping mall and advocate the view of cyberspace as a public space (O'Brien 2001). The rainbow metaphor recognizes the multiple usage patterns in retrieving and creating relevant content, encompasses conventional and new media, and emphasizes the interplay of social and technical dimensions in infrastructure development, helping to define which services are essential to whom. Most importantly, it helps to identify access gaps, in other words, those social segments likely to be left out by market forces acting alone, and hence emphasizes the need for their protection via collective public initiatives. The seven layers conceptualized for the rainbow metaphor are carriage, devices, software, content, service/ access provision, literacy, and governance; these also correspond to the important regulatory distinctions between carriage and content.

Table 2 lists the seven layers of the rainbow metaphor for the Canadian network (Clement and Shade 2000) and alternative conceptualizations for the Indian NSDI. The table illustrates how the rainbow metaphor, while emphasizing the

Layer	Characteristics for the following NSDI conceptualization:	
	Current (Canada)	Alternative (India)
Carriage	Multiplicity of networks, key role of the Internet, promotion of access at affordable costs	Universal connectivity through a wide range of telecommunication technologies and the Internet, reduction of inequities in service provision
Devices	Proliferation of device forms and increasing affordability, wireless connectivity, and use of PDAs and mobile phones	Affordable ICT devices based on needs, community-based models for sharing device resources
Software tools	Increased embedding of software; affordable, multilingual, interoperable, privacy-enhancing applications	Free and open-source applications in local languages
Content/services	Provision of a wide range of information to user groups; enabling information to speak to other information; affordability, reliability, cultural compatibility, freedom from censorship	Accommodation of a wide range of users based on participatory principles; reduction of government controls to make spatial data more freely available to civil society; identification of services not provided by market forces but required by civil society
Service/access provision	Access primarily through employers or educational institutions; need for access to affordable network providers	Need for government departments owning spatial data to become more user-friendly and marketing oriented; alternative intermediary institutions (e.g., NGOs) facilitating access and quality control of data; identification of central (provider-specific) and local (user-specific) databases and mechanisms for their interaction
Literacy/social facilitation	Inclusion of expertise based on formal and informal education relevant to everyday life	Building awareness and culture around spatial data; reform of university curricula to include II/SDI-specific education; key role of training institutions in the private sector for shorter-term capacity building
Governance	Multiplicity of stakeholders, inculcating democratic participatory process still a challenge	Inclusion of users, private sector, research organizations, and universities; bottom-up participatory process; access to SDI as not an end in itself but a means of addressing development and business concerns

Table 2. Rainbow metaphor characteristics.
Clement and Shade 2000.

various interconnected layers of the NSDI implementation effort, can also facilitate exploration of locally developed solutions. This reconceptualization provides more effective alternatives than those dictated by the information superhighway and marketplace metaphors. We look at several locally inspired technological initiatives in India from the last decade or so to demonstrate the usefulness of the rainbow metaphor.

Carriage: STD booths. The indigenous development of digital switching hardware and software for rural automatic exchanges (RAX) by the Centre for Development of Telematics was the key driver behind India's telecom revolution of the late 1980s and early 1990s. These designs were appropriate not only for the hot, humid, and dusty operating environment but also for the social settings. Subscriber trunk dialing (STD) facilitated by this infrastructure provides national and international telephonic connectivity to people through more than 700,000 STD "booths" that dot the urban and rural landscapes of India. These booths are owned and managed largely by private entrepreneurs. Chakravartty (2004, p. 242) comments, "Although a seemingly modest accomplishment by the technological standards of the day, the STD booth phenomenon has truly transformed the communications landscape in urban, and increasingly rural India."

This example points to the need to adopt locally supported technologies compatible with historically based social relations and customs.

Devices: Gyandoot. An innovative experiment called Gyandoot (meaning "purveyor of knowledge") was aimed at using ICT to empower local communities. It was initiated in 1999 in one of the most underdeveloped tribal areas of the country (Dhar district of Madhya Pradesh state). Communities were actively involved, for example, in providing locally relevant input for system design and service prioritization. The main objective of the Gyandoot project was to establish a distributed computer network in the district so as to provide online information to the local people on subjects and problems that are part and parcel of everyday rural life in India. It also facilitated access to district government departments for residents who in the past had to physically travel long distances to the district headquarters (Rajora 2002). In analyzing Gyandoot's success, Warschauer (2003a, p. 38) observed:

> While the number of users is a small percentage of the population, but the benefits of this project, such as improved government services, eventually ripple outwards to friends, families and co-workers. . . . the underlying approach—a combination of well-planned and low-cost infusions of technology with content development and educational campaigns targeted to social development—is surely a healthy alternative to projects that rely on planting computers and waiting for something to grow.

A lesson for the NSDI is that the ICT-based applications need to be socially sensitive and consonant with end user perspectives.

Software tools: GRAM. The development of GRAM (Geo-Referenced Area Management) (in local parlance, gram means "village") was initiated in the early 1990s by the Centre of Studies in Resources Engineering (CSRE), Indian Institute of Technology, Mumbai, to provide cheaper, easily understood GIS software consistent with the specific needs of India's rural sector for spatial applications. GRAM, now fully operational, does not provide some of the more sophisticated and high-end features included in commercial GIS packages. While catering to commonly used GIS features, GRAM also provides an excellent learning platform to potential future users of spatial technologies. Such indigenous development efforts need to be acknowledged and incorporated into NSDI thinking, which is currently being shaped by external vendors. Systems like GRAM also offer significant cost savings compared to commercial products.

Content/services: Bhoomi. The recent Bhoomi (meaning "land") project in the state of Karnataka in India is aimed at digitalizing land records, which provide ownership information. Authenticated copies of land ownership information are often needed by individual farmers for many reasons: for example, applying for a bank loan, obtaining an electricity connection, and so forth. Before these records were digitalized, the ownership certificates had to be obtained from the local patwari (a junior official in the land records department located at subdistrict level). In addition, these records were not regularly updated (such as incorporating transfer/sale deeds into the existing records). Warschauer (2003b) describes how the Bhoomi project helped verify land sales and update 20 million records. Copies of these records can now be purchased for about 30 U.S. cents without long waiting periods or the need to make several visits. The earlier unethical practice of the patwari extracting "unofficial payments" from the needy farmers has also been, by and large, done away with.

The Bhoomi initiative represents a simple, yet effective spatial application which caters to a massive local need, demonstrating that often the simplest information is the most valuable. The Indian NSDI initiative should be more oriented towards providing such uncomplicated but much needed services initially before moving on to more complex application fields, involving, for example, 3D modeling and simulations.

Service/access provision: GIS implementation in Anantapur district. The development and use of spatial technologies like GIS in India has generally been characterized by top-down approaches and a marked absence of user participation (Sahay and Walsham 1996, 1997), resulting in project failures (Walsham and Sahay 1999). More recently, bottom-up approaches drawing on the potential of ICT in local development contexts, mainly driven by political initiatives (like in Andhra Pradesh), have also been promoted. These approaches encourage the incorporation of indigenous domain knowledge into microlevel programs: for example, knowledge around wise land use and water harvesting in local settings, accumulated by farming communities through trial and error over generations (Mathias 1994).

In the Anantapur district of Andhra Pradesh, for example, the requisite spatial data was collected by local teams composed of technicians, nongovernmental organizations (NGOs), and farmers (using handheld GPS devices). The superior technical infrastructure and scientific knowledge available at premier national-level organizations (like the National Remote Sensing Agency) was then drawn upon to construct a geographically consistent GIS database according to prescribed rigorous standards of accuracy. The database and the accompanying applications were used to support local development plans. The use of this database improved land productivity as well as contributed to more efficient water harvesting in this historically drought-prone area (Puri and Sahay 2003).

An implication of this for the NSDI program is the need to incorporate bottom-up design approaches as well as to be able to realistically support microlevel needs of spatial data and its analysis. Embracing a judicious mix of top-down and bottom-up methods has also been emphasized in the SDI literature (Groot and McLaughlin 2000; Georgiadou et al. 2005).

Literacy/social facilitation: role of intermediary agencies. NGOs have come to play increasing advocacy and intermediary roles vis-á-vis governments and people. These agencies are generally able to effectively communicate with government officials because of their educational background, experience, and contacts with the media and can serve as "gateways" between people and officials. The increasingly significant role of local governmental agencies and NGOs as mediators between global challenges and local concerns of exclusion and marginalization also needs to be recognized (Beck et al. 2004). In the NSDI context, active attempts need to be made to identify and encourage technically competent and locally rooted NGOs and academics to act as mediators between scientific departments and end users such as district and subdistrict government agencies. This could potentially foster user participation as well as provide more effective communication channels to enhance mutual understanding among designers and users of technologies like the NSDI.

Governance. The above illustrations of relatively successful applications of ICTs in developmental scenarios underscore several common themes: (1) locally relevant hardware and software, (2) multiplicity of stakeholders (including end users) involved in design and implementation, (3) bottom-up and participatory processes, (4) involvement of end users based on respect for their knowledge and aspirations, and (5) ICT not as an end but as a means of addressing development concerns. These characteristics constitute the governance layer of the rainbow metaphor envisaged for the Indian NSDI.

CONCLUSIONS

This sociotechnical thinking inspired by the use of metaphors emphasizes the need to examine technological and organizational characteristics together rather than in mutual exclusion. Kling et al. (2000) point out that, in contrast to standard task-technology models, sociotechnical network approaches underscore work practices, conceptual understanding of the actors, and the realization that "[socio-technical] configurations interact with human activities, such as work" (p. 46). These approaches focus on improving work practices and understanding how collaborative learning takes place. Skill development takes precedence over improving efficiencies. Sociotechnical design approaches are not yet evident in the Indian NSDI case, and end user participation is nearly nonexistent. As a result, end user organizations remain limited in their ability to meaningfully apply spatial data in specific contexts, which is the *raison d'être* of the NSDI. The focus needs to shift from the D in SDI to the I. Since an NSDI is large, layered, and complex, and "because it means different things locally" (Star 1999, p. 382), its development cannot be top-down but must involve negotiations among different stakeholders to accommodate their competing interests. The rainbow metaphor can help NSDI planners to visualize why and how sociotechnical approaches can promote successful SDI design and implementation.

ENDNOTES

1. Information infrastructures are heterogeneous networks subsuming varied technologies, networks, and standards to support a diversity of information system application areas over time and space (Hanseth 2000). SDI is a special case of II specifically geared towards geographic information. A national SDI comprises all relevant geographic data of a country.

2. Public goods are nonrival in consumption and nonexcludable. Hence, they can elicit in consumers the temptation for free riding as well as unwillingness to reveal their level of demand. Markets are unable to supply nonexcludable goods. As a result, public goods are underprovided, unless the government intervenes. In the case of geographic data, government intervenes by producing the data itself.

3. The famous Taj Mahal, one of the Seven Wonders of the World, is located in Agra, India.

REFERENCES

Beck, E. 2002. P for political: Participation is not enough. *Scandinavian Journal of Information Systems* 14:77–92.

Beck, E., S. Madon, and S. Sahay. 2004. On the margins of information society: A comparative study. *The Information Society* 20:279–90.

Blavin, J. H., and I. G. Cohen. 2002. Gore, Gibson and Goldsmith: The evolution of metaphors in law and commentary. *Harvard Journal of Law & Technology* 16:265–85.

Castells, M. 1996. *The rise of the network society: The information age, economy, society and culture.* Oxford: Blackwell Publishers.

Chakravartty, P. 2004. Telecom, national development and the Indian state: A postcolonial critique. *Media, Culture & Society* 26:227–49.

Checkland, P. 1989. Soft systems methodology. In *Rational analysis for a problematic world,* ed. J. Rosenhead, 71–100. John Wiley & Sons.

Clement, A., and L. R. Shade. 2000. The access rainbow: Conceptualizing universal access to the information/communication infrastructure. In *Community informatics: Enabling communities with information and communication technologies,* ed. M. Gurstein, 32–51. Idea Group Publishing.

Clement, A., and L. R. Shade. 1998. The access rainbow: Conceptualizing universal access to the information/communications infrastructure. Information Policy Research Program working paper 10, University of Toronto, Faculty of Information Studies.

Clement, A., and L. R. Shade. 1997. What do we mean by "universal access?" *Social perspectives in a Canadian context.* Information Policy Research Program working paper 5, University of Toronto, Faculty of information Sciences.

DST. 2001. *National spatial data infrastructure strategy and action plan: discussion document.* ISRO-MMRMS-SP-75-2001, New Delhi, Government of India.

Drummond, H., and J. Hodgson. 2003. The chimpanzees' tea party: A new metaphor for project managers. *Journal of Information Technology* 18:151–58.

GAO. 1995. *Information superhighway: An overview of technological challenges.* Washington, DC.

General Accounting Office. 1995. GO/AIMD-95-23, Report to Congress. http://www.gao.gov/archive/1995/ai95023.pdf (accessed January 12, 2006).

Georgiadou, Y., S. K. Puri, and S. Sahay. 2005. Towards a potential research agenda to guide the implementation of spatial data infrastructures: A case study from India. *International Journal of Geographic Information Science* 19:1113–1130.

Georgiadou, Y., and R. Groot. 2002. Capacity building for geo-information provision: A public goods perspective. Paper presented during GISDECO, 6th Seminar on GIS and Developing Countries, Enschede, the Netherlands.

Giddens, A. 1991. *Modernity and self-identity.* Palo Alto, CA: Stanford University Press.

Gore, S. A. 2003. A Rose by any other name: Judicial use of metaphors for new technologies. *Journal of Law, Technology & Policy* 2:403–456.

Groot, R., and J. McLaughlin. 2000. Introduction. In *Geospatial data infrastructure: Concepts, cases, and good practice,* ed. R. Groot and J. McLaughlin, 1–12. Cambridge: Cambridge University Press.

Gurstein, M. 2004. Effective use and the community informatics sector: Some thoughts on Canada's approach to community technology/community access. *Communications in the Public Interest* 2:223–44. Ottawa: Canadian Centre for Policy Alternatives.

Hanseth, O. 2000. The economics of standards. In *From Control to Drift,* eds. C. U. Ciborra, K. Braa, A. Cordella, B. Dahlbom, A. Failla, O. Hanseth, V. Hepso, J. Ljungberg, E. Monteiro, and K. A. Simon, 56–70, Oxford: Oxford University Press.

Hirschheim, R., and M. Newman. 1991. Symbolism and information systems development: Myth, metaphor and magic. *Information Systems Research* 2:29–62.

Kling, R., H. Crawford, H. Rosenbaum, S. Sawyer, and S. Weisband. 2000. *Learning from social informatics: Information and communication technologies in human contexts.* Project report version 4.6b. Indiana University, The Center for Social Informatics.

Lakoff, G., and M. Johnson. 2003. *Metaphors we live by.* Chicago: University of Chicago Press.

Mathias, E. 1994. *Indigenous knowledge and sustainable development.* Working paper 53, Cavite 4118, International Institute of Rural Reconstruction, Y. C. James Centre, Philippines.

Morgan, G. 1986. *Images of organizations.* Thousand Oaks, CA: Sage Publications.

Mosco, V. 2003. *The digital sublime: Myth, power, and cyberspace.* Cambridge, MA: MIT Press.

Mosco, V. 1998. Myth-ing links: Power and community on the information highway. *The Information Society* 14:57–62.

National Research Council. 1994. *Rights and responsibilities of participants in networked communities.* Washington, DC: National Academy Press.

O'Brien, R. 2001. Research into the digital divide in Canada. Information Policy Research Program, Faculty of information Sciences, University of Toronto. http://www.web.net/~robrien/papers/digdivide.html (accessed February 15, 2006).

Organization for Economic Cooperation and Development. 2000. Materials from the special session on information infrastructures. In *National information infrastructure initiatives: Vision and policy design,* eds. B. Kahin and E. J. Wilson III, 569–612. Cambridge, MA: MIT Press.

Onsrud, H. J. 2004. Exploring the library metaphor in developing a more inclusive NSDI. *GeoData Alliance.* http://www.geoall.net/library_harlanonsrud.html (accessed June 22, 2006).

Ortony, A. 1979. Metaphor: A multidimensional problem. In *Metaphor and Thought,* ed. A. Ortony, 1–18. Cambridge, MA: Cambridge University Press.

Peyer, C., J. Carron, M. Thorndahl, M. Egger, S. O'Siochru, and Y. Steiner. 2005. *Who pays for the information society?* Challenges and issues on financing the information society. Lausanne: Bread for All. http://www.ppp.ch/cms/IMG/ Financing_IS.pdf (accessed March 16, 2006).

Puri, S. K., and S. Sahay. 2003. Participation through communicative action: A case study of GIS for addressing land/water development in India. *IT for Development* 10:179–99.

Rajora, R. 2002. Bridging the digital divide. New Delhi: Tata-McGraw Hill Publishing Company Ltd.

Rohrer, T. 1995. Conceptual blending on the information highway: How metaphorical inferences work. Philosophy Department, University of Oregon. http://philosophy .uoregon.edu/metaphor/iclacnf4.htm (accessed March 10, 2006).

Sahay, S., and G. Walsham. 1997. Social structure and managerial agency in India, *Organization Studies* 18:415–44.

Sahay, S., and G. Walsham. 1996. Implementation of GIS in India: Organizational issues and implications. *International Journal of Geographical Information Systems* 10:385–404.

Sawhney, H. 2003. Wi-Fi networks and the rerun of the cycle. Info: The Journal of Policy, *Regulation and Strategy for Telecommunications* 5:25–33.

Sawhney, H. 1996. Information superhighway: Metaphors as midwives. *Media, Culture and Society* 18:291–314.

Sawhney, H. 1992. The public telephone network: Stages in infrastructure development. *Telecommunication Policy* 16:538–52.

Sfard, A. 1998. On two metaphors for learning and the dangers of choosing just one. *Educational Researcher* 27:4–13.

Star, S. L. 1999. The ethnography of infrastructure. *American Behavioural Scientist* 43:377–91.

Streeter, T. 2003. The net effect: The internet and the new white collar style. School of Social Science (SSS) Workshop paper 14 (unpublished), Princeton University, Institute for Advanced Study, School of Science. http://www.sss.ias.edu/ papers/paper14.pdf (accessed February 3, 2006).

Task Force on Financial Mechanisms. 2004. The report of the task force on financial mechanisms for ICT for development. World Summit on the Information Society. http://www.itu.int/wsis/tffm/final-report.pdf (accessed March 9, 2006).

Walsham, G., and S. Sahay. 1999. GIS for district-level administration in India: Problems and opportunities. *MIS Quarterly* 23:39–65.

Warschauer, M. 2003a. Demystifying the digital divide. *Scientific American* (August):34–39.

Warschauer, M. 2003b. *Technology and social inclusion: Rethinking the digital divide.* Cambridge, MA: MIT Press.

White, P. B. 1996. Online services and "transactional space": Conceptualizing the policy issues. A paper presented at the 20th AIERI/IAMCR/AIECS Conference and General Assembly, Sydney, Australia. http://www.komdat.sbg. ac.at/ectp/White_P.htm (accessed December 5, 2005).

Winner, L. 1989. *Autonomous technology: Technics-out-of-control as a theme in political thought.* Cambridge, MA: MIT Press.

Geospatial Data Infrastructure for Sustainable Development of East Timor

TRACEY P. LAURIAULT AND D. R. FRASER TAYLOR

*GEOMATICS AND CARTOGRAPHIC RESEARCH CENTRE, DEPARTMENT OF GEOGRAPHY AND
ENVIRONMENTAL STUDIES, CARLETON UNIVERSITY, OTTAWA, CANADA*

ABSTRACT

East Timor became independent from Indonesia in August of 1999, and shortly
thereafter an Indonesia-backed militia destroyed critical natural resources and
infrastructures. A host of overseas development agencies assisted with humanitarian aid, reconstruction, and nation building. East Timor is unique, since
its Constitution includes sustainable development as a key principle. Accurate
and timely geospatial data can assist sustainable development decision making.
Geospatial data infrastructures (GDIs) are the intersectoral, cross-domain, interdepartmental consensus-making mechanisms by which a nation can manage its
geospatial data assets. Aid agencies compiled much data and rendered some maps,
but these were project specific, limiting their utility in other contexts. Currently
no formal body is coordinating these information resources. East Timor has many
data requirements, and an East Timor GDI will help meet the constitutional mandate for sustainable development. Seven GDIs models are examined, and elements
from them are selected for a proposed hybrid GDI for the nation. It is argued that
the major need is for institutional rather than technological development.

The Democratic Republic of East Timor became the United Nations' 191st member in 2002, shortly after its physical and social infrastructures were totally destroyed. Today a new nation state is being created with sustainable development as a core value.

Sustainable development is best achieved when decisions are made with reliable, accurate, relevant, and timely spatially referenced data. Geospatial data infrastructures (GDIs) are a good mechanism to manage these data (Masser 1998, p. 7; Lopez 1997). A GDI is a collection of technologies, policies, people, and institutional arrangements that facilitate the availability of and access to geospatial data (Groot and McLaughlin 2000). It directs the who, how, what, and why of geospatial data at various scales as the data are collected, stored, manipulated, and analyzed (Ezigbalike, Selebalo, Faiz and Zhou 2000). Sustainable development is "development which meets the needs of the present without compromising the ability of future generations to meet their own needs" (UNCED 1992) and "the reconciliation of society's developmental goals with its environmental limits" (NRC 2002). East Timor is balancing the provision of essential services to its citizens with trying to sustainably manage its natural resources.

The reconstruction of East Timor's information and communication technologies (ICTs) benefits from the well-developed ICTs of various aid and financial agencies operating in the country as well as those of its historical allies (Portugal, United States, Canada), neighboring countries (Australia, New Zealand, Malaysia, Japan), and regional and overseas development agencies. Unfortunately, there is little coordination of information among these actors. Consequently, East Timor has been mapped in an ad hoc fashion before and after the 1999 referendum by a host of international organizations to inform reconstruction, orient aid workers, locate resources, facilitate international communication, and support defence. Australian Defence and AusAID have assisted with the creation of a small geomatics (geospatial technology) unit, and no plans exist to manage these data or build the capacity to effectively utilize them throughout the government of East Timor. Aid agencies complain of a lack of data to inform their work, and there is no mechanism to collect and coordinate existing data.

East Timor's culture, geography, size, and population, combined with its unfortunate history, may in fact be its assets in terms of building a GDI. East Timor does not have the difficulties experienced by more established economies attempting to build GDIs, since it does not have to change existing practices and structures. It may be able to build consensus-making and collaborative work environments from the outset. A GDI is more than technology and geospatial data; "of equal importance are the individuals, institutions, and technological and value systems that make it a functional entity, one that serves as a basis for much of the business of a nation" (Mapping Sciences Committee 1993). GDI initiatives are beginning to be recognized by aid agencies as good governance strategies.

East Timor is located in a region where excellent GDI examples and considerable geospatial experience exist: the Permanent Committee on GIS Infrastructure for Asia and the Pacific (PCGIAP); the Australia and New Zealand Land Information Council (ANZLIC), which guides the activities of the Australian Spatial Data Infrastructure (ASDI); Malaysian Geospatial Data Infrastructure (MyGDI), formerly known as National Infrastructure for Land Information System (NaLIS); and the Japanese National Spatial Data Infrastructure (JNSDI).

Both Canada and Portugal have aid and historical ties to East Timor, and the Canadian Geospatial Data Infrastructure (CGDI) and the Portuguese Sistema Nacional de Informação Geográfica (SNIG) could provide assistance.

We argue that building a GDI at an early stage is a good governance strategy for East Timor, particularly if the national government is to meet its responsibilities to sustainably manage its territory. The new government needs data and information to fulfill its constitutional mandate, to make informed decisions, and to address social, economic, and environmental issues. We examine a number of GDI models and identify elements to be incorporated in a proposed GDI for East Timor. A GDI is a key mechanism for sustainable development and should be built sooner rather than later. The initial focus should be on institutional, organizational, and managerial components followed by technological components, and outside assistance will be required to help build indigenous capacity.

INDEPENDENCE OF EAST TIMOR

After centuries of Portuguese colonial neglect and a 25-year brutal Indonesian illegal occupation, a referendum supported by the UN Mission in East Timor (UNAMET) took place on August 30, 1999 (figure 1). In that referendum 78.5 percent of citizens voted in favor of autonomy, and immediately thereafter a premeditated Indonesian-backed campaign of violence, looting, and arson took place (UN-CCA 2000, UN and WB 1999, UN General Assembly 2000). Hundreds of civilians were killed, and all government buildings and records were destroyed (UNDP and the UNPF 2000), leaving limited institutional memory.

On October 25, 1999, the United Nations Transitional Administration in East Timor (UNTAET) was established with full legislative and executive authority and responsibilities for law, order, and security in order to begin the process of sustainable development (UN and World Bank 1999). Many aid agencies supported the humanitarian and reconstruction effort, but there was a lack of coordination. There was also a lack of local participation and very limited forward planning (UNTAET 2000a).

At independence East Timor was one of the poorest areas in Southeast Asia.

Every aspect of East Timor's physical infrastructure, institutions, social infrastructure, and communications and every sector of its economy had to be rebuilt.

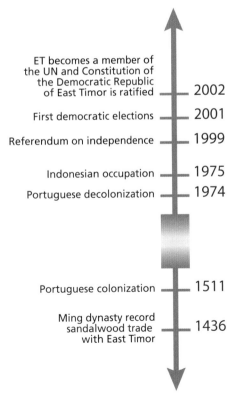

ET becomes a member of the UN and Constitution of the Democratic Republic of East Timor is ratified — 2002

First democratic elections — 2001

Referendum on independence — 1999

Indonesian occupation — 1975

Portuguese decolonization — 1974

Portuguese colonization — 1511

Ming dynasty record sandalwood trade with East Timor — 1436

Figure 1. Historical timeline.

Rebuilding meant balancing immediate humanitarian needs with long-term sustainability. Giving sustainable development priority during such a crisis situation is difficult to justify, and environmental mismanagement during reconstruction caused much destruction (Tais Timor 2000a; Sandlung et al. 2001; Phillips 1999).

CREATING THE GOVERNING STRUCTURES OF A NEW NATION

Constitution of the Democratic Republic of East Timor. Thirty thousand citizens participated in a series of 200 hearings held in 65 subdistricts to create the East Timor Constitution (UN Department of Public Information 2001). The Constitution includes "the development of the economy and the progress of science and technology" along with the protection of the environment and the preservation of natural resources (Constituent Assembly of East Timor 2002). Furthermore, section 61, Environment, states that "everyone has the right to a humane, healthy, and ecologically balanced environment and the duty to protect it and improve it for the benefit of future generations." The Constitution provides for the preservation of cultural heritage and the creation, promotion, and guarantee of the "effective equality of opportunities between women and men."

Governance. Governance is "the exercise of political, administrative and economic authority to manage a country's affairs at all levels. It comprises the mechanisms, processes, and institutions through which citizens and groups articulate their interests, exercise their legal rights, meet their obligations, and mediate their different interests" (UN-CCA 2000, p. 92). GDIs are mechanisms for a good governance strategy. In the East Timorese context, governance means "a high degree of simplicity, flexibility and adaptation to ongoing and dynamic change" (UN-NPDA 2002, p. 70) and a lean, merit-based civil service (UN and WB 1999, p. 19).

The new administrative structure consists of 13 districts (figure 2) and 65 subdistricts with appointed and elected public officials at each level. The traditional councils of elders, local leaders, and hamlet heads also remain part of an informal decision-making structure (UN-NPDA 2002). Stratified local male- and elder-centered hierarchical authority structures remain strong in rural areas (Barbero Magalhães 1994, UN-CCA 2000, Phillips 1999, Taylor 1999, Ormeling 1957), but allegiances based on lineage are being challenged, as women and youth want to be included in decision-making processes (UN-NPDA 2002).

Citizens have been integrated into all major decision-making positions in the administration (one-third of them women), and authority was transferred to the districts. The administrative Civil Service Academy has given introductory courses which include computer training, and a capacity development plan has been implemented to train senior and middle managers (UN Secretary General 2001; UN-NPDA 2002). There are few computer technicians, engineers, and social scientists. ICT is a new concept, and there remains a need to develop an information and technology culture. Computer literacy is low, although some returned refugees have had some exposure. There have been few opportunities to develop professional associations around ICT, standards, geomatics, or other technical issues related to GDIs. It will take time before the institutional and research capacity in this area can be built.

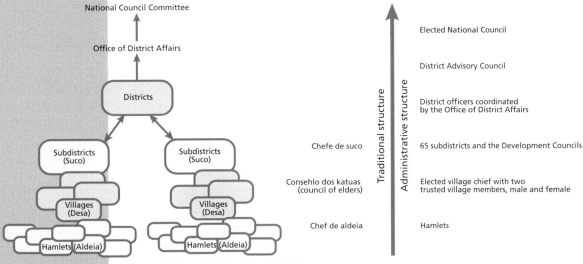

Figure 2. New administrative structure.
Compiled from UN-NPDA 2002, UNTAET 2002, Phillips 1999, UN and WB 1999, UN-CCA 2000.

The new government's goals are to enable the transition from a postconflict, donor-driven development process toward a sustainable democratic, independent, and economically viable society. National priorities are to ensure democratic, social, and macroeconomic stability; develop a stable state sector; alleviate poverty and unemployment; develop human resources; establish a legislative and regulatory environment; and establish and operate a physical infrastructure (UN-NPDA 2002) by and for East Timorese.

DATA AND INFORMATION REQUIREMENTS

Constitutional and regulative responsibilities. Rules, regulations, and responsibilities embedded in constitutional and regulative frameworks guide the actions of state institutions and the behaviour of citizens. East Timor is committed and constitutionally bound to sustainably manage its territory by protecting the environment and promoting sustainable behaviour and related determinants such as culture, heritage, health, security, social welfare, and the economy, which are all data-intensive activities. A key determinant of sustainable development is information, particularly geospatial data (UNCED 1992).

A series of regulations and sections in the Constitution address the environment, economy, society, data gathering, and the use of science and technology (table 1). They include promotion of the wise use of forests and protection of areas of inherent environmental, cultural, or historical value. A border regime was created to regulate the transport of goods, the quarantine of livestock, immigration, and the collection of taxes and tariffs, which are intensive national data collection activities. Borders remain a security problem requiring accurate maps to support international peacekeeping efforts. The provision of education, health, and social services and the management of business activities also require demographic data. Laws were created to ensure that a census is taken and to mandate electoral, business, and land registries and a tax regime. Ethical-conduct

Regulation or constitutional principle
Regulations
2000/9: Border regime 2000/15: Telecommunication 2000/17: Reduction of deforestation 2000/19: Protected places 2000/20: Revenue system 2001/3: Establishment of central registry 2001: Timor Gap Agreement 2002/4: Business registry
Constitutional principles

Section 6
  Economic and financial organizations
  Subsection f: Protection of the environment
  Subsection g: Preservation of cultural heritage
  Subsection d: Science and technology
Section 56: Social security and assistance
Section 57: Health
Section 59: Education and culture
Section 61: Environment
Section 139: Natural resources

Table 1. East Timor regulations and constitutional principles related to environmental protection.
Source: Constituent Assembly of East Timor 2002.

provisions were introduced to ensure that citizens in East Timor have the right "to truthful information and protection of their health, safety and economic interests" (Constituent Assembly of East Timor 2002).

Government departments and ministries. Regulative frameworks require institutions to operationalize laws, directives, and constitutional principles. A government administration carries out this function, and table 2 lists 36 new departments, units, and ministries. These require data and information to inform managerial decisions and to address crosscutting issues such as economy, health, education, and agriculture.

Environmental, social, and economic issues. The lack of information and data about East Timor (Sandlund et al. 2001; Phillips 1999; UN-CCA 2000; Ormeling 1957) make it "difficult to give an accurate, well-researched overview of the environmental problems" (UN-CCA 2000:88). The "lack of updated and reliable information regarding natural resources and biodiversity in East Timor" (Sandlund et al. 2001, p. 31) is acute, and there are few scientific sources and even less social science information. The destruction of government records only compounded the problem (Phillips 1999). The Department of Census and Statistics includes provisions to supply other departments with data; however, there is no mechanism to ensure that departments and ministries manage their data in a standardized way, nor is there a central agency that coordinates departments to ensure cooperation and collaboration on information needs.

Reconstruction, humanitarian, and aid efforts suffered from a lack of coordination between overseas development agency (ODA) projects and government departments. Many issues such as health and environment require an integrated management approach together with research in the physical and social sciences and appropriate institutions to conduct it (Sandlund et al. 2001).

Department or ministry	
Agricultural Affairs	Fisheries
Internal Administration	Foreign Affairs
Census and Statistics	Economic Affairs
Office of Communication and Public Information (OCPI)	Inspector General's Office
Political Affairs and Timor Sea	Border Service and Control Unit
Civic Education	Investment Promotion
Civil Service and Public Employment	Labour
Civil Registration Unit	Judicial Affairs
Civilian Police	Land and Property
Commerce, Industry, and Tourism	Mineral Resources Section, Department of Economic Affairs
Communications	Central Payment Office
Health Authority	Central Fiscal Authority
Division of Social Affairs	Power Services
Education	Social Services
Information Technology Post Telecommunication (ITPT)	Reconstruction
Environmental Protection Unit	Public Utilities (power, water, and sanitation)
Finance	Transport

Table 2. East Timor government departments and ministries.

The appendix lists issues, projects, and needs for a variety of subjects related to spatial data in East Timor. Currently, there are more issues than there are maps or datasets. Many issues are project and subject specific or follow the mandates of the organizations that fund them, and their data are not easily transferable to other contexts. Since projects are not integrated, information and data collected do not address the interrelated causes and effects of problems, nor are the benefits of sharing and integrating specialized knowledge realized. GDI framework data, clearinghouses, and working groups would help meet these data needs.

Ongoing data collection activities. Some data gathering and mapping projects have been implemented, important datasets are being compiled, mapping projects are ongoing, and potentially very useful maps exist. There is, however, a pressing need for their coordination, interoperability, and integration. Data on many topics are still missing (see appendix), and to date there is no central data clearinghouse, archive, library, or coordinating body to manage these.

A GEOSPATIAL DATA INFRASTRUCTURE FOR EAST TIMOR

Choosing a GDI model. East Timor has significant unmet data requirements. There is no central data- and information-coordinating body, and ICT capacity and related education levels are low. There is an opportunity to start building a GDI now, particularly since there is no bureaucratic entrenchment, and the initial administrative and steering functions of a GDI could be created as governmental units, departments, and ministries are being developed.

Seven GDI models were examined for applicability to East Timor: Sistema Nacional de Informação Geográfica of Portugal (SNIG), Canadian Geospatial Data Infrastructure (CGDI), Australia and New Zealand Land Information Council (ANZLIC), Japan National Spatial Data Infrastructure (JNSDI), the Malaysian Geospatial Data Infrastructure (MyGDI) (formerly National Land Information System [NaLIS]), Asia-Pacific SDI (APSDI), and global SDI (GSDI). None of these was found suitable for the current context in East Timor, but selected elements of each could be combined to form a hybrid GDI.

GDIs have distinct operational and implementation functions with coordinative, administrative, and managerial elements. Secretariats, advisory boards, steering committees, working groups (WGs), and partnership programs are managerial components, while mapping, standards, framework data, interoperability, metadata, and clearinghouses are technical. Some national GDIs (e.g., JNSDI and MyGDI) have their managerial elements as formal extensions of their national mapping organizations, while others (e.g., SNIG, ANZLIC, and PCGIAP) have separate institutions to carry out and guide these activities. An accord, directive, or order formally mandating the existence of a GDI can also guide and define the activities of committees.

The MyGDI and JNSDI models (figures 3 and 4) centralize decision making and restrict information to the public sector. Given East Timorese traditions, elements of a centralized decision-making structure need to start the process while collaboration and consensus building are nurtured. Malaysia has integrated geomatics and GDIs into education curricula at all levels and offers ongoing training programs. This could be a method to develop, coordinate, and integrate geomatics capacity in East Timor, particularly in its university and polytechnic. Information and data could be coordinated centrally but managed by departments and units. Data should be made available at no cost to the public, nongovernmental organizations (NGOs), and other government departments. The MyGDI and JNSDI models are restrictive in this respect.

East Timor is small, with distinct rural and urban economies. Most government offices, electrical services, and the limited ICT infrastructure are located in urban areas. Government authority and responsibilities have been redistributed to the 13 districts, 65 subdistrict offices, and villages (figure 2); however, offices are rudimentary, gathering data manually and sending it to urban offices for computer input. Rural areas have low literacy, numeracy, and education levels. For these reasons, the jurisdictional Australian model is not recommended. The ASDI model relies on advisory and policy direction from ANZLIC (figure 5) to coordinate state and territorial GDIs. The ASDI does not have a central data

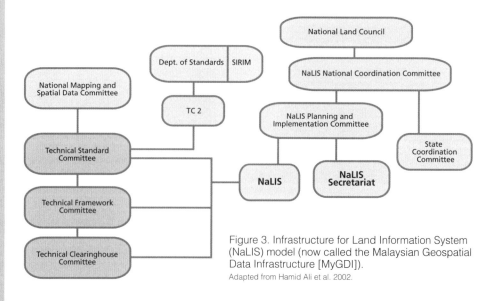

Figure 3. Infrastructure for Land Information System (NaLIS) model (now called the Malaysian Geospatial Data Infrastructure [MyGDI]).
Adapted from Hamid Ali et al. 2002.

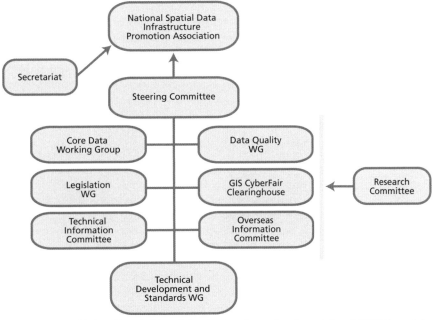

Figure 4. Japan National Spatial Data Infrastructure Promotion Association (NSDIPA) model.
Adapted from Masser 1998 and NSDIPA 2003.

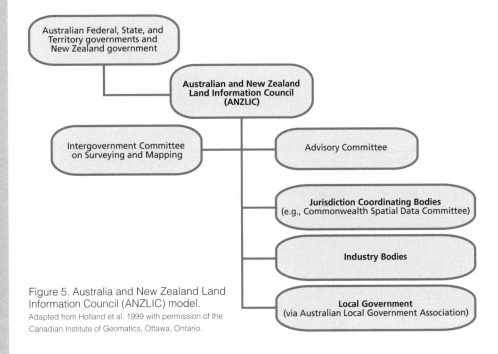

Figure 5. Australia and New Zealand Land
Information Council (ANZLIC) model.
Adapted from Holland et al. 1999 with permission of the
Canadian Institute of Geomatics, Ottawa, Ontario.

repository, but instead jurisdictional GDIs adhere to policies, standards, and mapping processes that enable the creation of national framework datasets and clearinghouses linked to a central hub. Rural offices in East Timor do not have the necessary physical infrastructure to house equipment, the electrical and telephone infrastructure to power equipment and transmit data, or the institutional, managerial, and technical capacity to coordinate a jurisdictional GDI. Local offices can contribute data, develop community mapping projects, learn surveying techniques, and provide local knowledge to inform spatial data and populate national thematic datasets. These offices are map and data end users and should be included in GDI development plans.

Given the ICT and academic capacity in East Timor, the GSDI model (figure 6) is also not suitable. It is an association of professionals, industry leaders, communities of interest (e.g., defence, hydrography, Open Geospatial Consortium [OGC]), and established international mapping, cartographic, and surveying associations. East Timor could benefit from the skill development, international outreach, and capacity-building support offered by members of the GSDI community. The APSDI is also a jurisdictional model (figure 7), but its UN-type administrative reporting structure might be cumbersome. The APSDI mechanisms of defining roles and responsibilities could be useful. GSDI and APSDI could assist East Timor with dataset development activities, support to attend meetings and conferences, and the PCGIAP skill development and training programs.

The Centro Nacional de Informação Geográfica (CNIG) in Portugal is a government-arm's-length research centre that guides the GDI activities of SNIG. This type of steering agency could be useful to East Timor and could be based in the Universitas Timor Timur (UNTIM). The funding could come from aid agencies, research-granting institutions, and government. Necessary courses, skills, and training programs could operate in the university, along with geomatics laboratories. Participants would include civil servants from all departments with geospatial data requirements; officials from districts, subdistricts, and villages; educators; professors; librarians; and NGOs. This cross section of stakeholders could collaboratively steer the development of a GDI for East Timor.

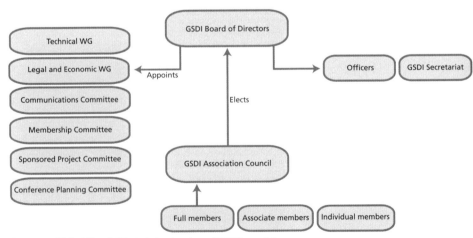

Figure 6. Global Spatial Data Infrastructure (GSDI) model.
Adapted from the GSDI Association Organizational Chart 2006 with permission.

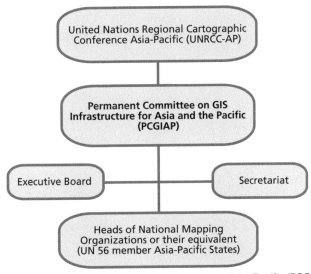

Figure 7. Permanent Committee on GIS Infrastructure for Asia and the Pacific (PCGIAP) model.
Adapted from Holland et al. with permission of the Canadian Institute of Geomatics, Ottawa, Ontario.

The Canadian model (figure 8) includes librarians, municipalities, information end users, and so on, along with geomatics experts. The CGDI includes application advisory nodes (i.e., WGs) and formal programs to develop partnerships and support from the private sector. The program advisory network is separate from operations and is a network consensus-making environment. In East Timor an informal social structure of problem resolution is based on consensus, and production is carried out by a massive network of familial and social ties. Building on an inclusive traditional decision-making social structure would ensure participation and align methods with cultural norms. WGs on a variety of themes such as health, agriculture, aid agency data could be part of the GDI.

Proposed East Timor GDI. An equivalent of Portugal's CNIG could guide application area WGs. This would ensure formal collaboration between academia, educators, and the public sector, which could result in mutually beneficial training opportunities. Executive-level membership from the civil service would ensure that decisions are implemented. Application WGs could focus on data needs and develop policies and standards similar to NaLIS. Crosscutting WGs on aid agency data coordination and strategic training could be formed. Other WGs could include health, agriculture, transportation, utilities infrastructure, natural resources, and social infrastructure. The research center could be responsible for analyzing WG needs and providing recommendations on data standards and policies. Once data requirements are identified and the civil service is more governance oriented, the GDI application WGs could evolve into technical ones dealing with standards, framework data, and related topics.

An East Timor GDI (figure 9) could be housed in the Information Technology Post Telecommunication department, which has some ICT capacity, or in the Office of Communication and Public Information, so as to build on its information and dissemination experience. The secretariat can provide administrative

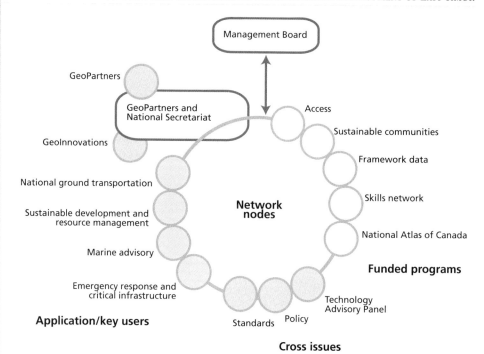

Figure 8. Phase 1, Canadian Geospatial Data Infrastructure (CGDI) organizational model.
Adapted from GeoConnections 2003 with permission.

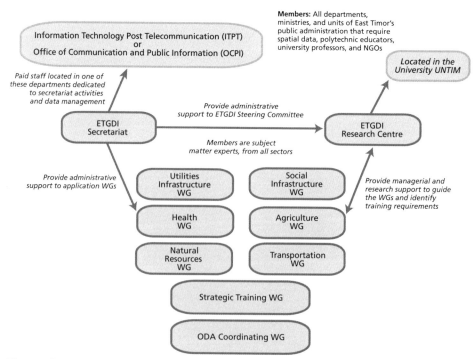

Figure 9. Proposed East Timor Geospatial Data Infrastructure (ETGDI) model.

support to application area WGs and the research/steering committee. The secretariat may temporarily store and disseminate data and ODA reports until a formal database, a clearinghouse, and national mapping organizations are established. The civil service is still restructuring, and an opportunity exists to formalize two new institutions: a GDI secretariat and a research center. This would ensure that the public sector has a formal mechanism to identify interdepartmental, cross-sectoral data needs and begin to coordinate activities.

A strategic-training WG could seek support from PCGIAP, GSDI, and neighboring-nation training programs, while funding could be acquired from aid agencies, research institutes, granting agencies, and international mapping associations. The public administration could cover the costs of civil servant training programs and pilot studies. The university can provide social science, agricultural, civil engineering, and legal knowledge, while the polytechnic can develop mapping, database, and cartographic technical training programs designed for students and public servants. Postsecondary institutions could collaborate on pilot studies such as community mapping, data gathering, surveying, and traditional knowledge in partnership with government departments. Private-sector involvement could be nurtured. Once some capacity has been built and students have graduated, research and development incubation centers could be created between public, private, and academic institutions to spark a private geomatics sector.

These recommendations are offered to stimulate thought and discussion among East Timorese officials grappling with geospatial data issues and decision-making processes. A final East Timor GDI model would have to be worked out in East Timor with officials from all jurisdictions and be formally funded and supported by the Council of Ministers. Few financial resources are directed at nonproject-specific data management activities, as these are not aid agency priorities. These ideas are offered as possibilities, should coordinating and managing spatial data become a central priority in East Timor and aid agencies recognize the possibilities of integrated data and information management as part of reconstruction and nation-building activities. GDIs are good governance initiatives, and building one for East Timor would contribute to sustainable development.

Building a GDI for East Timor. Building a GDI requires political support and commitment, appropriate terms of reference, business solutions along with national solutions, the management of aid funding, project sustainability, organisational change, public–private sector partnerships, and data (Land and McLaren 1998).

Political commitment ensures access to resources and the cooperation of officials, high-ranking politicians, and managers. The proposed East Timor GDI (figure 9) is contingent on broad-based support. Steering committees should represent all stakeholders. Political commitment to finance the operation of a secretariat requires resources for dedicated full-time staff, offices, computers, training, and a budget to support travel and facilitate stakeholder meetings. Staffing issues are critical, and NaLIS officials have identified this as an ongoing problem (Hamid Ali et al. 2002).

Convincing high-ranking politicians of the utility of a GDI is not easy; therefore, an informal forum to enable interested officials, geomatics staff, AusAID officials, and academics to discuss information and geospatial data issues, identify problems, and demonstrate the need for a coordinated effort should be created. It could enlist PCGIAP support to facilitate roundtable discussions on the benefits of GDIs, show examples from the region to officials and politicians, spark interest, and harness support from the Council of Ministers.

Once the utility of a GDI is demonstrated, a vision should be articulated (NRC 2002; Ezigbalike, Selebalo, Faiz, and Zhou 2000). The GSDI Cookbook (2000) includes references to GDI visions from around the world.

Well-defined terms of reference (TOR) and mandates guide decision making, and these should be based on local needs (Land and McLaren 1998; Masser 1998; Ezigbalike, Selebalo, Faiz, and Zhou 2000). They guide the work of steering committees, advisory boards, and WGs. PCGIAP members, chairs, data custodians, and sponsors have well-defined rules, roles, and responsibilities. Terms of reference in East Timor should specify the roles and mandates of departments, units, and ministries as well as regulations and laws, and formally include liaison functions with donors, the Cabinet, and departments. Key datasets should be developed in alignment with political opportunities (Land and McLaren 1998). The proposed application area WGs could identify critical datasets (see appendix). A focus on end users facilitates a service-oriented approach rather than a "government-needs" one (Land and McLaren 1998).

Until comprehensive business and project management plans have been designed, some datasets have been gathered, and policies have been tested, it is best to avoid the purchase of technology that could end up as "dusty boxes on desks." An approach that "focuses on broader human and organizational issues as a key component of capacity building rather than emphasizing technology and method" is considered more desirable in less developed countries (LDCs) (Britton 2000, p. 9), and aid efforts need to be coordinated (Hall 2000) to ensure that aid meets real needs. The following guiding principles (Ezigbalike, Selebalo, Faiz, and Zhou, 2000) are provided:

- Favor demand-led rather than technology-led projects
- Ensure that support is provided to local initiatives with a foreseeable integration into overall institutions
- Ensure autonomy by supporting small projects that sustain themselves with local finances and human resources
- Incorporate the "soft" components of implementation such as capacity building and social and organizational issues
- Support longer-term projects rather than short-term ones

The overseas development aid (ODA) coordinating WG (figure 9) could oversee some of these activities, and consultative support to build business plans could be provided by the agencies themselves.

Developing the institutional components will give East Timor time to gain some experience and carry out some agreed-upon pilot projects without being bound to the purchase of expensive technology. This means investing in people first (particularly managers) and developing human-resource plans. Partnership and training opportunities could be explored, and a capacity development plan could accompany a management plan to ensure that training meets identified data requirements and is in alignment with East Timor's long-term needs. Training strategies could involve secondments in other countries, attendance of conferences and workshops, short courses, internships, coaching and mentoring, study tours, university and polytechnic courses, visits by international researchers, and distance learning. Building on lessons learned is key, and information needs to be disseminated. Case studies, regional workshops, project evaluation, and reviews are good sources of this type of knowledge (Land and McLaren 1998).

A conscious effort is required to ensure that the newly created knowledge is retained and transferred. Retaining knowledge means embedding it in routines, technologies, and individuals and sharing it between groups, departments, divisions, and establishments (Argote and Ophir 2002). Moving personnel from one department, unit, or WG to another enables the transfer of tacit as well as explicit knowledge. The WGs and the research center are formal multidisciplinary knowledge transfer methods. Timorese traditional knowledge transfer practices should also be explored.

Sound data-sharing policies should be developed sooner rather than later, the most pressing problems should be tackled first, pressures for access that could produce premature decisions should be resisted, and partnership approaches to data sharing should be encouraged (King 1995). The geomatics unit in East Timor has explored cost recovery, while the government of East Timor remains committed to sharing data with citizens. This contradiction needs to be addressed so as to avoid confusion and ensure a consistent national approach to all data-related activities. Cost recovery is considered an impediment to knowledge transfer in cash-strapped environments.

A policy of inclusiveness and transparency is critical, and it is highly recommended that GDI activities include village, subdistrict, and district officials to ensure that data producers are involved. Incorporating local knowledge into data gathering could yield positive results.

CONCLUSIONS

East Timor has many geospatial data requirements and difficult social, economic, and environmental issues that require an integrated approach. The context in East Timor is not conducive to the full implementation of a GDI at present; however, administrative, steering, research, and WG functions of a GDI can be developed. This would give East Timor time to develop a vision as well as business and strategic training plans. Once some datasets have been created and integrated, experience has been accumulated, and learning has been codified, technological components can follow. East Timor should enlist outside consultative support to draft policy, technology, and training plans. Building the institutional and organizational components of a GDI first may seem simplistic; however, the literature suggests that the institutional components are the most difficult (O'Donnell and Penton 1997; Sherif in Pinto and Onsrud 1995; Pinto and Onsrud 1997; Hamid Ali et al. 2001). It will take time for ICT acculturation to grow and the data management cycle to take form, particularly given the nature of the economy and traditional, normative, and cultural structures.

An East Timor GDI is a long-term endeavor, and success will depend on ongoing support from all stakeholders and requires investments in science, technology, social sciences, cartography, planning, engineering, and knowledge management.

Relevant training could be incorporated into all levels of the education system. Regional and global GDIs will hopefully broaden their mandates and provide consultative, training, and political support.

The proposed East Timor GDI structure is a collaborative decision-making environment that will include consultations with citizens, data users, and data producers to ensure that needs are at the center. Small rural pilot projects are critical, as is providing training to rural officials on data-gathering methodologies and integrated analysis. Sustainable development should be informed by traditional land management and forestry practices. Participatory and community mapping pilot projects could effectively demonstrate the utility of data and maps and of the inclusion of local knowledge.

The proposed GDI organizational structure will be shaped by the cultural and normative structures of East Timorese society. At first, strong central leadership should reinforce a merit-based system, and slowly a shared-leadership and consensus-making approach to managing can be nurtured. Concerted effort will be required to overcome subtle but important social power dynamics.

East Timor must develop new GDI institutions sooner rather than later, even if it means only a few small WGs at first. As AusAID officials aptly point out (AusAID 2002), it is difficult to change established institutions. If the government drafted a charter, an accord, or a directive to mandate public service organizations to participate in an East Timor GDI and share their data, this problem could be reduced and full cooperation could increase. The civil service is still changing, and East Timor can hopefully enlist the support of PCGIAP and the GSDI Association to impart to the Council of Ministers the benefits of a GDI.

A GDI for East Timor would help ensure that required geospatial data are collected, managed, and disseminated so as to inform the work of reconstruction and sustainable development. Outside consultative support is required to develop business and training plans and facilitate collaboration. Other short-term projects that fit with the training and business plan could also be carried out with outside support. The resolutions of the 17th UNRCC-AP/12th PCGIAP meetings in Bangkok in 2006 recommended "that the Government of Timor Leste, member countries of PCGIAP, the UN Department of Peacekeeping Operations and other international efforts, collaborate as appropriate in developing this spatial data infrastructure so as to maximize its value" (PCGIAP 2007). This is very encouraging news indeed, and we hope that the recommendations in this article can be used to inform the development of an East Timor GDI.

Making the case for an East Timor GDI in the face of more pressing national needs is challenging, but recognizing the GDI's considerable potential for contributing to knowledge creation, an integral component of the fabric of society, lessens the challenge.

ACKNOWLEDGMENTS   Tracey P. Lauriault thanks Bella Galhos, who was once an East Timorese refugee in Canada, for inspiration.

Appendix: Sustainable development issues in East Timor requiring geospatial data and information	
Issue	Requirements
Agriculture, animal husbandry, fishing	
Agricultural rehabilitation Animal husbandry Coffee Fishing Irrigation Steep mountain slopes Importation of rice, eggs, sugar, flour Rainy season, dry season Self-sufficiency in fruit, vegetables, and beef Staples (corn, rice, cassava, gourds) Sugar cane production	Agricultural survey maps Climatic data (precipitation, temperature, etc.) Hydrographic maps Land capacity maps for agricultural production (slope, cover, bioregion, precipitation, etc.) Land classification and use maps Land allocation data and maps Range land data Soil survey maps Topographic maps Vegetation cover
Air quality	
Car emissions Lack of car maintenance Mosquito fumigation Old cars	Car registry with attributes Mosquito vector maps Wind maps
Biodiversity	
Limited knowledge	Ecological models Flora and fauna data Habitat and hunting ground maps Data and maps for protected places
Border control	
Citizenship Control Monitoring Quarantine Taxes Travel permits	Border regime and travel data Border survey maps Census data Defence maps National accounting system Registry of dangerous animal diseases, pests, or invasive plant species Remote sensing images Tax registry Travel document data
Buildings	
Building registry	Administrative land records Building permit data Cadastre Land management Legal land records
Economy	
All banks and credit unions destroyed Business registry Dili Market waste generation Distribution network has collapsed Household economic survey to alleviate poverty Local construction materials not used in housing aid (marble, bamboo, thatch, cement) Revenue system Tax collection	Census Inventory of natural resources Inventory of markets Land, air, and water transportation network maps Location of pathways of pollutant dispersal Location and capacity of waste sites Location of sources of pollution Mines, quarries, and minerals maps

Continued on next page

Appendix (continued)	
Issue	Requirements
Education	
Age of students High school Illiteracy New buildings Postsecondary Presence of girls High student/teacher ratio (60/1) Supplies Teacher population	Cadastre Demographic data (age profiles, population growth) Labor force survey Location of buildings Retention rate data School catchment areas
Elections	
Election registry Privacy Voter registration Voting stations	Demographic data District, subdistrict, and village maps
Emergency preparedness	
Disaster response	Emergency plans Evacuation sites Inventory of hazardous sites and reporting plan Land-use maps Meteorological data
Energy	
Diesel generators Generation capacity Hydroelectric stations Power lines Power station rehabilitation Removal of kerosene subsidy increased reliance on wood for cooking Solar energy Time of coverage unstable	Climatic data for solar and wind power generation Data on peak demand locations Demographic data Electrical system infrastructure and management maps Forestry maps Data on fuel consumption needs Hydrographic maps Location of infrastructure
Forestry	
Community-based forestry program Fuelwood Loss of fuel subsidy increased demand for fuel, causing conflicting priorities with reforestation programs Mangroves Palm wine Population dislocation increased demand for forestry products in urban areas Sandalwood trade Slash-and-burn agriculture Soil erosion Tamarind	Demographic data Endangered species Forest management sites Forest product inventory Forest inventory Fuel demand studies Land-use maps Plantation maps Road network Soil survey Topographic maps Vegetation cover maps

Continued on next page

Appendix (continued)	
Issue	**Requirements**
Health	
AIDS Asthma Communicable-disease surveillance system Community health centres Dengue fever Diet improvement program Environmental health impact assessment Epidemiological bulletins Family planning First aid posts Food insecurity Gastroenteritis Health posts Health programs Hospital services Human resource development Lack of district and subdistrict health services Lack of understanding of water sanitation Lower respiratory tract illness Malaria Maternal mortality Mobile health clinics Mosquito fumigation moved insects inside National health surveillance health network National blood program National tuberculosis program and diagnostic centres Need for new laboratories, pharmaceutical drug program Obstetric care Pit latrines and sewage Population density Tuberculosis Vector-borne-disease control working group	Cadastre Demographic data Distribution of disease vectors Fumigation sites Health indicators Labor force survey Location and capacity of institutions and catchment areas Location of water sanitation services Maps of mosquito vectors Mortality rates Social and natural vectors Water quality data
ICT	
Access to land for towers and wires, right of persons to appeal law Archiving Basic telecommunication services Computerization of functions such as census Effective wireless system is in place Internet sites Internet ISP Internews Linking schools and hospitals to the Internet Newspaper distribution Postal service Power lines Printing consortia Radio bands, stations Radio service being used to encourage hook-up and boil-water advisories Records management system Rehabilitation of the infrastructure Satellites Server cables, hubs, routers Television broadcasting Transmission towers	Cell phone tower location analysis maps Communication network Data on energy needs Infrastructure location data with attributes Land-use data Location and capacity of post offices Reach of radio waves Readership data
Labor	
ICT training Lack of trained health personnel Lack of trained teachers Lack of civil engineers Lack of computer technicians, network specialists, programmers, dBASE managers Lack of occupation profiles Shortage of mid- to top-level management	Demographic data Occupation profile statistics Education profile statistics Inventory of education and training programs Labor forecast data

Continued on next page

Appendix (continued)	
Issue	Requirements
Land	
Land titles Property rights	Cadastre Land-use data Land allocation data
Mines and minerals	
Marble Oil and gas reserves (Timor Gap Agreement) Phillips petroleum pipeline Resource exploitation and exploration Salt	Hydrographic data Infrastructure maps and management plan Location and capacity of mines and quarries Analysis of resource needs Shipping data Survey map
Population	
Registry	Census Registry
Sensitive, protected places and cultural sites	
Biodiversity Coral reefs Designated areas Endangered species Governance is required Historical sites Lake Irralalarus flora and fauna Mangrove areas Wetlands	Flora and fauna maps Inventory of protected areas Location of historical and cultural sites Location of sites or issues that threaten protected places Management plan Special research project on Lake Irralarus Transportation network maps Wetlands maps
Shelter	
Housing shortage in Dili Shelter kits displaced local economy and their transport damaged roads Traditional building materials (bamboo, palm leaf roofs)	Cadastre Distribution of shelter kits Geodemographic data Location of traditional materials Data for needs analysis
Soil erosion	
Agricultural runoff Deforestation Demand for fuelwood Flash floods Clogging of irrigation systems Mountain slopes Mud slides Siltation Watershed management	Hazards and vulnerable-sites data Hydrological data Irrigation management data Land-use data Land classification Soil survey Topographic map Transportation network
Tourism	
Air travel Beaches Ecotourism	Coastline data Data on tourist preferences Decision support for land use for facilities Location of historical sites Location of protected places Location of facilities Location of places of interest Transportation network

Continued on next page

Appendix (continued)	
Issue	Requirements
Transportation	
Bridge reconstruction and repairs Bridges destroyed by flash floods Motor vehicle registry Only 11 East Timorese civil engineers Only 55 percent of roads paved Only 12 buses Port improvement Rehabilitation of navigation systems Rehabilitation of roads	Coastline data Disaster relief data Elevation data Fleet management data Flight route maps Geological maps Hydrographic data Infrastructure location maps and management schedules Labour force survey Pavement and road management maps Ridership preferences and needs Road network maps with attributes, transit route maps
Treaties	
Convention on International Trade in Endangered Species (CITES)	Location of sites that fall under treaties
Urban issues	
Crime and social unrest Lack of urban planning Population dislocation to urban areas increased demand for fuelwood from surrounding areas Population density is a vector of disease and causes stress High migration from rural to urban areas Sewage	Cadastre Census Crime data and programs Epidemiological maps Location of infrastructure capacity and services Location, capacity, and waste flow of sewage facilities Migration data Waste management and dump plans
Waste management	
Garbage truck fleet management Garbage smell, transportation in the rain, toxic smoke, lack of a collection system Road network	Inventory of waste sites, capacity, type of waste
Water	
Access to fresh water Agricultural runoff causing eutrophication Coastal zone Domestic sewerage, pit latrines Drainage canals Drinking water wells were poisoned, and dead bodies were dumped in them Flash floods Hydroelectricity Irrigation rehabilitation Lake Irralalarus Purification River management Sanitation services Siltation Soil erosion Stagnant water Water supply rehabilitation Water pumps were stolen Watershed management	Geophysical data Irrigation and water diversion needs Land-use maps Location of water users Location of sewage treatment facilities Location of water bodies and courses with flow and condition data Location and capacity of dams Meteorological data, precipitation Mosquito vector data Water consumption data Groundwater mapping

Sources: UNPDA 2002, ETTA 2000, ICRC 2001, WHO 2000 and 2001, Internews 2002, Phillips 1999, Sandlund et al. 2001, Tais Timor No. 8 2000, TaisTimor August 2000b, TimorAid 2001, UN Secretary General 2001, UN-CCA 2000, UNDP 2002, United Nations and World Bank 1999, UNTAET 2000a, UNTAET 2000/27, UNTAET 2000b update, UNTAET 2001.

Argote, Linda, and Ron Ophir. 2002. Intraorganizational learning. In *The Blackwell companion to organizations,* eds. Joel A. C. Baum Malden, 181–207. Malden: Blackwell Publishers.

Australian Agency for International Development (AusAID), 2002, Interim Capacity Building Program East Timor (CAPET, Agriculture and Landuse Mapping and GIS Development and Training Project GIS Extension 2002, Facsimile communication Vanessa Loney, Program Officer East Timor Section, received August 15, 2002.

Australian Defence and ESRI, 2003, Project Toposs. http://www.ead.com.au/downloads/ead/presentations/Project_TOPOSS.pdf (accessed May 10, 2003).

Australia New Zealand Land Information Council (ANZLIC), 1996, Spatial Data Infrastructure for Australia and New Zealand, discussion paper. http://www.anzlic.org.au/ (accessed September 18, 2002).

Barbedo Magalhães, Antonio. 1994. Lesson 1: Population Settlements in East Timor and Indonesia. Introductory course on Indonesia and East Timor. http://www.uc.pt/timor/CURSO1A.HTM (accessed October 09, 2001).

Britton, James M. R. 2000. GIS Capacity in the Pacific Island Countries: Facing the Realities of Technology, Resources, Geography and Cultural Difference. *Cartographica* 37 (4): 7–19.

Canadian Geospatial Data Infrastructure. 2002. http://www.geoconnections.org.

Constituent Assembly of East Timor. 2002. The Constitution of the Democratic Republic of East Timor. www.timor-leste.org (accessed August 13, 2002).

ETTA. 2000. Mission Statement of the Department of Census and Statistics. http://www.goc.east-timor/old/statistics/main.php (accessed November 18, 2000).

Ezigbalike, Chukwudozie, Qhobela Cyprian Selebalo, Sami Faiz, and Sam Z. Zhou. 2000. Spatial data infrastructures: Is Africa ready? Proceedings of the 4th Spatial Data Infrastructure Conference, Cape Town, South Africa.

GeoConnections. 2003. About GeoConnections, Government of Canada. http://www.geoconnections.org/english/about/index.html (accessed April 12, 2003).

GERTiLFA.UTL Regressa a Timor. 2001. http://www.utl.pt/internacional/GERTiL2.htm.

Global Spatial Data Infrastructure. 2000. SDI Cookbook. http://www.gsdi.org/oubs/cookbook/chapter08.html#spatial (accessed November 20, 2001).

Global Spatial Data Infrastructure. 2003. GSDI Internet Site. http://www.gsdi.org (accessed April 24, 2003).

Global Spatial Data Infrastructure. 2006. GSDI Association Organizational Chart. http://www.gsdi.org/Association%20Information/bylaws%20and%20organization%20chart/GSDI-OrgChart-July03.pdf (accessed March 18, 2006).

Groot, Richard, and John McLaughlin, eds. 2000. Geospatial Data Infrastructure, Concepts, Cases and Good Practice. Oxford: Oxford University Press.

Hall, Brent. 2000. The Information Age, Capacity Building, and the Use of Spatial Information Technologies in Developing Countries. *Cartographica* 37 (4): 1–5.

Hamid Ali, Dato, Ahmad Fauzi Nordin, and Mohamad Kamalis Adimin. 2002. Malaysian Spatial Data Infrastructure Initiatives and Issues. www.fes.uwaterloo.ca/crs/gp555/ pdfs2003/malaysian_sdi.pdf (accessed April 28, 2003).

Holland, Peter, Alister Nairn, Graham Baker, Bob Irwing, Peter Boersma, Brian Burbidge, and David Robertson. 1999. The Development of a Spatial Data Infrastructure in the Asia-Pacific Region with Special Reference to Australia. *Geomatica* 53 (1): 47–55.

International Committee of the Red Cross. 2001. East Timor: The ICRC hands over Dili Hospital to UNTAET. ReliefWeb. www.reliefweb.int (accessed September 2002).

International Federation of the Red Cross. 2001. East Timor: Focus on Institutional and Resource Development Appeal 01.42/2001, Programme Update 2. ReliefWeb. www.reliefweb.int (accessed August 8, 2002).

Internews. 2002. Internews East Timor. http://internews.org/regions.seasia/easttimor.htm (accessed August 13, 2002).

King, John Leslie. 1995. Problems in public access policy for GIS DataBases: An economic perspective. In *Sharing Geographic Information*, 255–77. New Brunswick: Centre for Urban Policy Research.

Land, Nick, and Robin McLaren. 1998. Key success factors in implementing LIS projects in developing countries and countries in transition. Proceedings of the XXI International F.I.G., Congress, Brighton, U.K.

Lopez, Xavier. 1997. The dissemination of spatial data: A North American and European comparative study on the impact of government information policy. University of California, Berkeley.

Mapping Sciences Committee. 1993. *Toward a Coordinated Spatial Data Infrastructure for the Nation*. Washington, DC: National Academy Press.

Masser, Ian. 1998. *The First Generation of National Geographic Information Strategies*. Canberra. http://www.gsdi.org (accessed September 12, 2001).

National Infrastructure for Land Information System. http://www.nalis.gov.my.

National Research Council. 2002. *Down to earth: Geographic information for sustainable development in Africa*. Washington, DC: National Academy Press.

National Spatial Data Infrastructure Promotion Agency. 2003. Internet site of the Japanese National Spatial Data Infrastructure Promotion Agency. http://www.nsdipa.gr.jp/ENGLISH (accessed April 28, 2003).

O'Donnell, J. Hugh, and Cyril R. Penton. 1997. Canadian perspective on the future of national mapping organizations: Framework for the world. *GeoInformation International* 214–25.

Ormeling, J. 1957. *The Timor Problem*. The Hague, Netherlands: J. B. Wolter, Groningen and Martinus Nijhoff.

Permanent Committee on GIS Infrastructure for Asia and the Pacific. 2003. http://www.gsi.go.jp/PCGIAP/pcmemb.htm (accessed April 24, 2003).

Permanent Committee on GIS Infrastructure for Asia and the Pacific (PCGIAP). 2007. Resolutions of the 17th UNRCC-AP/12th PCGIAP of the 12th Meeting of the Permanent Committee on GIS Infrastructure for Asia and the Pacific. Bangkok, Thailand. http://www.gsi.go.jp/PCGIAP/unrcc/bangkok/12th%20PCGIAP%20Report%20all.pdf.

Phillips, David L. 1999. Social and Economic Conditions in East Timor, Fafo Institute for Applied Social Science, Oslo. http://www.fafo.no/pub/929.htm (accessed September 18, 2002).

Pinto, Jeffrey K., and Harlan J. Onsrud. 1995. Sharing Geographic Information Across Organizational Boundaries: A Research Framework. *Sharing Geographic Information* 44–64. New Brunswick: Centre for Urban Policy Research.

Pinto, Jeffrey K., and Harlan J. Onsrud. 1997. In Search of the Dependant Variable: Toward a Synthesis in GIS Implementation Research. In *Geographic Information Research Bridging the Atlantic,* eds. Massimo Craglia and Helen Couclelis, 129–45. London: Taylor & Francis.

Sandlund, Odd Terje, Ian Bryceson, Demetrio de Carvalho, Narve Rio, Joana da Silva, and Maria Isabel Silca. 2001. Assessing Environment Needs and Priorities in East Timor: Issues and Priorities. UNOPS and Nina-Nikku.

Sistema Nacional de Informação Geográfica (SNIG). 2003. http://snig.igeo.pt.

Tais, Timor. 2000a. Rains Bring Death, Destruction to Timor Island. *Tais Timor Newsletter* 1 (8) (May 29–June 11). http://www.gov.east-timor.org/old/news/East_Timor_Update (accessed September 18, 2002).

Tais, Timor. 2000b. News Briefs. *Tais Timor Newsletter* 1 (July 24–August 6, 2000). http://www.gov.east-timor.org/old/news/East_Timor_Update (accessed September 18, 2002).

Taylor, John G. 1999. *East Timor: The Price of Freedom.* London: Zed Books.

Timor Aid. 2001. About East Timor. http://www.timoraid.org/timortoday/html/about_east_timor.htm (accessed September 18, 2002).

UN and World Bank. 1999. Overview of external funding requirements for East Timor. www.worldbank.org/html/extdr/offrep/eap/etimor/donorsmtg99/unwb.pdf (accessed August 9, 2002).

UN Commission on Environmental and Development. 1992. Information for decision-making. Chapter 40, section 40.1, introduction, http://www.unep.org/Documents/Default.asp?DocumentID=52&ArticleID=90 (accessed April 4, 2002).

UN Committee on Economic Development. 2002. Johannesburg Summit, frequently asked questions, http://www.johannesburgsummit.org/html/basic_info/faqs_sust_dev.html#sd1 (accessed September 3, 2002).

UN Common Country Assessment. 2000. Building blocks for a nation. *East Timor Common Country Assessment.* www.unagencies.east-timor.org/08_publication/publication.htm (accessed October 10, 2001).

UN Department of Public Information (UNDPI). 2001. Constitutional hearings draw 30,000 East Timorese, UN Mission Says. *ReliefWeb,* www.reliefweb.int (accessed September 18, 2002).

UN Executive Board of the UNDP and the UNPF. 2000. Assistance to East Timor, third regular session 2000, New York. www.undp.org/execbrd/html/advance4.htm (accessed August 8, 2002).

UN General Assembly. 2000. General Assembly, fifty-fourth session agenda item 96, question of East Timor security council fifty-fifth year. Distr. General A/54/726, S/2000/59. http://www.unhchr.ch/huridocda/huridoca.nsf (accessed October 10, 2001).

UN National Development Program (UNDP). 2001. East Timor work force to gain new skills. *ReliefWeb,* www.reliefweb.int (accessed August 8, 2002).

UN National Development Program (UNDP). 2002. US$5 million project to tackle poverty in rural East Timor. *ReliefWeb,* www.reliefweb.int (accessed August 8, 2002).

UN National Development Program (UNDP). 2002b. BP Foundation aids action against poverty in East Timor. *ReliefWeb,* www.reliefweb.int (accessed August 8, 2002).

UN National Planning and Development Agency (UNPDA). 2002. East Timor-governance and public sector management capacity development-program overview. www.undp.east-timor.org/teks%20frame/undp/pubs.htm (accessed May 18, 2003).

UN Secretary General. 2001. Progress report of the secretary general on the United Nations transitional administration in East Timor (S/2001/719). www.reliefweb.int (accessed August 11, 2002).

UN Transitional Administration in East Timor (UNTAET). 2000. Regulation No. 2000/9: On the establishment of a border regime for East Timor. www.un.org/peace/etimor/untaetR/r-2001.htm (accessed November 12, 2001).

UN Transitional Administration in East Timor (UNTAET). 2000. Regulation No. 2000/17: On the prohibition of logging operations and the export of wood in East Timor. www.un.org/peace/etimor/untaetR/r-2001.htm (accessed November 12, 2001).

UN Transitional Administration in East Timor (UNTAET). 2000. Regulation No. 2000/19: On protected places. www.un.org/peace/etimor/untaetR/r-2001.htm (accessed November 12, 2001).

UN Transitional Administration in East Timor (UNTAET). 2000. Regulation No. 2000/27: On the temporary prohibitions of transactions in land in East Timor by Indonesian citizens not habitually resident in East Timor and by Indonesian corporations. www.un.org/peace/etimor/untaetR/r-2001.htm (accessed November 12, 2001).

UN Transitional Administration in East Timor (UNTAET). 2000b. External review of the humanitarian response to the East Timor crisis. *ReliefWeb*, www.reliefweb.int (accessed August 8, 2002).

UN Transitional Administration in East Timor (UNTAET). 2001. Regulation No. 2001/4: On the replacement of Regulation 2000/4 on registration of businesses. www.un.org/peace/etimor/untaetR/r-2001.htm (accessed November 12, 2001).

UN Transitional Administration in East Timor (UNTAET). 2001. Regulation No. 2001/8: On the registration of motor vehicles in East Timor. www.un.org/peace/etimor/untaetR/r-2001.htm (accessed November 12, 2001).

UN Transitional Administration in East Timor (UNTAET). 2001. Regulation No 2001/3: On the establishment of the Central Civil Registry for East Timor. www.un.org/peace/etimor/untaetR/r-2001.htm (accessed November 12, 2001).

UN Transitional Administration of East Timor (UNTAET). 2001. Regulation No. 2001/15: On the establishment of an authority for the regulation of telecommunications in East Timor. www.un.org/peace/etimor/untaetR/r-2001.htm (accessed November 12, 2001).

UN Transitional Administration in East Timor (UNTAET). 2001. Regulation No. 2001/20: To Amend Regulation No. 2000/18: On a revenue system for East Timor. www.un.org/peace/etimor/untaetR/r-2001.html (accessed November 12, 2001).

UN Transitional Administration in East Timor (UNTAET). 2001a. Education fact sheet. UNTAET Press Office. http://iiasnt.leidenuniv.nl:8080/DR/2002/01/DR_2002_01_03/14 (accessed September 18, 2002).

UN Transitional Administration in East Timor (UNTAET). 2002a. East Timor: Border demarcation teams complete survey. *ReliefWeb*, www.reliefweb.int (accessed August 8, 2002).

World Health Organization (WHO). 2000. East Timor health sector situation report *ReliefWeb*, www.reliefweb.int (accessed August 8, 2002).

World Health Organization (WHO). 2001. WHO's Contribution to health sector development in East Timor. *ReliefWeb*, www.reliefweb.int (accessed August 8, 2002).

A Framework for Comparing Spatial Data Infrastructures: An Australian–Swiss Case Study

CHRISTINE NAJAR,[1] ABBAS RAJABIFARD,[2] IAN WILLIAMSON,[2] AND CHRISTINE GIGER[1]

INSTITUTE OF GEODESY AND PHOTOGRAMMETRY, SWISS FEDERAL INSTITUTE OF TECHNOLOGY (ETH), ZURICH, SWITZERLAND,[1] AND CENTRE FOR SPATIAL DATA INFRASTRUCTURES AND LAND ADMINISTRATION, DEPARTMENT OF GEOMATICS, UNIVERSITY OF MELBOURNE, VICTORIA, AUSTRALIA[2]

ABSTRACT

Spatial data infrastructures (SDIs) facilitate data sharing and rely on effective management of data, metadata, and Web services. A framework for evaluating SDI initiatives has been developed, and the national SDIs of Australia and Switzerland are compared on the basis of these criteria.

Many countries are developing spatial data infrastructures (SDIs) to better manage and utilize their spatial data by taking a perspective that starts at the local level and proceeds through state, national, and regional levels to the global level. The emergence of different SDIs at and between these levels has brought attention to the SDI hierarchy (figure 1).

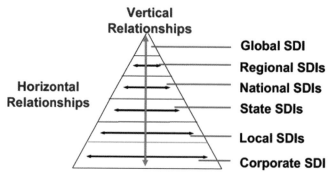

Figure 1. Hierarchy of SDI levels.
Reprinted from Rajabifard et al. 2000 with permission of Taylor and Francis Books.

Since SDI development, maintenance, and operation require huge investments, indicators for judging SDI performance and determining cost/benefit ratios are needed. Stakeholders are interested in monitoring SDI development and impact on the geoinformation market.

Although nearly all of the SDI initiatives use the same (or similar) basic technology and standards, they are hard to compare because of their different legal and organizational backgrounds. Best-practice solutions do not seem to exist because the organizational framework plays an important role in an SDI's success and cannot easily be transferred from one institution or country to another. Clear means of determining the developmental status of an SDI or measuring its impact on geoinformation availability and usage are lacking.

We offer a framework for comparing SDI initiatives using a set of clearly defined indicators, which are as much as possible independent of the organizational backgrounds of the SDIs. We present the results of ongoing research on SDI and metadata management activities in Switzerland and Australia. The framework can simplify characterization of complex enabling platforms and diverse peculiarities and can be used for any national SDI initiative.

We review current SDI monitoring and evaluation methodologies, discuss problems of interoperability on the national level, and apply proposed key indicators to case studies in Switzerland and Australia.

National SDIs. Many governments and organizations have recognized the economic, social, and environmental benefits of SDIs. In 2002 more than half the countries of the world had national SDI initiatives (Borrero et al. 2002; Crompvoerts et al. 2003). By 2005, most countries had national SDI initiatives (Warnest et al. 2005).

SDI development is a long-term process which needs long-term investment and the consideration of organizational and technical issues. The national level is of special significance to SDIs, as this is where juridical, political, and administrative decisions are made for a country and guidance is given for local levels.

As Masser (1999) points out, successful national SDIs are increasingly composed of three elements: identification of core datasets for a wide range of users, development of metadatasets, and a coordination framework to develop the infrastructure.

Although stakeholders understand the need for core spatial data, they are not always aware of the need for adequate metadata. Metadata is needed to describe and label the spatial datasets and thus make them findable for search engines. Furthermore, metadata management, standardization, and modeling are key factors for interoperability within an SDI (Giger et al. 2003).

Web services support the user in processing, accessing, and visualizing data. A worldwide assessment of spatial data clearinghouses found that one of the main factors for success is the inclusion of Web services, and the latest definitions of clearinghouse put more emphasis on the inclusion of services (Crompvoets et al. 2004).

Data, Web services, and metadata are important components of SDIs but not the only ones. Comparative approaches for examining their roles can help identify best practices and targets for improvement (Williamson et al. 2005). SDIs can be compared on the basis of technical, institutional, and conceptual factors.

Comparisons of SDIs. SDI initiatives appeared in several countries many years ago more or less simultaneously, whereas other countries have only just begun planning their SDIs. SDI success can be assessed in different ways. Establishing a general framework and focusing on metadata, spatial data, standards, and Web services makes it easier to find a common denominator. Methods for comparing SDIs on the basis of institutional, political, and financial factors are discussed below.

Rackham and Rhind (1998) compare the UK SDI (National Geospatial Data Framework) with international initiatives by looking at inclusion of the public and private sectors within the NGDF Board, formal political support, and the emphasis on fostering services and facilitating business rather than data.

For comparing the national SDIs of the Netherlands and United States, Kok et al. (in press) created an "organisational maturity matrix" and looked at four organizational "context shaping" components: leadership, vision, communication channels, and self-organization.

Another indicator, SDI readiness, is based on the e-readiness index, which can be defined as the degree to which a country is prepared to participate in the networked world (Group@IMRB et al. 2003), and on factors identified in previous studies (Giff et al. 2002; Kok et al. 2005; Crompvoets et al. 2004). An SDI readiness index uses qualitative factors such as organization, information and data availability, people, access network, and financial resources, as well as decision criteria like political vision-commitment-motivation, institutional leadership, and umbrella legal agreements. Based on a model using fuzzy-compensatory logic, the SDI readiness index compares SDI progress over time within a country. This framework was used to evaluate the Cuban SDI (Delgado Fernandez et al. 2005).

The comparison of national spatial clearinghouses throughout the world by Crompvoets et al. (2004) looked at the year of first implementation, number of data suppliers, monthly number of visitors, number of Web references (Alta Vista and Google), languages used, frequency of Web updates, metadata accessibility, number of datasets, most recently produced dataset, use of maps for searching, registration-only access, and metadata standard.

The land administration framework developed by D. Steudler (2004) combines technical, institutional, and political factors and takes into account the different stakeholders. We extend this framework to SDIs.

Interoperability of SDIs on the national level. To find suitable indicators of SDI success, objectives have to be clearly defined. On the national level, many different organizations (private and public) are involved in producing and maintaining spatial datasets. Often they work as islands or are only partly interconnected. The ideal SDI connects these sources via the Internet and provides users with Web services to find, process, and analyze the spatial data on the spot.

The sharing of datasets using a common platform faces the challenge of interoperability. Figure 2 shows how institutions can connect to a national SDI on the Web. Many data providers manage a great variety of datasets that differ in scale, quality, topic, format, method of acquisition, purpose, and model. This heterogeneity is due partly to missing technical regulations and standardization and partly to institutional obstacles, for example, communication between national agencies or between departments within an agency. Most organizations are simultaneously data providers and users.

Although many problems of access to data will likely decline, problems of data incompatibility or unsuitability for reasons of scale, coverage, or methodology are much less tractable (Bayfield et al. 2005). The exchange of data is possible only with accurate, standardized metadata. Web services of national clearinghouses can search for ASCII-based metadata.

FRAMEWORK FOR COMPARISON OF SDIs

A comparison of national SDIs on the basis of data, metadata, and Web services will lead to a better understanding of best practices and help improve the interoperability of an enabling platform and the integration of data and metadata in common datasets and models (Giger and Najar 2003). Using a case study approach, we compare the national SDIs of Australia and Switzerland, focusing on the important components and related processes especially for metadata and Web services.

The geographic, historic, and political context needs to be understood; objectives need to be clearly defined; the roles of stakeholders and coordinating agencies need to be delineated; and government policy on data sharing (intellectual property rights, privacy issues, and pricing) needs to be take into account (table 1).

Table 2 lists the indicators for the evaluation related to data-metadata, Web services, and standards. Data capture and updating procedures (indicator 1) can be standardized by ISO 9001 certification, for example. Besides standardization, these procedures need to be well-documented.

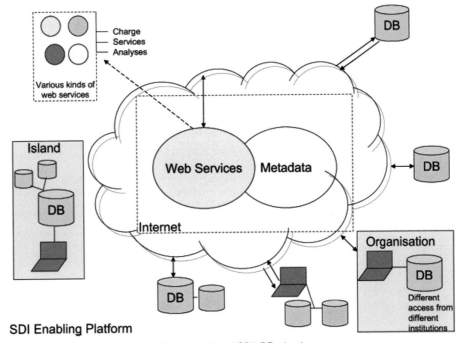

Figure 2. Horizontal interoperability in a national SDI. DB, database.

Characteristic
History
Objectives
Components
Global context
Coordinating agency
Stakeholders
Institutional partnerships
Government mandate
Data-sharing policy
National standards

Table 1. General characteristics of a national SDI.

Component	Indicator	
	Technical	Organizational
Data and metadata	1. Data capture process 2. Definition of core datasets 3. Data format and conceptual model 4. Data management 5. Data quality and accuracy 6. Common modelling language and tools 7. Harmonization of data and metadata	8. Custodianship 9. Data sharing and partnerships agreements 10. Business models 11. Coordinating arrangements
Web services	12. Application profile 13. Clearinghouse and geoportal	14. Clearinghouse organization
Standards	15. Interoperability	16. Organizational arrangements for standardization

Table 2. Indicators for comparing SDIs on the basis of Web services and data management.

Standardization is also important for core datasets (sometimes called reference data) (indicator 2), especially if these are acquired from different organizations. Core datasets are the basic geographic data and should have the highest priority. INSPIRE's Reference Data and Metadata working group has defined three functional requirements for core datasets:

- Provide an unambiguous location for user information
- Enable the merging of data from various sources
- Provide a context for understanding the information

Core datasets need to be clearly defined and well documented.

Integration of data from heterogeneous sources in a network environment requires a neutral format and a conceptual model (indicator 3).

Data management (indicator 4) relies on standardized update cycles and standardized, computer-processable metadata. The infrastructure should be available 24 hours a day, 7 days a week, and updates should be standardized, frequent, and documented. The relationship between data and metadata should also be standardized and documented. SDIs must also effectively manage large databases (e.g., photogrammetric imagery).

Data quality (indicator 5) standards (e.g., update frequency, precision, actuality, reliability) need to be documented for specific applications.

Conceptual models described in a standardized modeling language (e.g., UML) are needed. Tools for importing a model, testing its quality, and checking for mistakes should be provided (indicator 6).

The value of standardized metadata increases if, for example, the metadata is updated simultaneously with the spatial data. Harmonization standards are needed, and data models and metadata models must be coordinated (indicator 7).

Horizontal interoperability can be measured by homogeneous metadata–spatial data sets and catalogues, and it requires that participating organizations have clear guidelines and a focused vision. Indicators 8 to 11 address horizontal interoperability from an organizational point of view.

Well-documented, accessible, checkable, and standardized guidelines for licensing, regulations for custodianship, restrictions on data use and reproduction, and juridical parameters (indicator 8) are needed.

Partners can enter into specific data-sharing agreements or contracts. Clear rules for commercial use, reproduction, and pricing (indicator 9) should be well-documented, accessible, checkable, and standardized.

The basic functioning of an SDI must be secured. This might be possible with a business model that defines a minimal infrastructure that can be financed by a neutral organization (e.g., a government agency) (indicator 10).

Clearly defined and documented coordinating arrangements for participating organizations (indicator 11) ensure continuous workflows and quality control. For example, if a new organization wants to participate in the SDI, it needs to be provided with a set of rules.

In an ideal SDI, horizontal and vertical interoperability will be supported by diverse Web services based on standards set by the Open Geospatial Consortium (ISO/TC211, CEN/TC 287) and the World Wide Web Consortium (3WC) (indicator 12).

A clearinghouse is a specialized, complex Web service for sharing spatial data, and a geoportal is its access point on the Internet (indicator 13). A search engine for finding and retrieving spatial data is a basic function for an SDI.

Data accessibility relies on clearinghouse policies, institutional arrangements, and agreed-upon technical standards (indicator 14). A neutral organization or committee can be officially entrusted to make important decisions for the clearinghouse.

The tranfer of functions accompanying the transfer of data must be standardized (indicator 15). Standardization responsibilities need to be assigned to appropriate organizations (indicator 16).

AUSTRALIA AND SWITZERLAND

Australia and Switzerland are both highly developed countries administered by a federal structure, but their cultural and organizational frameworks differ.

Australian SDI. Australia is the world's sixth largest country in land area and the only country covering an entire continent. It's one of the least populated countries in the world and one of the most urbanized, due to approximately 85 percent of the population residing in urban areas along the eastern and southeastern coastlines. Australia is a Commonwealth comprising six states and two territories.

Australian SDI (ASDI) has been coordinated by Australia and New Zealand Land Information Council (ANZLIC) since 1986, with the aim of making Australia's spatially referenced data, products, and services available and accessible to all.

ANZLIC comprises 10 members representing the Australian government, the New Zealand government, and each of the state and territory governments of Australia. Each member represents a spatial information coordinating structure for whole-of-government within their jurisdiction.

ANZLIC, the Spatial Information Council, the Intergovernmental Committee for Surveying and Mapping (ICSM), and Public Sector Mapping Agencies Australia Ltd. (PSMA) are the key stakeholders of the ASDI.

Swiss SDI. Switzerland is a small but heterogeneous country, with 4 official languages, 26 cantons, and 7 million people. A patchwork of political, legal, and technical issues needs to be harmonized for the national SDI to function effectively.

On the 15th of June 2001, the Swiss Federal Council gave the interdepartmental GIS Coordination Group (COGIS) an official mandate to come up with a plan for a national SDI. This project is embedded in the e-geo.ch initiative, which offers an e-government framework for cooperation among public agencies and the private industry.

The COGIS center is administratively attached to the Swiss Federal Office of Topography (the national mapping agency) but is practically independent.

Table 3 summarizes the comparison of the Australian and Swiss SDIs.

Based on the fact that Australia is 183 times larger than Switzerland and contains vast areas of desert, the data capture methods are very different. Both countries have well-established rules and regulations for cadastral data capture, and most of the spatial data is digitalized. The Swiss federal mapping agency is certified by ISO 9001. Therefore, the capturing and updating of the federally acquired cadastral data is standardized through the printing of maps. However, data capture procedures at lower levels of the federal system are not certified, and applications other than cadastral data capture are not well documented or standardized. In Australia on the other hand, national standards are provided by ANZLIC and are based on ISO 9001. They are recommended to the states for adoption. Most states are certified according to ISO 9001 (indicator 1 in table 3).

Australia has defined core datasets organized in five themes: primary reference, administration, national environment, socioeconomic environment, and built environment. Yet, each state in Australia defines its own core datasets. Switzerland is currently inventorying basic spatial data of national interest for upcoming legislation on geoinformation (indicator 2 in table 3).

Data format normally depends on the GIS system used, but Switzerland has defined an additional standardized format, INTERLIS, for data transfer between different GIS systems and models. INTERLIS is mandatory for cadastral data and is becoming more widely used for other data. In both countries the data is stored in a decentralised way, with the custodians of the data also in charge of the metadata and the updating.

In Australia, different jurisdictions have established different solutions, data models, and processes for their digital cadastral data systems. The Intergovernmental Committee on Survey and Mapping provides overall coordination of cadastral standards and promotes data harmonisation (Dalrymple et al. 2003). Switzerland established a national data model for cadastral surveying in the early 1990s, and every canton is obliged to follow it, with the option of extending it to other uses. INTERLIS provides several modeling tools, for example, for checking compliance with the model (www.interlis.ch). In both countries, metadata and spatial data are saved in separate files or databases (indicator 3 in table 3).

The Australian Spatial Data Directory (ASDD) and the Environmental Data Directory (EDD) are the two main national metadata cataloges in Australia. The latter is maintained by the Department of the Environment and Heritage and contains biological survey data, documentation of species, vegetation data, and biological nomenclature. The former is maintained by Geoscience Australia on behalf of ANZLIC and contains 40,000 metadata records on 25 distributed nodes (24 public and one private) covering various topics. The two metadata catalogues are not linked to each other. Update cycles for spatial data are determined at the jurisdictional level (e.g., each state determines its own update cycle for cadastral data).

In Switzerland, Geocat.ch is the main national metadata catalogue. It was launched on the Internet in 2004 and offers metadata in English, German, and French on various topics. Geocat.ch is well documented and is available 24 hours a day, 7 days a week. However, it covers mainly metadata and has no special provisions for large datasets like photogrammetry.

Indicator	Switzerland[a]	Australia[a]
Data and metadata		
Technical		
1. Data capture process		
Well documented	+/-	+/-
Standardized	+/-	+/-
Accessible	+/-	+/-
Verifiability	+/-	+/-
2. Definition of core datasets		
National	++	+
State		
3. Data format and conceptual model		
Format	+	+/-
Model	+	+/-
4. Data management		
Availability	+	+
Standardized update cycles	+	+
Consistency	+/-	+/-
Provision of large datasets	-	-
5. Data quality and accuracy	+	-/+
6. Common modeling language and tools		
Accessibility	++	++
Usability	+/-	-/+
Usefulness	++	-/+
7. Harmonization of data and metadata		
Coordination	+	+/-
Common models	+	+/-
Organizational		
8. Custodianship		
Well documented	+/-	++
Standardized	+/-	++
Accessibility	+/-	+
Verifiability	+/-	+/-
9. Data sharing and partnership agreements		
Well documented	+/-	+
Standardized	+/-	+
Accessibility	+/-	+
Verifiability	+/-	+/-
10. Business models	++	+
11. Coordinating arrangements		
Definition	+/-	+
Documentation	+/-	+
Web services		
Technical		
12. Application profile	++	+
13. Clearinghouse and geoportal	+	+
Organizational		
14. Clearinghouse organization	++	+/-
Standards		
Technical		
15. Interoperability standards	+	+/-
Organizational		
16. Organizational arrangements for standardization	++	+
[a] ++, very good; +, good; +/-, first steps in a positive direction; -, not so good; -, bad.		

Table 3. Comparison of the Australian and Swiss national SDIs.

On the technical side, both countries are trying to harmonise historical metadatasets. Various metadata profiles have been launched in Australia (ANZLIC metadata profile versions 1.0 and 2.0) and Switzerland (SIK-GIS, CDS) in the last 10 years. Switzerland adopted ISO 19115 for its national profile in 2004, and Australia is in the process of doing so. Both countries are developing freeware and open-source metadata entry tools supporting ISO 19115 (indicator 4 in table 3).

In both Switzerland and Australia, the main Web service is the OGC (Open Geospatial Consortium) Web Map Server (WMS). In Switzerland, a few SOAP (simple object access protocol)-based Web services were developed for metadata, Swiss names, and geocoded addresses. These basic services were coordinated by KOGIS (Coordination, Geoinformation, and Services Division). Other specialized Web services (e.g., for geodetic transformations) are offered by national agencies (e.g., the national mapping agency swisstopo has services for conversion of national map coordinates [Swiss grid] to geographical coordinates [WGS84 datum, etc.] [http://www.swisstopo.ch/en/online/calculation/index]), but these are not coordinated by the national SDI and thus not linked directly (e.g., by a common portal).

Geoscience Australia offers similar Web services, for example, conversion of static GPS data into Geocentric Datum of Australia (GDA) and International Terrestrial Reference Frame (ITRF) coordinates. Other services are offered by other national agencies. As in Switzerland, WMS is the most commonly used standard in Australia. In Switzerland the application profile for Web services already exists, and in Australia it is being developed on the national level by ANZLIC (indicator 12 in table 3).

Neither country has a national clearinghouse as such, but the metadata catalogue serves the function of a search engine. Geoscience Australia, the national mapping agency, lists institutions offering maps. ANZLIC has coordinated an assessment of services provided by the Australian SDI (indicator 13 in table 3).

The role of KOGIS as the coordinating agency for provision of basic services in Switzerland has been questioned. The upcoming legislation on geoinformation will codify federal standards for modeling, acquisition, and exchange of spatial data; regulate copyright and privacy issues; and assign responsibilities for coordination of spatial data. Each Australian state has its own clearinghouse, which determines policies, the institutional framework, and technical standards, while ANZLIC provides national standards and policies for all jurisdictions to implement.

Our indicators apply easily to the two national SDIs, even though the cultural, organizational, and legal contexts differ significantly between Switzerland and Australia. The indicators are still not suitable for assessing SDIs from a cost–benefit point of view, but they do help identify best practices and targets for improvement or for cooperation between countries and institutions. For example, Switzerland can offer technical solutions, whereas Australia can provide excellent expertise on organizational issues.

CONCLUSIONS

We have offered a framework for comparing SDIs on the basis of Web services and data and metadata management, taking a holistic approach encompassing technical and organizational factors.

The SDIs have to reflect the complex organizational, institutional, and technical patchwork of the federal systems of Switzerland and Australia by including local and national representatives and experts.

Australia has a longer history of SDI development and therefore has more results, such as the two metadata cataloges, with which it tested and developed different technologies (Z39.50, three metadata standards). The SDI is very well documented and has clear guidelines, for example, for adding metadata to the ASDD metadata catalogue. However, the existence of two independent metadata catalogues confuses users and complicates interoperability.

Switzerland has a strong data modeling tradition, especially in cadastral datasets. The common modeling (conceptual schema) language INTERLIS, which can be used with a variety of free tools, is an important contribution to interoperability. All SDI components in Switzerland have to provide user interfaces in at least two languages: French and German. The multilingual solutions may be applicable to other countries.

Both countries struggle with inhomogeneous and inconsistent metadata. Common metadata–spatial-data models are especially feasible in Switzerland, where the nationally standardized model for cadastral data, for example, allows integration of different layers. Both countries recognize that free metadata acquisition tools must be offered by the government to help coordinate and facilitate metadata management. In Switzerland these tools are provided by KOGIS (geocat.ch) using the Swiss profile of the ISO 19115 standard. In Australia this work is in progress.

ACKNOWLEDGMENTS

We acknowledge the support of the Institute for Geodesy and Photogrammetry, Swiss Federal Institute of Technology Zurich (ETH); Centre of Coordination of Geo-information (KOGIS); the University of Melbourne; and the members of the Centre for Spatial Data Infrastructures and Land Administration at the Department of Geomatics, University of Melbourne.

REFERENCES

Australian Spatial Data Directory. http://asdd.ga.gov.au/asdd (accessed June 2006).

Balfanz, D. 2002. Automated geodata analysis and metadata generation. SPIE conference on visualization and data analysis (VDA). San Jose, CA.

Bayfield, N. G., J. Conroy, R.V. Birnie, A. Geddes, J. L. Midgley, M. D. Shucksmith, and D. Elston. 2005. Current awareness, use and perceived priorities for rural databases in Scotland. *Land Use Policy* (22):153–162.

Borrero, S. 2002. The GSDI Association: State of the art. GSDI-6 Conference. Budapest, Hungary.

Bulterman, D. C. A. 2004. Is it time for a moratorium on metadata? *IEEE* 11 (04): 10–17.

CEN-Technical Committee 287 (CEN/TC 287). http://www.cenorm.be (accessed June 2006).

Clarke, D., O. Hedberg, and W. Watkins. 2003. Development of the Australian Spatial Data Infrastructure. In *Developing spatial data infrastructures: from concept to reality,* eds. R. A. Williamson, M.-E. F. Feeney. New York: Taylor & Francis.

Crompvoets, J., and A. Bregt. 2003. World status of national spatial data clearinghouses. *URISA Journal* 15(Access and Participatory Approaches [APA] 1).

Crompvoets, J., A. Bregt, A. Rajabifard, and I. Williamson. 2004. Assessing the worldwide developments of national spatial clearinghouses. *Int. J. Geographical Information Science* 18 (7): 665–89.

Dalrymple, K., I. Williamson, and J. Wallace. 2003. Cadastral systems within Australia. *The Australian Surveyor* 48 (1): 37–49.

Delgado Fernandez, T., K. Lance, M. Buck, and H. Onsrud. 2005. Assessing an SDI readiness index. GSDI-8 Conference. Kairo, Egypt.

Environmental Data Directory (EDD). http://www.deh.gov.au/erin/edd (accessed June 2006).

Frank, A. U., E. Grum, and B. Vasseur. 2004. Procedure to select the best dataset for a task. GIScience 2004. Maryland: Springer-Verlag.

Geocat. www.geocat.ch (accessed June 2006).

Giff, G., and D. Coleman. 2002. Spatial Data Infrastructures Funding Models: A necessity for the success of SDI in emerging countries. FIG XXII International Congress. Washington, DC.

Giger, C., and C. Najar. 2003. Ontology-based integration of data and metadata. 6th AGILE Conference on Geographic Information Science. Lyon, France.

Group@IMRB. 2003. E-Readiness assessment of central ministries and departments (draft report). India.

Hunter, G., and M. Goodchild. 1995. Dealing with error in a spatial database: A simple case study. *Photogrammetric Engineering & Remote Sensing* 61 (5): 529–37.

INSPIRE Reference Data and Metadata Group (RDM). http://inspire.jrc.it/ir/index.cfm-9 (accessed May 2006).

INTERLIS. http://www.interlis.ch (last accessed June 2006).

ISO/TC 211 International Organization for Standardization. http://www.isotec211.org (accessed November 2005).

Kok, B., and B. van Loenen. 2005. How to assess the success of national spatial data infrastructures? *Computers, Environment and Urban Systems* 29: 699–717.

Masser, I. 1999. All shapes and sizes: The first generation of national spatial data infrastructures. *International Journal of GIS* (13): 67–84.

Najar, C., C. Giger, F. Golay, C. Moreni, and M. Riedo. 2004. Geo Web Services: An NSDI-embedded approach. 7th AGILE Conference on Geographic Information Science. Heraklion, Greece.

Najar, C., C. Giger, F. Golay, C. Moreni, and M. Riedo. 2004. Geodata does not always equal geodata: Meet the challenge of a GDI in the Swiss Federal System. 10th EC GI and GIS Workshop. ESDI: State of the Art. Warsaw, Poland.

Onsrud, H., G. Camara, J. Campbell, and N. Sharad Chakravarthy. 2004. Public commons of geographic data: Research and development challenges Geographic Information Science. Maryland: Springer-Verlag.

Open Geospatial Consortium. http://www.opengeospatial.org (last accessed June 2006).

Rackham, L., and D. Rhind. 1998. Establishing the UK national data framework. SDI'98. Ottowa, Canada.

Rajabifard, A., F. Escobar, and I. P. Williamson. 2000. Hierarchical reasoning applied to spatial data infrastructures. *Cartography Journal* 29 (2).

Rajabifard, A., M.-E. F. Feeney, I. P. Williamson, and I. Masser. 2003. *Developing spatial data infrastructures: From concept to reality.* New York: Taylor & Francis.

Steudler, D. 2004. A framework for the evaluation of land administration systems. University of Melbourne, Australia.

Warnest, M., A. Rajabifard, and I. Williamson. 2005. A collaborative approach to building Nntional SDI in federated state systems: Case study of Australia. GSDI-8 Conference. Cairo, Egypt.

Williamson, I., D. Grant, and A. Rajabifard. 2005. Land administration and spatial data infrastructures. GSDI-8 Conference. Cairo, Egypt.

Williamson, I., A. Rajabifard, and M.-E. F. Feeney. 2003. Developing spatial data infrastructures: From concept to reality. New York: Taylor & Francis.

World Wide Web Consortium (2WC). http://www.w3.org (accessed December 2005).

Institutions Matter: The Impact of Institutional Choices Relative to Access Policy and Data Quality on the Development of Geographic Information Infrastructures

BASTIAAN VAN LOENEN AND JITSKE DE JONG

DELFT UNIVERSITY OF TECHNOLOGY, DELFT, THE NETHERLANDS

ABSTRACT

Access to geographic information is critical for the development of geographic information infrastructures (GIIs). Two access policy options are dominant: open access and cost recovery. Cost recovery policies are generally thought to be associated with high-quality datasets, while open-access policies are thought to be associated with poor-quality data. However, how data collection is organized is more critical for the quality and the use of the dataset than access policies are. We compared parcel and large-scale topographic datasets in five jurisdictions of comparable sizes, population densities, and socioeconomic levels and argue that institutional choices affect GII development.

Access policies have been deemed critical for geographic information infrastructures (Borgman 2000 p. x; Masser 1999, p. 81; Tosta 1999, p. 23). Two access policy options have been dominant: open access and cost recovery. Cost recovery policies have been thought to be associated with high-quality datasets (GITA 2005; Lopez 1998, p. 79), while open-access policies have been thought to be associated with poor-quality data (GITA 2005; Aslesen 2002).

These findings are based on examples from the United States and the European Union. According to the Geospatial Information and Technology Association (GITA) survey, the cost recovery policies of the UK Ordnance Survey are justified by the quality of its products, which "far exceed the quality, in terms of accuracy and timeliness, of most products given away in the United States" (GITA 2005; see also Lopez 1998, p. 79). In addition, the experiences of academics in the United States suggest that access policies become more restrictive as the level of information detail increases. Figure 1 shows the assumed relationship between access policy and data quality.

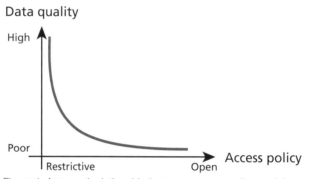

Figure 1. Assumed relationship between access policy and data quality.

We sought to test the hypothesis that the quality and the access policy of a dataset are related: datasets with restrictive access policies have excellent quality, while datasets with open access policies have poor quality. The hypothesis assumes that access policy is decisive for the quality of the dataset and that the different access policies of European and U.S. jurisdictions may account for the differences in data quality. The hypothesis conforms to the economic theory predicting that quality of information is related to the price and use restrictions.

We report the findings of a study (Van Loenen 2006) comparing parcel and large-scale topographic datasets in five jurisdictions of comparable sizes, population densities, and socioeconomic levels (the Netherlands, Denmark, the German state of North Rhine–Westphalia; and in the United States, Massachusetts and the Metropolitan region of Minneapolis and St. Paul).[1]

In the open-access model, information held by a government can be accessed by those outside the government for a price not exceeding the cost of reproduction and distribution (marginal cost of dissemination), with the imposition of as few restrictions as possible. The information is available to all (nonexclusively) on a nondiscriminatory basis (see also NRC 1997, p. 15). Acceptable restrictions include national security, trade secrets, and privacy.

Although the open-access model may initially have been enacted to control government, "[it] fosters a process for adding value to raw government information resources" (Lopez 1998, p. 58). This spin-off effect promotes the use of the information, which results in higher (income, company, or value-added) tax revenues going to the government. The increased use of geographic information resulting in innovative solutions to societal needs encourages GII development and the information economy (Pira et al. 2000; Weiss and Pluijmers 2002).

However, open policies make government entities responsible for the collection of geographic data fully dependent on general budgets. This is a precarious position, especially in an economic climate of recession. For example, the U.S. national mapping agency, U.S. Geological Survey (USGS), has suffered from significant real budget reductions, causing it to scale back updates of the 1:24,000 map series (NRC 2003, p. 22). The diminished quality of the data may result in diminished use.

Cost recovery approaches seek profits from the sale of information to support the development and maintenance of the datasets (Lopez 1998, p. 43; Onsrud 1992). Information collection, maintenance, and dissemination are not fully supported by public funds, and the costs must be covered through other means. The agency is forced to generate income from the sales of information or products or through the provision of services. The cost recovery model assumes that sufficient income will be generated for the creation and maintenance of the dataset.

Access to information may therefore be restricted to cope with the financial conditions established by government rules. In practice this entails a charge for the information at more than the marginal cost of dissemination, and restrictions are imposed on the use of government information through copyright and database rights. Further, use restrictions are often imposed through contractual or licensing provisions. Government expertise may be used to respond to private requests for specific geographic products.

In several cost recovery models, individual government agencies are in control of their budgets, making them independent of the fluctuating budgets of the national governments (Onsrud 1998, p. 146). As a consequence, government agencies are able to offer (access to) accurate, consistent, standardized databases with sufficient coverage (Aslesen 2002). Therefore, the cost recovery model may provide sustainable funding to individual government agencies, allowing them to maintain their information collection activities over time (Onsrud 1992). This is especially true in instances where "income is generated through actions required by statute" (Pira et al. 2000, p. 44; see also Coopers Lybrand 1996).

The cost recovery model may be summarized as follows: "[it] benefits end-users who are interested and able to acquire high-quality geographic information, directly from government" (Lopez 1998, p. 58). The continuous availability of high-quality datasets supports GII development.

A GII may be defined as a framework continuously facilitating the efficient and effective generation, dissemination, and use of needed geographic information within a community or between communities (after Kelley 1993). The framework consists of six interdependent components: (framework) datasets, institutional framework, technology, standards, financial resources, and human resources. Interaction between the components is a condition for the further development of the infrastructure.

In the early stages, GII development is data-centric: data creation and integration, reduction of duplication, effective use of resources, and creation of a base from which to expand the productivity of the geographic information sector and the geographic information market (Rajabifard et al. 2003, p. 101, 107; Rajabifard et al. 2002, p. 14). After a dataset has achieved sustainable quality in meeting the needs of primary users, value-added use of the information is the driver (Rajabifard et al. 2003; cf. Masser 2000; Van Loenen 2006).

Users of datasets. Users of the GII "will probably be the most mentioned group and yet actually the least considered" (McLaughlin and Nichols 1994, p. 72). This, however, does not imply that all potential user groups or applications need to be identified. It does mean that the user community has to be considered as part of the total infrastructure, and that real, rather than purely academic requirements have to be met (McLaughlin and Nichols 1994, p. 72).

We distinguish four user groups:

1. Primary users: the data collector and major users
2. Secondary users: incidental users with similar purposes as the primary user
3. Tertiary users: users that use the dataset for other than primary purposes
4. End users: users that use the end product, such as a map

Primary users are those that use the dataset in line with the initial purpose of information collection on a continuous basis. They are typically members of a government organization that has collected and processed the information. Secondary users may be found in government, the private sector, or academia. They use the information for similar purposes but only incidentally. Tertiary users are typically found in the private sector. They add value to the dataset. A tertiary use may be the integration of several topographic datasets into one layer for a jurisdiction, the linkage of a dataset with several thematic layers, provision of user-friendly access to the dataset (e.g., adding search facilities, explanation, help desk), or intermediary services for distributing the dataset. End users are citizens, decision-makers, and others that use the end product of geographic information, for example, a map, an animation, or an answer to a question.

Technical and nontechnical characteristics of framework datasets. Framework datasets are a key component of GIIs. The value of a framework dataset rests on its "coverage, the strengths of its representation of diversity, its truth within a constrained definition of that word, and on its availability" (Longley 2001, p. vii). A user decides to use the dataset or not to use it on the basis of its technical and nontechnical characteristics (Van Loenen 2006). Although each user category and within each category each user may have specific needs for framework data characteristics, the technical and nontechnical data characteristics described below are assessed to satisfy most users.

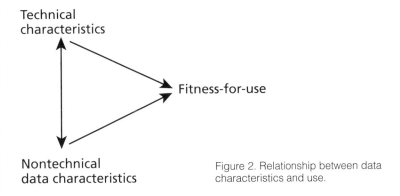

Figure 2. Relationship between data characteristics and use.

Technical data characteristics are commonly referred to as "data quality." From a technical perspective, an ideal framework dataset includes full jurisdictional coverage; a jurisdiction-wide uniform data model; adherence to open data formats; current, accurate, clearly demarcated coverage area; comprehensive metadata; and a high level of interoperability with other datasets. In the ideal dataset, each of these characteristics is sustainable over time (Van Loenen 2006).

Nontechnical data characteristics include access policy (price and use restrictions) and ease of access (e.g., access through a clearinghouse). From a nontechnical point of view, an ideal framework dataset is available, findable, easily assessable, and readily accessible (i.e., the dataset can be accessed immediately from a one-stop shop with few legal restrictions at an affordable price) (Van Loenen 2006). The access policy should be transparent and generic throughout the government.

Together, the nontechnical and technical characteristics determine the dataset's fitness for use (figure 2).

RESEARCH METHODOLOGY

To assess the impact of access policies on the technical characteristics and use of a (framework) dataset, a multiple-case-study design was used. The case studies were selected on the basis of their similarities to the Netherlands in socioeconomic level, jurisdiction size, population density, and government type. In maximizing variance in access policies, both literal and theoretical analogies were sought. We also assumed that jurisdictions of comparable geographic sizes and population densities have similar needs for geographic datasets. Therefore, jurisdictions comparable to the Netherlands in geography were chosen: the Netherlands, Denmark, the German state of North Rhine–Westphalia, the U.S. state of Massachusetts, and the U.S. metropolitan region of Minneapolis–St. Paul (Metro region).

We focused on parcel and large-scale topographic datasets for four reasons:

1. These datasets are important for local levels of GIIs (Rajabifard et al. 2000).
2. The high level of detail at the local level can be used as the basis for higher GII levels.

3. These datasets are relatively expensive to collect, process, and maintain.

4. They have barely been addressed in access policy research.

The case study was conducted in 2004.

The findings for parcel information are organized in three categories: technical characteristics, nontechnical characteristics, and use.

Technical characteristics. The technical data characteristics were judged to be advanced for the Danish and Dutch parcel datasets. The Netherlands and Denmark have full digital parcel information coverage. In each country, one national public organization has the responsibility for parcel information. The datasets adhere to one national data standard; have a legal basis, comprehensive content, full topology, and generally no gaps or overlaps; and are fully harmonized and current for their purposes. In addition, the Danish dataset has comprehensive metadata documentation.

The Metro dataset also meets the technical data requirements of a GII. However, it's less comprehensive than the Danish and Dutch datasets. Because its parcel information comes from seven counties, the Metro dataset is not as consistent with respect to positional accuracy and content. However, Metro has full digital parcel information coverage, and all county datasets adhere to the U.S. parcel information standard, although individual counties have modified it to meet their specific needs. The Metro dataset has comprehensive metadata documentation.

North Rhine–Westphalia has the most comprehensive parcel dataset, including both parcel information and full topographic detail. Its 54 counties have created their own parcel datasets, with different technical characteristics. The state mapping agency has, together with the counties, initiated the integration of these 54 datasets into one statewide dataset. However, as of 2006, this dataset did not cover the entire jurisdiction of North Rhine–Westphalia, and some parts were still in analogue format. This inconsistency led to poor scores on technical data characteristics. With full digital coverage, the North Rhine–Westphalian dataset could be comparable to those of Denmark and the Netherlands. As with the Metro dataset, differences between some of the North Rhine–Westphalian component datasets may lead to gaps or overlaps in the integrated dataset.

The Netherlands, Denmark, Metro, and North Rhine–Westphalia datasets have legal foundations.

The Massachusetts' dataset has been judged inadequate in all technical categories. A significant percentage of the 351 component datasets are not in digital format and/or do not adhere to a standard data model. The datasets vary so much that harmonized, statewide parcel information coverage is not expected soon.

Nontechnical characteristics. The open-access policies of Massachusetts state government are conducive to GII development. However, obtaining parcel information covering the entire state is difficult because of the need to contact the 351 dataset sources.

The other four jurisdictions have restrictive access policies. The Danish and Dutch parcel datasets, for example, cost €2,400,000 and €3,500,000, respectively, and cannot be resold without prior consent of the data provider. However, the datasets can be acquired more easily than the datasets in Massachusetts. To obtain the dataset for all of Denmark or the Netherlands, only one organization needs to be contacted. The Metro integrated parcel dataset is available to MetroGIS participants; others need to contact each of the seven counties. The North Rhine–Westphalia datasets can potentially be made available from one contact point, but currently the 54 data providers need to be contacted separately. In none of the cases is access immediate.

Table 1 lists the findings for parcel information. Table 2 lists the overall scores.

Characteristic	Score[a]				
	DK	NL	NRW	MA	Metro
Technical					
Internal					
Content	+	+	++	-	0
Horizontal positional accuracy	+	++	++	-	-
Currency	++	++	++	-	++
Structure	0	+	--	--	+
Quality consistency throughout the (integrated) dataset	+	+	--	--	-
Average	+	+	0	-	0
External					
Digital coverage (vector format)	++	++	+	-	++
Number of datasets for jurisdiction coverage	++	++	-	--	+
Standard adherence	+	+	-	-	+
Data model	+	+	-	-	+
Metadata documentation	++	-	--	--	++
Quality assurance	++	++	++	-	+
Average	+	+	-	-	+
Nontechnical					
Access policy					
Legal access	-	-	-	++	-
Financial access[b]	- (++)	-	- (++)	++	- (++)
Average	- (+)	- (+)	- (+)	++	- (+)
Accessibility					
Publication	++	-	- -	- -	++
Points to contact for maximum coverage of jurisdiction	++	++	-	- -	+
Acquisition procedure	+	+	+	+	+
Time between request and access	+	+	+	+	+
Average	+	+	-/ 0	-	+

a DK, Denmark; NL, Netherlands; NRW, North Rhine–Westphalia; MA, Massachusetts. ++, ideal; --, far from ideal.
b Free access for designated user groups.

Table 1. Findings for parcel information.

Characteristic	Score[a]				
	DK	NL	NRW	MA	Metro
Technical characteristics	+	+	0/ -	-	0/ +
Nontechnical characteristics	0	0	0/ -	+	0

a Users were mainly primary and secondary. DK, Denmark; NL, Netherlands; NRW, North Rhine–Westphalia; MA, Massachusetts.

Table 2. Overall scores for parcel information.

Use. Parcel information in all five cases was found to be used primarily by state and local governments and secondary users such as real estate managers, notaries public, utilities, architects, and engineering companies. We found few tertiary users creating value-added products. Some potential users indicated that use restrictions do not allow this; others indicated that the prices are too high to create value-added products. In Massachusetts, where prices and use restrictions are intended to promote reuse, the investment needed to integrate the 351 datasets into one is too high for the creation of value-added services.

FINDINGS FOR TOPOGRAPHIC INFORMATION

As with parcel information, the findings for topographic information are organized in three categories: technical characteristics, nontechnical characteristics, and use.

No one central organization is responsible for the collection and processing of large-scale topographic information in any of the five jurisdictions. Instead, data is collected cooperatively, and costs are shared. The Netherlands has strong cooperation between the public and private sectors, North Rhine–Westphalia has strong cooperation within the public sector, Denmark has cooperation within the public sector and between the public and private sectors, while in the Metro region and in Massachusetts cooperation is limited.

Technical characteristics. The technical characteristics of the European topographic datasets were found to be reasonably good. The consistency of quality needs to be improved, but each dataset is potentially sufficient for use as a framework layer in a GII.

Full coverage and the integration of local topographic datasets distinguish the Dutch dataset. The metadata documentation, however, does not meet the standards of a GII. The Danish datasets in combination also cover Denmark completely. However, users interested in full coverage need to integrate the individual datasets themselves and face the likelihood of geometric and/or semantic incompatibility. Several Danish datasets have good metadata documentation, while others have none. As with parcel information, North Rhine–Westphalia has the most comprehensive dataset for topography. However, as of 2004, this dataset did not cover the entire jurisdiction of North Rhine–Westphalia, and some parts were still in analogue format. This inconsistency led to poor scores on technical data characteristics. With full digital coverage, the North Rhine–Westphalian dataset could be comparable to those of Denmark and the Netherlands.

The Massachusetts and Metro region datasets have poor technical quality (e.g., data models are not harmonized, and data is collected in an ad hoc fashion) and large-scale topographic information for parts of the jurisdiction.

Nontechnical characteristics. Because of the involvement of utilities, all five jurisdictions have generally restrictive-access policies. The large-scale basemap of the Netherlands costs between €1,000,000 and €2,000,000. In North Rhine–Westphalia the large-scale topographic dataset costs €3,400,000. The Danish equivalent sells for approximately €5,000,000. Redistribution is generally prohibited.

The Netherlands topographic dataset is published only on the provider's Web site and is sufficiently accessible: only one point needs to be contacted. As with parcel information, the North Rhine–Westphalia topographic datasets can potentially be made available from a central location, but currently the 54 information providers need to be contacted separately. Only a few of the 54 datasets are included in the state's clearinghouse. In Denmark, the 68 data sources have to be contacted separately, but some of them may be found through the national clearinghouse. In the Metro region, county datasets are sufficiently accessible, and comprehensive metadata is available online in the region's clearinghouse. In Massachusetts, large-scale topographic information is difficult to find and is not provided by the state's clearinghouse. Consequently, the 351 local authorities need to be contacted separately. However, in many instances local governments rely on topographic data provided by the utilities.

Because of incomplete coverage, the quality of the Massachusetts and Metro region datasets is judged to be poor.

Table 3 lists the findings for topographic information. Table 4 lists the overall scores.

Characteristic	Score[a]				
	DK	NL	NRW	MA	Metro
Technical					
Internal					
Content[b]	+	+	++	-- (+)	-- (+)
Horizontal positional accuracy	+	++	++	- -	- -
Currency	+	++	++	- -	- -
Structure[b]	-	-	- -	-- (+)	-- (0)
Quality consistency throughout the (integrated) dataset	-	-/ 0	- -	- -	- -
Average	0	0/ +	0	- -	- -
External					
Digital coverage (vector format)	++	++	0	- -	- -
Number of datasets for jurisdiction coverage	-	++	-	- -	- -
Standard adherence	0	+	-	- -	- -
Data model	0	+	-	- -	- -
Metadata documentation	+	- -	- -	- -	- -
Quality assurance	-	-	++	- -	- -
Average	0/ +	+	-	- -	- -
Nontechnical					
Access policy					
Legal access	-/ - -	-/ - -	-	-- (++)	-
Financial access[c]	-	-	- (++)	- (++)	-
Average	-	-	- (+)	- (++)	-
Accessibility					
Publication[c]	-	- -	- -	- -	N/A (++)
Points to contact for maximum coverage of jurisdiction	-	++	-	N/A	N/A
Acquisition procedure	+	++	+	+	+
Time between request and access	+	+	+	+	+
Average	0	+	-/ 0	N/A	N/A

a DK, Denmark; NL, Netherlands; NRW, North Rhine–Westphalia; MA, Massachusetts. ++, ideal; --, far from ideal; N/A, not applicable.
b Individual datasets.
c Free access for designated user groups.

Table 3. Findings for large-scale topography.

Characteristic	Score[a]				
	DK	NL	NRW	MA	Metro
Technical characteristics	0	0/+	0/ -	- -	- -
Nontechnical characteristics	- /0	0	- /0	- -	0
Users	Primary and secondary	Primary and secondary	Primary and secondary	Primary and secondary	Primary and secondary
a DK: Denmark; NL, Netherlands; NRW, North Rhine–Westphalia; MA, Massachusetts.					

Table 4. Overall scores for large-scale topography.

Use. Uses in the five jurisdictions were similar. The primary users of large-scale topographic information are local governments and utilities. Secondary users (those that use the data incidentally) are engineers and planning agencies. Tertiary users are few or nonexistent. Restrictive policies and pricing, incomplete datasets, and lack of transparency may be the causes (the lack of transparency may also explain the findings of duplicate information collection in four cases). On the other hand, the products may be so large-scale and application-specific that they are of limited value to outside users.

SUMMARY OF FINDINGS

Both technical and nontechnical characteristics may affect the use of parcel information. All five jurisdictions had significant levels of primary and secondary uses but few value-adding activities, possibly because of restrictive-access and cost-recovery policies.

We had hypothesized that the technical quality of a dataset and its access policy are related: that technically excellent datasets have restrictive-access policies and technically poor datasets have open-access policies.

The case studies yielded conflicting findings. We identified several technically advanced datasets with less advanced nontechnical characteristics: the topographic dataset in the Netherlands and the parcel datasets of Denmark, the Netherlands, and the Metro region. We also identified technically insufficient datasets with restrictive-access policies: the topographic datasets in Denmark, North Rhine–Westphalia, Massachusetts, and Metro region. Thus, cost recovery does not necessarily signify excellent quality. We did not obtain sufficient information to draw any conclusions about open-access policies.

Although the links between access policy and use and between quality and use are apparent, we did not find convincing evidence for a direct relation between the access policy and the quality of a dataset (figure 3).

INSTITUTIONAL SETTING IS CRITICAL

The institutional setting of a jurisdiction affects the way data collection is organized (e.g., centralized versus decentralized control), the extent to which data collection and processing are incorporated in legislation, and the extent to which legislation requires use within government.

We have focused on the organization of data collection. The differences in data characteristics may be explained by the different way each jurisdiction has

Data quality

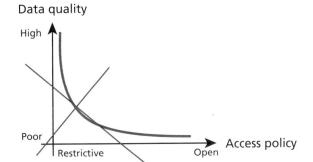

Figure 3. No convincing evidence was found supporting a causal relationship between access policy and data quality.

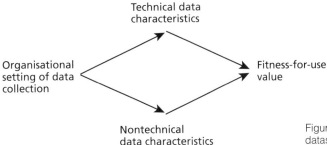

Figure 4. Factors determining dataset use.

organized its data collection. We found a direct link between the institutional setting and the characteristics of the datasets.

In jurisdictions where information collection was centralized in a single public organization, datasets (and access policies) were more homogenous than datasets that were not controlled centrally (such as those of local governments). Ensuring that data is prepared to a single consistent specification is more easily done by one organization than by many.

The parcel datasets of the Netherlands and Denmark are good examples of central control and high technical quality. They cover the entire jurisdictions in digital format and have harmonized content, accurate information, one standard data model, and no gaps or overlaps. Each has a consistent restrictive access policy. If information collection is not centralized or at least under the control of a central body (as in the case of surveys and mapping projects subcontracted to private companies by a central government agency), the likelihood of heterogeneous data characteristics increases and the fitness of the data for (tertiary) use diminishes (figure 4). For example, if information is collected by 300 local entities and each entity adheres to a different information model and policy, cross-jurisdictional users (such as tertiary users) need to find 300 datasets, contact 300 organizations, navigate 300 access policies, and ultimately integrate 300 datasets differing in content, currency, accuracy, and format. This would require significant investments, which may be difficult to recover. The 351 parcel datasets in Massachusetts, the 68 *Tekniske Korte* in Denmark, and the 54 *Automatisierten Liegschaftskarten* in North Rhine–Westphalia are examples of decentralized heterogeneous datasets.

**OVERCOMING
INSTITUTIONAL
BARRIERS**

The disadvantages of decentralized information collection can be redressed by institutional reform that would centralize responsibility for a specific dataset and formalize data collection arrangements (e.g., legislation), including the maintenance of the data over time.

Centralizing responsibility for a dataset has the advantage of establishing one point of contact with far-reaching authority that would enforce standards, execute agreements, and improve data quality. Access policies of central government agencies can be improved relatively easily. Moreover, a central organization would allow for investment in research and development activities specifically aimed at improving the dataset and developing the GII. In addition to contributing to a product-based GII strategy, a central organization is better able to appropriate resources for GII development. It can also coordinate activities efficiently and effectively. Increased trust may also accelerate the pace of GII development, especially with support from the highest levels of the central organization.

Based on the above factors, centrally organized datasets seem preferable to those that are not managed centrally. However, even centrally managed datasets can hinder GII development since it is not directly depending on developments in society. For example, the central agency may not readily adopt open standards or may be unwilling to provide metadata documentation. In a decentralized setting such practices may be corrected by best practices of some individual datasets. In a centrally managed situation, which in practice is a one-of-a-kind dataset, such best practices may not be stimulated. Central management may prevent individual datasets from having flexible policies and procedures and may stifle change. On the other hand, information societies increasingly communicate across borders, and organizations that operate centrally may learn from the best practices of other organizations.

Central governments may be unable or unwilling to counteract strong feelings of local independence in order to enforce institutional reform for the sake of GII development. Nevertheless, Dutch and Danish municipalities have been forced to merge (Tweede Kamer 1998–1999; Christoffersen 2005) in order to increase efficiency through the economies of scale, to organize knowledge better within the public sector, and to better address complex societal issues. GIIs will likely benefit from institutional reforms aimed at centralizing government functions.

The transformation of voluntary efforts into formal arrangements (e.g., legislation) can also be considered institutional reform. Institutional arrangements for collecting geographic data, maintaining its quality, and ensuring its accessibility are a way to promote the development of a GII. "In order to function as a foundation framework datasets should have guaranteed qualities, and central control over these qualities should exist" (Philips et al. 1999; see also NRC 1995, pp. 25, 27). To ensure data quality, standards and requirements can be legislated. For example, the Dutch government has assumed responsibility for the collection, processing, and dissemination of framework information for a wide range of framework datasets, the so-called national base registers (Stroomlijning Basisgegevens 2004). Government agencies are required to use this information and to provide feedback. The Dutch large-scale basemap (GBKN) is an example of a voluntary effort of public and private organizations, which is about to become part of the national base registries. This will result in a legislation-based dataset with guaranteed qualities, central control, and mandatory use. However, inclusion of data requirements in legislation may be against the interests of users, since they will be tied to the standard of a specific time frame and may not meet future needs. Legislation should therefore be flexible to allow for necessary changes to meet future needs.

For parcel information collection that's not centrally organized, strong cooperation between public information providers may facilitate GII development. We found that centrally organized datasets had better technical quality than decentralized ones. Information sharing and other forms of cooperation need to be initiated where centralization is lacking. A common understanding of the needs of all primary and secondary users makes successful cooperation more likely. Champions are key to the success of the effort (Craig 2001; Rietdijk 2000, p. 222).

The topographic dataset of the Netherlands and the integrated parcel dataset of Denmark offer examples of cooperative efforts in creating one jurisdiction-wide homogeneous dataset. Although they are not centrally organized, the datasets of North Rhine–Westphalia and Metro are integrated and harmonized (to some extent) by the central organizations *Landesvermessungsamt* and MetroGIS, respectively. In Massachusetts, the centrally operating organisation, MassGIS lacks the resources to accomplish comparable results. Technological developments allowing organizations to exchange information without losing their autonomy (Onsrud 1990) may aid cooperation.

CONCLUSIONS

Technical and nontechnical data characteristics are important factors on the basis of which users decide to use or not to use a dataset. We hypothesized that the technical and nontechnical components were independent of the specific institutional setting, but we found that the way information is organized has major implications for a dataset. The institutional setting can affect access policy, accessibility, technical quality, and consequently, the type and number of users.

The direct link between the institutional setting and the technical and nontechnical characteristics of datasets suggests that choices in institutional settings many years (or even centuries) ago may affect GII development.

ENDNOTE

1. Research reported here was conducted as part of Bastiaan van Loenen's PhD dissertation, Delft University of Technology.

REFERENCES

Aslesen, L. 2002. The U.S. experience. Data policy and legal issues position paper for Infrastructure for Spatial Information in Europe (INSPIRE). http://inspire.jrc.it/reports/position_papers/inspire_dpli_pp_v12_2_en.pdf.

Borgman, C. L. 2000. *From Gutenberg to the global information infrastructure; Access to information in the networked world.* Cambridge, MA: The MIT Press.

Christoffersen, H. 2005. The structural reform: A reform of the organisation and management of local authorities in Denmark. http://www.akf.dk/dk2005/summary/strukturreform.htm.

Coopers Lybrand. 1996. Economic aspects of the collection, dissemination and integration of government's geospatial information. A report arising from work carried out for Ordnance Survey by Coopers and Lybrand. http://www.ordnancesurvey.co.uk/oswebsite/aboutus/reports/coopers/index.html.

Craig, W. J. 2001. Spatial data infrastructure in Minnesota: Institutional mission and individual motivation. International Symposium on Spatial Data Infrastructure. University of Melbourne. http://www.metrogis.org/documents/presentations/craig_2001.pdf.

GITA. 2005. Free or fee: The governmental data ownership debate. http://www.gita.org/resources/white%20papers/Free_or_fee.pdf.

Kelley, P. C. 1993. A national spatial information infrastructure. Proceedings of the 1993 Conference of the Australasian Urban and Regional Information Systems Association. Adelaide, Australia.

Longley, P. A., M. F. Goodchild, D. J. Maguire, and D. W. Rhind, eds. 2001. *Geographic information systems and science.* Chichester, England: John Wiley & Sons Ltd.

Lopez, X. R. 1998. *The dissemination of spatial data: A North American-European comparative study on the impact of government information policy.* London: Ablex Publishing Corporation.

Masser, I. 1999. All shapes and sizes: The first generation of national spatial data infrastructures. *International Journal of Geographical Science* 3:67–84.

Masser, I. 2000. What is a spatial data infrastructure? Proceedings of the 4th Global Spatial Data Infrastructure conference. http://gsdidocs.org/docs2000/capetown/masser/masser.ppt.

McLaughlin, J., and S. Nichols. 1994. Developing a national spatial data infrastructure. *Journal of Surveying Engineering* 120:62–76.

NRC (National Research Council). 1995. *A Data Foundation for the National Spatial Data Infrastructure.* Washington, DC: National Academy Press.

NRC (National Research Council). 1997. *Bits of power: Issues in global access to scientific data.* Washington, DC: National Academy Press.

NRC (National Research Council). 2003. Weaving a national map: Review of the U.S. *Geological Survey concept of the national map.* Washington, DC: The National Academy Press.

Onsrud, H. J. 1990. The cadastral mapping challenge for surveyors in the U.S.A. Proceedings of the FIG XIX conference. http://www.oicrf.org.

Onsrud, H. J. 1992. In support of cost recovery for publicly held geographic information. *GIS Law* 1:1–7. http://www.spatial.maine.edu/~onsrud/pubs/Cost_Recovery_for_GIS.html.

Onsrud, H. J. 1998. The tragedy of the information commons. In *Policy issues in modern cartography,* ed. D. R. F. Taylor, 141-158. Oxford: Elsevier Science. http://www .spatial.maine.edu/%7eonsrud/pubs/tragedy42.pdf.

Phillips, A., I. Williamson, and C. Ezigbalike. 1999. Spatial Data Infrastructure Concepts. *The Australian Surveyor* 44 (1): 20–28. http://www.geom.unimelb.edu.au/research/publications/IPW/SDIDefi nitionsAusSurv.html.

Pira International Ltd., University of East Anglia, and KnowledgeView Ltd. 2000. Commercial exploitation of Europe's public sector information. Final report for the European Commission Directorate General for the Information Society. ftp://ftp.cordis.lu/pub/econtent/docs/commercial_final_report.pdf.

Rajabifard, A., I. P. Williamson, P. Holland, and G. Johnstone. 2000. From local to global SDI initiatives: A pyramid of building blocks. Proceedings of the 4th Global Spatial Data Infrastructure Conference. http://www.gsdi.org/docs2000/capetown/abbas.rtf.

Rajabifard, A., M.-E. F. Feeney, and I. P. Williamson. 2002. Directions for the future of SDI development. International Journal of Applied Earth Observation and Geoinformation 4:11-22. http://www.geom.unimelb.edu.au/research/publications/IPW/ITC%20Journal%202002.doc.

Rajabifard, A., M.-E. F. Feeney, and I. P. Williamson, and I. Masser. 2003. National SDI initiatives. In Developing spatial data infrastructures: From concept to reality, eds. Ian Williamson, A. Rajabifard, and M.-E. F. Feeney, 95–110. London: Taylor & Francis.

Rietdijk, M. 2000. Gemeentelijke vastgoedinformatievoorziening en regelgeving. Dissertation. Delft: Delft University Press.

Stroomlijning basisgegevens. 2004. De stelselnota: Op weg naar een werkendstelsel van basisregistraties. http://www.stroomlijningbasisgegevens.nl/pdfs2/Stelselnota_versie_14_4.doc.

Tosta, N. 1999. NSDI was supposed to be a verb: A personal perspective on progress in the evolution of the U.S. national spatial data infrastructure. In Integrating information infrastructures with GI technology, ed. B. Gittings, 13–24. Philadelphia: Taylor & Francis.

Tweede Kamer. 1998–1999. Gemeentelijk herindelingsbeleid (26331, nr. 1). Den Haag: Sdu Uitgeverij.

van Loenen, B. 2006. Developing geographic information infrastructures: The role of information policies. Dissertation. Delft: Delft University Press. http://www.library.tudelft.nl/dissertations/dd_list_paged/dd_metadata/index.htm?docname=088301.

Weiss, P., and Y. Pluijmers. 2002. Borders in cyberspace: Conflicting public sector information policies and their economic impacts. http://www.spatial.maine.edu/GovtRecords/cache/Final%20Papers%20and%20Presentations.

Legal Framework for a European Union Spatial Data Infrastructure: Uncrossing the Wires

KATLEEN JANSSEN AND JOS DUMORTIER

KATHOLIEKE UNIVERSITEIT LEUVEN, LEUVEN, BELGIUM

ABSTRACT

The development of a spatial data infrastructure (SDI) not only comprises technical aspects but also is supported by economic, social, organizational, and legal measures. The legal framework for a European Union (EU) SDI consists of two kinds of information policies: those that promote and those that hinder the availability of spatial data. Three types of policies promote spatial data availability, each with a different purpose: access, reuse, and sharing. In the European legal framework, these types of policies correspond, respectively, to three major legal directives: the directive on environmental information, the directive on reuse of public sector information, and the INSPIRE initiative. Among the policies that hinder the availability of spatial data are those dealing with privacy, liability, and intellectual property. Intellectual property rights in particular endanger the availability of spatial data for access, reuse, and sharing and can pose a considerable threat to the development of the EU SDI. Many European public agencies use their intellectual property rights on spatial data to gain additional funding for their activities. We address the relationship between the three legislative initiatives and look at the influence of intellectual property rights on the availability of spatial data.

The development of a spatial data infrastructure (SDI) not only comprises technical aspects but also is supported by economic, social, organizational, and legal measures. In many cases, these take more time and effort to establish than technical aspects, as the flexibility, inquisitiveness, and resilience of the human beings behind the technology are put to the test. Recently, more attention has been given to these human aspects of SDIs (Onsrud et al. 2004; Wehn De Montalvo 2003; Warnest et al. 2003), but numerous issues prevent SDIs from fully exploiting all their technical possibilities, with a large number of these issues being of a legal nature (e.g., access and commercialization, privacy, liability, security, etc.). In order to avoid the usual time gap between the emergence of new technologies and the development or adaptation of legislation to keep up with these technologies, the legal questions should be treated alongside the technical, economic, and organizational matters and not as an afterthought.

In this article we look at the legal framework for the availability of spatial data in a European Union (EU) SDI. The centerpiece of this framework is the INSPIRE (Infrastructure for Spatial Information in Europe) initiative of the European Community (Commission of the European Communities 2004), which was approved at the beginning of 2007. INSPIRE, however, is surrounded by other existing policies on the availability of information, which may be complementary to it or may cause conflicts. These policies can be divided into two categories.

On the one hand, certain information policies make spatial information, as part of a wide range of information (e.g., public sector information or environmental information), available to interested parties. These interested parties can be the public sector, the citizens, or companies. In the European Community, the most important rules are laid down in the 2004 Directive on Re-use of Public Sector Information (PSI directive) and the 2004 Directive on Public Access to Environmental Information. The PSI directive, the directive on environmental information, and INSPIRE each have their own purpose and target group, but they all aim to increase the availability and dissemination of spatial data.

On the other hand, the EU SDI is also regulated by legislation that can impede the availability of spatial information. These stem from concerns about privacy, intellectual property ownership, security, and liability (figure 1). The fact that most of the spatial data is held by the public sector also has an impact on these restrictions.

Facilitating spatial data sharing by reconciling these competing policies is an enormous challenge (Onsrud et al. 2004). Moreover, the conflicting legal policies are underpinned by questions concerning the relationship between the public and private sectors in the provision of spatial data and services and the role of public agencies in the information society and on the information market. Public agencies are in many cases the sole or the most important producers of specific spatial datasets, which in the language of competition law translates into a dominant position or a monopoly. Depending on their activities, they may be subject to European Union or national competition rules.

We focus on the elements of these information policies that currently are most discussed and that need to be addressed urgently to create a solid and well-functioning EU SDI. We will address the relationship between the three legislative initiatives that promote availability of spatial information and attempt to delineate their scopes. We will also look at the influence of intellectual property rights on the availability of spatial data and point out the need for clear references to the appropriate legal frameworks in any SDI discussion.

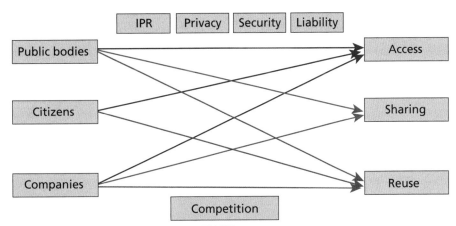

Figure 1. Information policies affecting the EU SDI.

INFORMATION POLICIES PROMOTING THE AVAILABILITY OF SPATIAL DATA

Spatial data are used by public agencies, the commercial sector, scientific researchers, community interest groups, and individual citizens for wide-ranging purposes (Onsrud et al. 2004). Depending on the type of user, but mostly on the purpose for which the spatial data is used, the process of obtaining data is addressed by different information policies. The scopes of the three European directives overlap to a considerable extent. The PSI directive applies to all public sector documents, the directive on environmental information addresses environmental information, and INSPIRE deals with spatial data. As shown in figure 2, a significant proportion of spatial data relates to the environment and is created or collected by the public sector. We will focus on this subset of data.

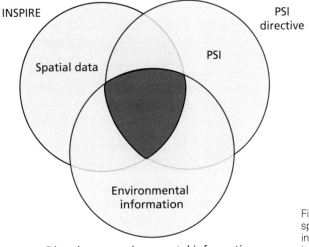

Figure 2. Relationship between spatial data, public sector information, and environmental information.

For each type of data, another distinction can be made between access, sharing, and reuse, based on the type of use. Thus, for our data subset, if a natural or legal person intends to obtain a document or information in order to exercise his democratic rights or obligations, we refer to this as access. Hence, access mainly involves democratic or political purposes and is generally outlined in freedom of information legislation and, in particular, in the directive on environmental information and the corresponding implementation legislation in the EU member states. INSPIRE also contains a number of elements that relate to access.

The second type of use policy is reuse. Our concept of reuse is based on the definition in the PSI directive: use for commercial or noncommercial purposes outside of the public task. We propose that the difference between access and reuse, in terms of purpose, could be described as follows. Reuse entails data processing beyond mere access to the document in order to learn its content. For example, researchers need to perform analyses on the data, in many cases combining it with other data. Merely accessing the data will not be enough. Academics and nonprofit organizations being the exception, reuse often has an economic goal. Mostly, it entails the creation of commercial products and services, often combining spatial data with other data layers, such as navigation systems, weather services, real estate services, and so on. However, it could also refer to the use of spatial data by the private sector for determining business strategies such as determining locations for new company branches by analyzing spatial data in combination with consumer statistics, environmental regulations, public transport arrangements, and so forth. No new products or services are created, but the data undergoes some form of processing, for purposes beyond the "checking up" on the government.

The third type of use policy, data sharing, refers to delivering or obtaining spatial data for the purpose of performing a public task. Generally, sharing involves the exchange of data between public agencies. However, if private companies are entrusted with the provision of a public service, some of their dissemination or acquisition of data may also fall under data sharing.

If we define the terms of access, reuse, and sharing based on the purpose of the use (and not on the type of user), the different users of spatial data, who can roughly be divided into public bodies, citizens, and companies, may all have access to, reuse, or share spatial data, depending on the goals of their activities (figure 3). For instance, a public body that disseminates or receives public spatial data for the purpose of performing a public task is sharing data. If it uses acquired public spatial data for creating and selling an information service on the market, this constitutes reuse. When a company requests data on, for example, environmental permits or cadastral units to find a new location for its headquarters or to create a service offered to other companies or the wider public, this also constitutes reuse, while asking for copies of the permits to check whether the government followed the rules falls under access.

The distinction between these three types of use of spatial data in the EU SDI should help determine with which information policies and regulations one must comply. In the European Union, access to environmental spatial data is covered by the directive on environmental information (which lays down the principles of the Aarhus Convention [UNECE 1998]) and chapter IV of the INSPIRE directive.

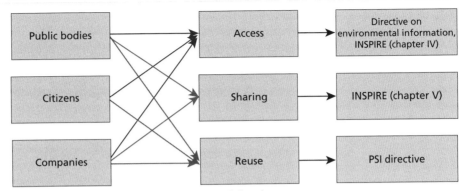

Figure 3. Types of use and corresponding legal directives.

Chapter V of the INSPIRE directive deals with the sharing of spatial data between public agencies. Reuse of public spatial data is addressed by the PSI directive.

As conditions and rules differ between these directives, a user must determine under which legal policy he has to formulate his request and which conditions he will have to comply with. In an ideal world, the distinction would be clear-cut, and no overlaps would exist. However, the ideal is already somewhat shattered by INSPIRE, which also contains some stipulations on access.

INSPIRE. INSPIRE lays down general rules for an EU SDI for performing public tasks that have an impact on the environment. It focuses on data sharing but also provides for public access to spatial data and spatial services. For now, INSPIRE addresses only the environmental sector, but it can be extended in the future to other policy domains, such as agriculture, transport, and energy. The directive specifies obligations for the EU member states and the public agencies on metadata, interoperability, network services, data sharing, monitoring, and reporting.

The basic principles of INSPIRE are as follows (INSPIRE DPLI Working Group 2002):

- Data should be collected once and maintained at the level where this can be done most effectively.
- It must be possible to seamlessly combine spatial information from different sources across Europe and share it between many users and applications.
- It must be possible for information collected at one level to be shared between all the different levels (detailed investigations, general strategic purposes, etc.).
- Geographic information needed for good governance at all levels should be abundant and widely available under conditions that do not restrain its extensive use.
- It must be easy to discover which geographic information is available, fits the needs for a particular use, and under what conditions it can be acquired and used.
- Geographic data must become easy to understand and interpret because it can be visualized within the appropriate context and selected in a user-friendly way.

PSI directive. The commercial value of spatial data did not escape the attention of public agencies. In many of the EU member states, public agencies are well aware that they can make money by offering their spatial data to the public at cost recovery or market prices. Many of them consider the sale of their data a necessary supplement to their government funding, even though opponents of this policy have argued that data that was already paid for by tax money should not be sold back to taxpayers at cost recovery or market prices. Some have argued that making public data available at lower prices or free of charge would actually create more income for the state, because of the expanded use of the data and the resulting increases in tax revenue (Pluijmers 2002; Weiss 2002; PIRA 2000).

Some public agencies not only sell their raw data but also create their own commercial products based on those data. In doing this, they become not only the suppliers of the private sector but also its competitors. This could create tension between the public and private sectors, as a public agency that sells commercial products and services will not be eager to pass on its data resources to competing providers. Since public agencies often have a monopoly, or at least a dominant market position, distortion of competition becomes a genuine risk. Temptation will be great for the public agency to sell at high prices to the competition or to eliminate the competition by charging very low prices to the consumers for the end products.

For these reasons, the European Commission felt that the different views and practices on making PSI available to the private sector impeded the development of the European information industry. Envious eyes were cast on the American information market, which was thought to benefit from easy and cheap availability of PSI, at least on the federal level (Commission of the European Communities 1998, 2001). Harmonization of the legislations of the EU member states should contribute to the development of the European market and information society. A European directive was voted for, creating a framework for the reuse of PSI, with spatial data being one of the most important categories.

The PSI directive lays down a set of minimum rules for public agencies to make their data available to the private sector. Member states are not obligated to disseminate data for reuse but are encouraged to do so under specified conditions. These conditions include time limits, available formats, fees, and transparency. The directive also makes sure that public agencies comply with the rules of fair competition. If a public body creates value-added products or services on the basis of its own documents for commercial activities outside the scope of its public tasks, the same charges and conditions should apply to the supply of the documents as those for other users. Otherwise, the activities of the public agency would lead to a distortion of market competition.

Directive on environmental information. The directive on environmental information aims to guarantee the right of access to environmental information held by or for public agencies and to ensure that environmental information is progressively made available to the public. The involvement of the European Community was an exception to its usual position that access issues should remain the sole competence of the EU member states. The Community considered the issue as too closely related to environmental matters, for which it does have competence.

The directive ensures free-of-charge on-site viewing of environmental information while allowing public agencies to charge a reasonable fee for supplying the information. As a general rule, the charges may not exceed the actual costs of production. However, when a public agency makes its environmental data available commercially in order to guarantee continued collection and publication of such information, market rate charges are allowed. This stipulation is intended to satisfy the concerns of public agencies regarding their income and regarding possible abuse of the received data by the citizen. However, market prices limit citizens' opportunities to stay informed about the activities of the government, in favor of providing the public agencies with the necessary resources for their activities.

As a side note, we can also mention a typical problem that legal policies encounter when they are confronted with new technologies. Any legal framework for an SDI should avoid technology dependency and use terminology that can be interpreted in the light of future developments. The directive on environmental information failed to do that, even though it is only a few years old. It distinguishes between making information available for on-site consultation and the supplying of information. How do we categorize viewing services on the Internet: as consultation or supplying? Making a document available for consultation allows the user to learn its content but not to retain a copy, while supplying a document provides the user with his own copy which he can consult afterwards. From a purely technical standpoint, viewing an Internet page leaves a cache copy on the hard drive, even if it is only temporary. However, a more teleological and perhaps common-sense interpretation of consultation—where the difference between on-site and remote consultations is outmoded (as it is merely a matter of physical presence)—would make remote viewing services a form of consultation (Janssen 2005). This openness to interpretation illustrates the importance of formulating legal measures as much as possible in a technology-neutral manner.

Relationship between the three directives. The three directives specify conditions for the availability of information. For instance, charges for access to information should, as a rule, not exceed the cost of production, charges for reuse can include a reasonable return on investment, and charges for sharing have to be kept to the minimum required to ensure the necessary quality and supply of spatial datasets and services together with a reasonable return on investment, while respecting the self-financing requirements of public authorities supplying spatial datasets and services. Public research institutions have to provide access to environmental information, but they do not fall under the scope of the PSI directive. The PSI directive includes transparency requirements for reuse (such as clearly specified conditions and charges), while the INSPIRE directive does not contain any transparency requirements.

The distinction between the legal policies for access, sharing, and reuse should be clear (figure 3). However, some incongruities will always remain, due to situations or requests for data the purpose of which may be hard define. For instance, in the example mentioned above, if a company CEO requests information on the building permits for a piece of land, is he intending to exercise his democratic rights as a citizen to know whether the public agency in question followed the rules in awarding the permit, or is he acting for his company and planning to use the information to determine the location of the next office branch?

Even a clear purpose does not guarantee the use of the proper procedure, however. Misuse or abuse is always possible, and public agencies cannot check up on how their data end up being used. Often the type and quantity of the requested data will be a first indication of the plans of the applicant but not always. The apprehension for this kind of abuse has led public agencies to take precautionary measures. Some of these are understandable, such as the INSPIRE idea to make spatial data available in a form that would prevent reuse for commercial purposes. Measures like these do not limit access or sharing, so they are not harmful to the SDI. However, if the concern for possible abuse leads public agencies to limit access and sharing or charge high prices, the development of an EU SDI is in danger. On the one hand, rules and procedures should be as harmonized as possible for all uses; on the other hand, public agencies should not use the possibility of abuse as an excuse to limit the availability of spatial data more than necessary.

INFORMATION POLICIES LIMITING THE AVAILABILITY OF SPATIAL DATA

Several legal policies limit the dissemination of spatial data. The privacy of the individual and the security of the nation have to be protected against the publication of certain data, and liability considerations lead public agencies to very strictly control what happens with their data. These limitations seem to be a lot less contested, however, than the restrictions stemming from intellectual property rights (IPR), a subject that has been debated extensively in the development of INSPIRE.

IPR. The goal of IPR is to stimulate innovation and the dissemination of information by rewarding authors who create original works for their effort and enabling them to control the use of their work by others (Laddie et al. 2000). In general, IPR require some level of originality (e.g., creativity for copyright, novelty for patents). However, with the advent of information technology and computer-generated works, the traditional level of originality has become harder to reach. This includes spatial data and databases. Spatial data constitute a factual representation of reality, and their value lies in their accuracy and interoperability with other data; spatial databases are valued mostly for completeness, user-friendliness, and interoperability. These elements limit the level of originality and the eligibility of data and databases for copyright protection. Yet the considerable investments required for the creation and collection of data and the ease with which the data can be copied increase the risk of misappropriation.

This is one of the reasons that the role of IPR has increasingly been shifting from the promotion of innovation in science and art towards the protection of the financial interests of the rightholders. IPR are being used as a tool against misappropriation, to protect effort and investment. For instance, a 1996 European Union directive on the legal protection of databases protects a database if its maker shows that there has been a substantial investment in either the obtaining, verification, or presentation of the contents to prevent extraction or reutilization of the whole or a substantial part of the database. This so-called *sui generis* right has been subject to criticism (Hugenholtz 1998; Maurer et al. 2001; David 2000) and has been curtailed by a number of judgments of the European Court of Justice in 2004 (European Court of Justice 2004). The Court made it clear that the resources used in the "creation" of the data did not count toward a substantial investment in the database and that only the resources used for obtaining preexisting data in order to assemble the database were relevant. The impact of the Court's decision may be considerable and should be examined. In addition, as the database right only protects nonsubstantial parts of a database from copying only if such copying conflicts with the normal exploitation of the database or unreasonably prejudices the legitimate interests of the database, the protection may not be as strong as is often assumed.

IPR of public agencies. The question of whether data are protected by IPR might also be influenced by whether they are held by the public sector. The Berne Convention (article 2.4), the most important international treaty on copyright, states that member states can themselves grant protection to official texts of a legislative, administrative, or legal nature and to official translations of such texts. It does not contain any definition of "official text," nor does it refer to any other documents of the public sector, such as reports, brochures, databases, and so on. These works can all be protected by copyright if they fulfill the conditions of originality and expression. Many national copyright acts exclude official texts from copyright protection but do not exclude other government works, such as reports, brochures, audiovisual material, etc. Notable exceptions are the United States and Austria, which exclude all works that are made public for "official use." The *sui generis* database right is not excluded for public agencies in any EU member state. In some countries, of which the most obvious example is the United Kingdom, IPR are used extensively as a source of funding for government agencies which have to finance the updating or expansion of their original government-funded datasets themselves.

IPR on spatial data. Because misappropriation risks and funding limitations worry public sector data producers and can potentially hinder the availability of information, IPR have been one of the major points of discussion in the development of an EU SDI and in the elaboration of the information policies mentioned above. Under the directive on environmental information, EU member states may refuse public access to data on the basis of IPR, and limiting public access on the basis of IPR has been one of the major debates in the development of INSPIRE. The European Commission and Parliament agreed that the IPR of public agencies should not limit public access to INSPIRE services (Commission 2004; European Parliament 2005, 2006), while the Council voted to allow EU member states to prevent public access to spatial data on the basis of IPR (Council of Ministers 2006). The final version of the directive holds only a small concession to Parliament: public access to discovery services cannot be restricted on the basis of IPR, but the other services can.

One of the Council's reasons for including IPR was to make the INSPIRE directive consistent with the directive on environmental information, which contains the same list of exceptions. Of course, consistency should be applauded, but it should not be the main reason for limiting public access to information. Moreover, the reasons for including this limitation in the directive on environmental information do not seem sufficient to justify this restriction of access. Public agencies may perceive a threat for two reasons: (1) access to spatial data may be misused to circumvent the rules and charges for reuse, and (2) potential revenue from selling data products or services may be lost. Both concerns can be allayed without denying access to spatial data. Other safeguards against misuses, such as the use of digital rights management (DRM), can be implemented.

Despite its bad reputation, stemming from its sometimes overenthusiastic application in the music industry, DRM may be of considerable value for the spatial data environment. DRM not only is a means of restricting the availability of information, but it can also be used for the electronic management and marketing of digital content (INDICARE 2004). If DRM is used to manage and track the use of spatial data, and to ensure compliance with use conditions, limiting access would be an excessive restriction of the public's right to obtain information from government. In addition, keeping in mind the public agencies' need for resources, limiting the access of the public out of concern for misuse may be counterproductive. What is not available obviously cannot be charged for, and anyone who is interested in actually buying the data will want to see if the data fits his purpose first. Thus, limiting public access to spatial data not only needlessly restricts the public's right to information but also goes against the public agencies' own interests.

Interestingly enough, the PSI directive's restrictions on reuse based on IPR seem to be less far-reaching than the INSPIRE directive. The PSI directive excludes only documents for which third parties hold IPR. In addition, the recitals of the directive encourage EU member states to exercise their copyright in a way that facilitates reuse. Thus, IPR limitations on reuse of public sector spatial data may be lower for commercial reusers than for citizens and for public agencies performing their public tasks.

Charges for spatial data. Charges for spatial data are closely linked to IPR and were debated at length during the formulation of the PSI directive. Should data that is collected at the expense of the taxpayer be freely available to the private sector, or should the burden on all the taxpayers be reduced by making the few parties that reuse the information pay for the expenses they cause? As mentioned above, some have claimed that making information available free of charge (or at the cost of dissemination) would lead to more income for governments because of the higher tax revenues, while others point to the necessity of public agencies to be self-sufficient and try to put cost recovery policies into perspective (Pluijmers 2002; Weiss 2002; PIRA 2000; Longhorn and Blakemore 2004). In the end, a very broad margin was given to the EU member states, allowing them to charge anywhere from no fee at all to an amount equivalent to the combined costs of collection, production, reproduction, and dissemination, plus a reasonable return on investment (article 6 of the PSI directive). The margin is somewhat understandable. After all, why should public agencies risk losing resources in giving away their information for free when the information industry intends to make a profit from reusing it?

More problematic, however, is charging for access to spatial data and for data sharing. The directive on environmental information makes it possible for public agencies to charge for the supply of information. The charges should be reasonable, as a general rule not exceeding actual costs of producing the material in question, except when a public agency makes the information available on a commercial basis, and this is necessary to guarantee the continuation of the collection and publication of such information. In that case a market-based charge is also possible. For INSPIRE, charging for spatial data was the most contested issue after IPR, both for access of the public to network services and for the sharing of spatial data between public agencies. Once again, the Council had a more restrictive view than the Commission and the European Parliament, wanting to charge the public not only for downloading but also for viewing data, if the charges would be essential to the maintenance of the datasets and services (Council 2006). The Commission and the Parliament wanted to keep access of the citizen to viewing services free of charge (Commission 2004; Parliament 2005, 2006). In addition, the Council wanted public agencies to be able to use licenses and require payment for sharing data with other public agencies, while the Parliament wanted those payments to not exceed the cost of collection, production, reproduction, and dissemination. Not including such a limit would have the bizarre effect that charges for reuse by the information industry would be limited to a reasonable return on investment, while prices for data sharing would not be restricted at all. This of course does not automatically mean that sharing spatial data would cost more than reusing it, but the possibility would exist. The final version reached a compromise: public access to viewing services should not be charged for, unless such charges secure the maintenance of spatial datasets and corresponding data services, and charges for data sharing should be kept to the minimum required to ensure the necessary quality and supply of spatial datasets and services, together with a reasonable return on investment, while respecting the self-financing requirements of public authorities supplying spatial datasets and services.

The Council's point of view on charging was influenced by a number of public agencies that are very large data producers and users, such as national mapping agencies and meteorological agencies. They feared that opening up their spatial databases to the entire European Union would create untenable pressure on their budgets, preventing them from maintaining data quality and breadth. Hence, one of the biggest complaints about INSPIRE is the current lack of knowledge of the exact costs and benefits of an EU SDI and whether the SDI might be sustainable under the model that the Commission and the Parliament proposed (Longhorn and Blakemore 2005). The same question was raised during the drafting of the PSI directive and has remained unanswered. Without a clearer view of the sustainability of the EU SDI as a whole, and not only the financial viability of existing data producers, deciding on the proper legal framework might be very difficult, especially in light of the fact that spatial data are very low on the political agenda and funding from the central government treasury is unlikely.

CONCLUSIONS

The legal framework for the EU SDI consists of two types of information policies: policies that stimulate the availability of spatial data and policies that hinder it. To "uncross the wires" between these different policies, a clear view of what part of the SDI is being addressed is crucial: access, reuse, or sharing of spatial data? Not merely a matter of terminology, this distinction also determines what legal framework is applicable and what consequences this has for the users and suppliers of the spatial data. Any discussion of access should not involve the PSI directive, while debates on the INSPIRE directive have nothing to do with the information industry creating added-value products based on spatial data held by the public sector. Mixing the terms and the issues creates confusion and needlessly hinders the development of the SDI by creating unrealistic expectations and demands on the public agencies. In addition, over- or underestimating the impact of information policies that limit the availability of spatial data may also lead to unwanted complications. For example, the extent to which spatial data and databases are protected by IPR and the extent to which IPR will influence the availability of spatial data are not entirely clear. Until these matters are sorted out, discussions between policy makers will be filled with uncertainty and knee-jerk reactions to perceived threats.

Ideally, all these information policies would fit seamlessly together, or even be harmonized into one framework, where sharing, reuse, and access would all have to comply with the same rules and IPR would be used to ensure a perfect balance between maximum availability of spatial data and the sustainability of the SDI. This would also eliminate the need to define the scope of the public task, which distinguishes reuse and sharing and depends on ever-changing political views. Because a harmonized framework may prove to be impossible, or in any case unrealistic in the very near future, we should look for the largest common denominator between the legal directives for access, reuse, and sharing and devise a framework where the first layer contains the common conditions for access, sharing, and reuse and additional layers or terms of use can be added according to the purpose for which the spatial data are required. Finding this legal common denominator is essential for an efficient and effective EU SDI.

ADDENDUM

After the manuscript was accepted for publication, a compromise was reached on INSPIRE and the directive became effective on 15 May 2007. The full text of INSPIRE is posted at http://www.ec-gis.org/inspire/directive/l_10820070425en00010014.pdf.

REFERENCES

Berne Convention for the Protection of Literary and Artistic Works of September 9, 1886. http://www.wipo.int/treaties/en/ip/berne/trtdocs_wo001.html.

Blakemore, Michael, and Roger Longhorn. 2005. Inspired by regulation, integration, confrontation and confusion. GIS@development Magazine, February. http://www.gisdevelopment.net/magazine/years/2005/feb/perspective.htm.

Commission of the European Communities. 1998. Public sector information: a key resource for Europe, COM(98) 585 final. http://europa.eu.int/information_society/policy/psi/docs/pdfs/green_paper/gp_en.pdf.

Commission of the European Communities. 2001. eEurope 2002: creating a EU framework for the exploitation of public sector information.COM (2001) 607 final. http://europa.eu.int/information_society/policy/psi/docs/pdfs/eeurope/2001_607_en.pdf.

Commission of the European Communities. 2004. Proposal for a directive of the European Parliament and of the Council establishing an infrastructure for spatial information in the Community (INSPIRE). COM (2004)516 final. http://www.ec-gis.org/inspire/proposal/EN.pdf.

Commission of the European Communities. 2005. DG Internal Market and Services working paper. First evaluation of Directive 96/9/EC on the legal protection of databases. http://europa.eu.int/comm/internal_market/copyright/docs/databases/evaluation_report_en.pdf.

Council of the European Union. 2006. Common Position adopted with a view to the adoption of a Directive of the European Parliament and of the Council establishing an infrastructure for spatial information in the European Community (INSPIRE). http://register.consilium.eu.int/pdf/en/05/st12/st12064-re02.en05.pdf.

David, Paul. 2000. The digital technology boomerang: new intellectual property rights threaten global "Open Science." http://www-econ.stanford.edu/faculty/workp/swp00016.pdf.

European Court of Justice. 2004. Fixtures Marketing Ltd v Svenska AB, C-338/02; Fixtures Marketing Ltd v Organismos Prognostikon Agonon Podosfairou EG, C-444/02; Fixtures Marketing Ltd v Oy Veikkaus Ab, C-46/02; British Horseracing Board Ltd v William Hill Organization Ltd, C-203/02.

European Parliament. 2005. Legislative resolution on the proposal for a directive of the European Parliament and of the Council establishing an infrastructure for spatial information in the Community (INSPIRE). http://www.europarl.europa.eu/sides/getDoc.do?pubRef=-//EP//NONSGML+TA+P6-TA-2005-0213+0+DOC+PDF+V0//EN.

European Parliament. 2006. Legislative resolution on the Council common position for adopting a directive of the European Parliament and of the Council establishing an Infrastructure for Spatial Information in the European Community (INSPIRE). http://register.consilium.europa.eu/pdf/en/06/st10/st10372.en06.pdf.

European Parliament and Council of the European Union. 1996. Directive 96/9/EC of 11 March 1996 on the legal protection of databases. OJ L 77, 27 March 1996, 20.

European Parliament and Council of the European Union. 2003. Directive 2003/4/EC of 28 January 2003 on public access to environmental information and repealing Council Directive 90/313/EEC. OJ L 41, 14 February 2003, 26.

European Parliament and Council of the European Union. 2003. Directive 2003/98/EC of 17 November 2003 on the re-use of public sector information. OJ L 345, 31 December 2003, 90.

Hugenholtz, Paul. 1998. Implementing the Database Directive. In Intellectual Property and Information Law. Essays in Honour of Herman Cohen Jehoram, ed. Jan Kabel and Gerard Mom, 183–200, The Hague: Kluwer Law International.

INDICARE. 2004. Digital rights management and consumer acceptability. A multi-disciplinary discussion of consumer concerns and expectations. http://www.ivir.nl/publications/helberger/INDICAREStateoftheArtReport.pdf.

INSPIRE DPLI Working Group. 2002. Data Policy & Legal Issues Working Group position paper. http://www.ec-gis.org/inspire/reports/position_papers/inspire_dpli_pp_v12_2_en.pdf.

Janssen, Katleen. 2005. INSPIRE and the PSI directive: public task versus commercial activities?. http://www.ec-gis.org/Workshops/11ec-gis/papers/303janssen.pdf.

Laddie, Hugh, Peter Prescott, and Mary Vitoria. 2000. *The modern law of copyright and designs*. London: Butterworths.

Longhorn, Roger, and Michael Blakemore. 2004. Revisiting the valuing and pricing of digital geographic information, *Journal of Digital Information* 4 (2), 1–27.

Maurer, Stephen M., Paul B. Hugenholtz, and Harlan J. Onsrud. 2001. Europe's database experiment, *Science*, 294, 789–90.

Onsrud, Harlan J., and Xavier Lopez. 1998. Intellectual property rights in disseminating digital geographic data, products, and services: conflicts and commonalities among European Union and United States approaches. In *European geographic information infrastructures: opportunities and pitfalls,* ed. Ian Masser and Francois Salge, 153–67, London: Taylor & Francis.

Onsrud, Harlan, Barbara Poore, Robert Rugg, Richard Taupier, and Lyna Wiggins. 2004. The future of the spatial Information infrastructure. In *A Research Agenda for Geographic Information Science,* ed. Robert B. McMaster and E. Lynn Usery, 225–55. Boca Raton: CRC Press.

Pira International Ltd., University of East Anglia, and Knowledge Ltd. 2000. Commercial exploitation of Europe's Public Sector Information. http://europa.eu.int/information_society/policy/psi/docs/pdfs/psd.pdf.

Pluijmers, Yvette. 2002. The economic impacts of open access policies for public sector spatial information. http://www.fig.net/events/fig_2002/fig_2002_abs/Ts3-6/TS3_6_pluijmers_abs.pdf.

Rajabifard, Abbas, and Ian P. Williamson. 2003. Anticipating the cultural aspects of sharing for SDI development. www.geom.unimelb.edu.au/research/publications/IPW/Spatial%20Sciences-2003-Abbas.pdf.

UNECE. 1998. Convention on access to information, public participation in decision-making and access to justice in environmental matters. http://www.unece.org/env/pp/documents/cep43e.pdf.

Warnest, Matthew, Abbas Rajabifard, and Ian Williamson. 2003. Understanding inter-organizational collaboration and partnerships in the development of national SDI. http://eprints.unimelb.edu.au/archive/00001112/01/URISApaper%5FWarnest.pdf.

Weiss, Peter. 2002. Borders in cyberspace: conflicting public sector information policies and their economic impacts. http://www.weather.gov/sp/Borders_report.pdf.

Wehn de Montalvo, Ute. 2003. In search of rigorous models for policy-oriented research: a behavioral approach to spatial data sharing. *Urisa Journal* 15, 19–28.

GeoDRM: Towards Digital Management of Intellectual Property Rights for Spatial Data Infrastructures

MOHAMED BISHR,[1] ANDREAS WYTZISK,[2] AND JAVIER MORALES[2]

INSTITUTE FOR GEOINFORMATICS (IFGI), UNIVERSITY OF MÜNSTER, GERMANY,[1] AND INTERNATIONAL INSTITUTE FOR GEO-INFORMATION SCIENCE AND EARTH OBSERVATION (ITC), ENSCHEDE, NETHERLANDS[2]

ABSTRACT

Web services are the building blocks of the modern spatial data infrastructure (SDI). Ubiquitous sharing and exchange of geospatial content and services is hampered by potential infringement of the intellectual property rights (IPR) of providers and producers. Proper management of IPR is a serious challenge that we face with the establishment of geospatial information commons and marketplaces. Recently, the Open Geospatial Consortium (OGC) established a geospatial digital rights management (GeoDRM) working group. The general objective of the working group is to define trusted infrastructures to protect rights to digital geospatial content. The GeoDRM working group also aims at facilitating the adoption of DRM as a technology for dissemination and management of IPR in the geospatial domain. In this article we present a novel approach to GeoDRM by providing a GeoDRM architecture based on our research into the current nongeospatial DRM technology. The adoption of DRM technology into the geospatial domain will enable the management and licensing of IPR on geospatial assets while keeping the effort of developing unique GeoDRM technology at a minimum. We also address the essential information model that is required for GeoDRM digital licensing to function. The model is based on our previous analysis and research into DRM information models for nongeospatial assets. This information model provides a technology-neutral information view of the requirements of digital licensing of geospatial assets. The GeoDRM policy of a given organisation relies on both a technical framework and a legal framework. This article does not address legal frameworks, because they are dependent on the organisational context in which GeoDRM technology is implemented. GeoDRM is in the early stages of development, and we propose a research agenda that we believe is necessary to further GeoDRM development and adoption within the geospatial community.

INTRODUCTION

As technology pushes various industries toward the digital frontier, many types of content are becoming available solely in digital format, and geospatial data is no exception. Copyright-protected geospatial content used to be sold on paper sheets but is now available in digital format. As a result, such content can now be used by a variety of users and devices: from car navigation systems, handhelds, and mobile phones all the way to corporate applications and business-related geospatial functions. Digital geospatial datasets moving across computer networks can be easily copied, transformed, or incorporated into new value-added products and services. Geospatial-data producers and owners are faced with the challenge of controlling the dissemination of their digital geospatial assets downstream in the geospatial value chain.

Onsrud et al. (2004) argue that the desire of producers for proper intellectual property rights (IPR) management is driven not by monetary gain but rather by the moral rights associated with the use of spatial data. The author's rights to have his assets used in an appropriate manner and the recognition gained from that are important issues for the establishment of the public geoinformation commons.

Digital rights management (DRM) is a popular term for a field that came into being in the mid-1990s. Two basic definitions of DRM exist: a narrow one and a broad one. The narrow definition of DRM focuses on persistent protection of digital content. It allows the distributor of data to control how the data is used, and by whom, in accordance with predetermined rules and agreements. The broad definition of DRM encompasses everything that is required to define, manage, and track rights on digital content. In addition to persistent protection, the latter definition also includes business rights or contract rights and access tracking. DRM solutions in the broad sense are potentially capable of tracking access to and usage of operations on content. Information about usage is often inherently valuable to content providers, even if they do not charge for the usage of content (Rosenblatt and Dykstra 2003).

Recently, higher awareness of IPR problems within the spatial data infrastructure (SDI) community gave rise to the discussion of the challenges surrounding the management and dissemination of data. Open Geospatial Consortium (OGC) formed a geospatial DRM working group (GeoDRM WG) with the mission of adopting the work done in the area of data ownership and digital rights management to the geospatial community. The working group addresses the lack of GeoDRM capability as a barrier to wider adoption of Web-based geospatial technologies. In this article we define the general components of the GeoDRM framework with a focus on the technical aspects of GeoDRM. We describe the GeoDRM architecture as a high-level architecture of loosely coupled services that manage digital licensing functionality for GeoDRM based on the study of current digital licensing infrastructures. We also discuss the GeoDRM information model, which is a technology-independent information view of the elements needed for a GeoDRM architecture to function. Both the architecture and the information model are based on proven and available DRM technology.

GeoDRM is defined as a set of technologies and legal frameworks that are fit for a certain organisational need, enabling rights-managed geospatial networks (e.g., SDIs), where all rights over geospatial assets are specified by licensors and any licensee would be "trusted" to honour the licensor's conditions within and beyond the network's trusted environment (e.g., remote clients). In this definition, trust is not synonymous with "enforcement of digital licenses," which in turn might or might not be part of a certain GeoDRM framework. In addition, a legally binding framework of licenses and licensing policies that are mapped to digital equivalents must support GeoDRM. This makes a GeoDRM license a legal tender that must be respected.

An asset is defined in Webster's dictionary as "an item of value owned." In the realm of service-oriented architecture (SOA), an organisation can own the rights to a service rather than digital content. Hence, the value attribute may apply to services as well. In this sense, we use the term asset to refer to both digital geospatial datasets and services in an SDI.

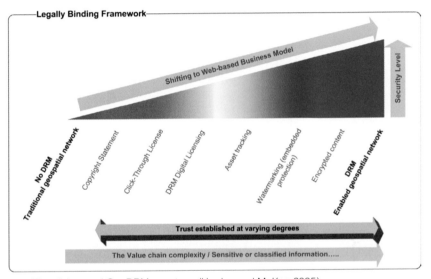

Figure 1. The elaborated GeoDRM spectrum (Vowles and McKee 2005).

Vowles and McKee (2005) provide an interesting perspective on GeoDRM as a spectrum of technologies. Figure 1 shows an elaborated version of this GeoDRM spectrum:

- As business models change according to organisational needs, so do the GeoDRM schemata used to disseminate IPR. SDIs like INSPIRE opt for click-through licensing for most of the data themes covered by the initiative, while Ordnance Survey UK Master Map already has click-through licensing and is opting for digital licensing and tracking of its assets.
- The modern SDI relying on SOA provides a stimulus for more sophisticated techniques of IPR management.
- For the SDI to support higher security models and access rights, more sophisticated GeoDRM technologies are needed.

- The legal framework is inseparable from the technical framework in GeoDRM, and together they form the essential GeoDRM policy. GeoDRM implementations will use a combination of the technical and legal tools to achieve the desired level of IPR management within an organisation.

- Trust is an essential component of GeoDRM and is an explicit reflection of the moral aspects of IPR law. In a geospatial data-sharing environment, a number of institutions explicitly admitted to sharing data freely with people they know and trust, while making it difficult for others outside their circle of trust to gain access (Harvey and Tulloch 2004). Additionally, Harvey (2003) argues that trust is important for geospatial data sharing. As geospatial information and maps are replacing experiential knowledge (e.g., "I trust that this map is accurate and will base my decisions on it"), establishing trust in geospatial data-sharing environments is essential. GeoDRM assures parties of each other's identities, and the preservation of their IPR increases trust in the data-sharing environment.

To illustrate how the GeoDRM spectrum can be implemented, geospatial digital watermarking can be used to track misuse of digital assets by identifying the sources of infringement. Thus, a combination of digital licensing and watermarking can assist geospatial-data providers in an SDI with monitoring the usage of their assets for the correct purpose. Once a watermarked dataset is found in a usage scenario where it shouldn't have been used (i.e., a map watermarked for user X being used by user B), the providers can trace back the source of infringement and minimise the damage to their intellectual property. This ability to combine technology (digital watermarking) with the correct legal actions to achieve the optimal degree of IPR management across the SDI is among the future research challenges for GeoDRM.

GeoDRM ARCHITECTURE

In this section, we present an overview of the proposed GeoDRM architecture. We first provide a brief description of DRM services by examining the general DRM reference architecture and the digital licensing infrastructures that are common amongst most DRM systems. We then show that, from a digital licensing perspective, DRM systems and GeoDRM are not fundamentally different, since they both manage licensing of digital assets, no matter how different those assets might be. We also discuss an approach for adopting current digital licensing and DRM technology into the geospatial domain to leverage already established and tested technology.

DRM. Rosenblatt et al. (2002) introduced a general high-level DRM reference architecture that is widely adopted in the DRM domain. The DRM reference architecture includes the most relevant components in DRM systems. It includes a publisher service that provides the original asset and provides facilities for delivery to the client. The licensing server is the administrative hub of the digital licensing server. It brokers negotiations between vendors and consumers and grants licenses. The licensing server generates the digital license, and the publisher attaches it to the asset being sent to the client. The recipient or the client has a controller service which receives the user's request to exercise certain rights on the content. The controller acquires the user's identity information and obtains a license from the license server. It then retrieves the encryption keys from the license, decrypts the content, releases it to the rendering application, and executes the license terms.

For management of digital licensing, more advanced architectures have been proposed. We follow the digital licensing infrastructure model of Thompson and Jena (2005). The digital licensing infrastructure enables machines to negotiate and issue licenses to protect assets and to regulate how assets and licenses are sold or used. The infrastructure also enables asset holders to track and monitor compliance with terms and conditions of use. The components of the digital licensing infrastructure are described in the context of the GeoDRM architecture below. This digital licensing infrastructure is designed to manage licensing of content and services in a variety of settings. We believe that GeoDRM would require such complex licensing capability.

We have established that the GeoDRM architecture for digital licensing should not be fundamentally different from that of other DRM, since digital licenses are what is being managed. Although geospatial licenses would differ to a degree from other types of digital licenses, from a broad perspective the differences between the two architectures will likely be minimal. Hence, by building on existing DRM technology, we leverage DRM into GeoDRM by creating a GeoDRM architecture for digital licensing of geospatial datasets.

GeoDRM. Figure 2 illustrates a formal UML (Unified Modeling Language) component diagram of the high-level GeoDRM architecture. The diagram combines the functionalities of DRM architectures (Rosenblatt et al. 2002) and digital licensing infrastructures (Thompson and Jena 2005). The GeoDRM service acts as the workflow hub of the GeoDRM system and coordinates the interactions between services, while the GeoDRM client component has the responsibility of managing licensing-related client side interactions with the GeoDRM service. The GeoDRM service can be located in the geoinformation publisher's environment, but it could also be a third-party trusted license provider service.

The data discovery service will provide one or both of the following solutions:

- **Extended CSW.** An OGC Catalogue Service for the Web (CSW) (OGC 2002) specification supports the registry and discovery of geospatial information resources. A CSW plays the role of a directory in the open distributed Web service environment, and it also allows data and service providers to register their capabilities using metadata, which users can query to discover information of interest. The metadata defined in the CSW could be extended so that users can search for data based on GeoDRM-related keywords (e.g., to see if the extracting feature is permitted). This will play a central role in the discovery process, where service metadata must express relevant GeoDRM metadata criteria (e.g., service requires prelicensing or licensing occurs after the data is released).
- **Product catalogue.** A provider's product catalogue would specify several licensing schemata mapped to assets so that a negotiation service (described below) can negotiate terms with users. Each license would be dynamically generated for a selected product.

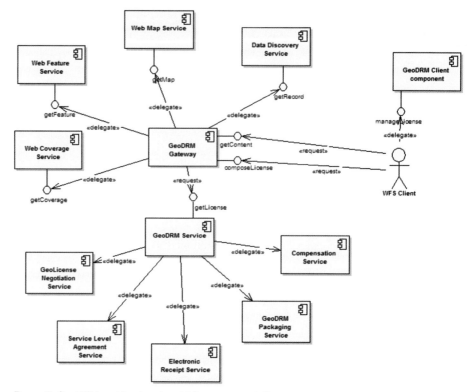

Figure 2. GeoDRM architecture in a UML component diagram.

Other GeoDRM specialised services are as follows:

- The **GeoDRM packaging service** packages a geospatial dataset file or stream with its license and, if requested, encrypts the stream of files at run time before dispatching the package as a geospatial object.

- The **service level agreement service** defines an optional section of a license. For certain service types (i.e., mission-critical or emergency) some licenses would contain penalties and service level parameters. A vendor might agree to pay a penalty for performance below some quantified level of service or acknowledge legal liabilities.

- The **compensation service** defines a compensation scheme and a method of payment. License servers would use compensation models such as pay-me-now perpetual or term-limited licenses (the consumer purchases a license for use of a given object over a specified time), pay-per-user licenses (the license server generates license keys based on the number of stand-alone applications purchased), and pay-per-use or subscription licenses (the consumer purchases units of usage of some asset, so a license key is generated every time the service is invoked).

- The **electronic receipt service** generates a persistent log entry (e-receipt) for each purchase or use of an asset. This e-receipt records which services were used and the form of payment. The service can generate e-receipts for geospatial Web services usage, client interactions, calls to a geospatial database, or datasets purchased over the Internet.

- The **geolicense negotiation service** provides offers and counteroffers between vendors and consumers until they reach an agreement (translated into a digital license). Negotiation can be manual, semiautomatic, or automatic; the service in the manual case would act as a facilitating medium of contact between the two parties involved.

In the GeoDRM architecture, we define two main use categories: (1) protecting and publishing an asset and (2) searching for and acquiring an asset.

In data publishing, providers perform the following tasks:

1. Define license terms and pricing schemes for services and content published in an SDI.
2. Create the appropriate Right Expression Language (REL) profile schema to support custom needs by means of management tools.
3. Publish the REL profile schema on the GeoDRM system.
4. Create metadata using the SDI metadata profile which has an extension to search datasets based on right descriptions as specified by the REL licenses.

In asset search and acquisition, users (asset consumers) perform the following tasks:

1. The user accesses the catalogue service on the SDI and searches for certain data offered by vendors who can provide licenses suitable for the user's needs.
2. The catalogue returns to the user the needed data.
3. The user interacts with the GeoDRM system where a data provider is registered to acquire the license and data.

Once users locate the required data, they send a request to the GeoDRM gateway, which in turn forwards it to the GeoDRM service. The GeoDRM service interacts with the services that hold the data to generate the custom license for the requested asset. The actual sequence of interactions between GeoDRM clients, the GeoDRM system, and the data provider may vary from one implementation to the other. The following three cases show different interaction flows from data suppliers to users.

Case 1 (figure 3) is most suitable for situations where data acquisition from a certain service requires users to have a valid usage license. In this scenario, possession of a license is a prerequisite to acceptance of data requests.

Case 2 (figure 4) is most suitable for users who already have the data but need to upgrade or modify their license terms. For example, a user may have rights to a dataset for noncommercial use and requests a license for commercial use.

In case 3 (figure 5) users interact with the GeoDRM gateway, which ensures a proper workflow, to request both data and a license from a provider according to the provider's business model. Also the Web Feature Service (WFS) is assumed to be extended to allow for relevant GeoDRM functionality. For example, each dataset sent must have minimal instructions for where to obtain licenses (e.g., the licensing server URI [Uniform Resource Identifier]).

Irrespective of the specific implementations of the three cases above, separating the license and the data is important. In earlier DRM systems the data and the

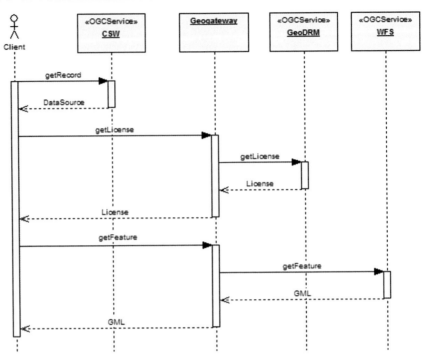

Figure 3. Case 1: license first and then release data.

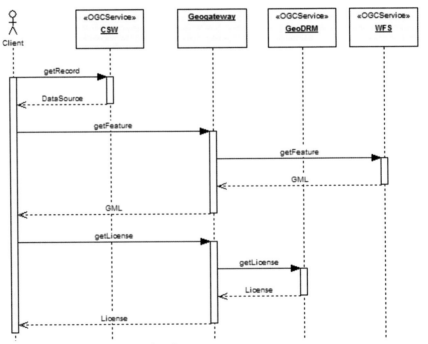

Figure 4. Case 2: data first and then license.

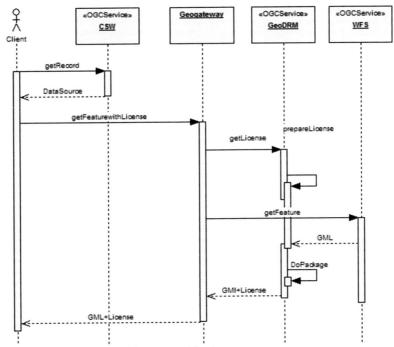

Figure 5. Case 3: data and license combined.

license were hardwired together (Rosenblatt 2002). This model would not have enabled cases 1 and 2. By decoupling licenses from data, GeoDRM technology allows users to negotiate unique terms instead of relying on a static license schema.

GeoDRM INFORMATION MODEL

A license data model is specified by Right Expression Language (REL). This means that the structure of the language schema and its semantics would determine the exact implementation of the data model. However, the language needs to be extended for GeoDRM, and a conceptual information model needs to be developed. The GeoDRM information model described below is independent of any REL or technology implementations.

Figure 6 illustrates the general GeoDRM license information model. A license can be assigned to various asset types (individual features, feature attributes, or feature types). The elements of the GeoDRM information model are as follows:

1. Rights information: a definition of types and meaning of the permissions granted in a license (this includes spatial- and nonspatial-data permissions and how they are expressed).

2. Rights constraints: a definition of the types and meaning of spatial- and nonspatial-data constraints that may apply to each permission (principal or resource).

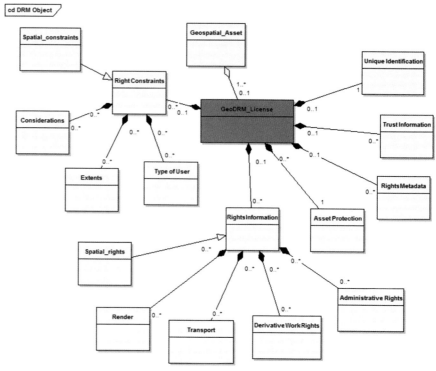

Figure 6. GeoDRM information model.

3. Unique identification (1) provides necessary unique identities to geospatial-asset holders and licensees in an open distributed environment like the Web and (2) enables licenses to refer to assets being managed.

4. Rights metadata: general information to enable the search and discovery of assets in a Web environment based on certain GeoDRM criteria (e.g., rights).

5. Trust information: authentication of identity and verification that the license has not been modified in any significant manner; trust in the social sense when parties are well-defined.

6. Asset protection: mechanisms to embed information in the assets in order to detect misuse and identify infringing parties.

Rights information. Rights in the broad sense, as identified in DRM, are actions or a class of actions (or activities) that a subject may perform on an asset (Contentguard 2001; Iannella 2002; Park and Sandhu 2004). Therefore, rights enable users to access assets in a particular mode, such as reading or writing (Park et al. 2004). Rosenblatt et al. (2002) identify render rights, transport rights, and derivative rights as the three comprehensive categories, with rights within the categories (playing, extracting, etc.) varying according to content and applications.

In addition to the three types of rights mentioned by Rosenblatt et al., issuance rights (the rights to issue grants of other rights) and delegation rights (the rights to delegate a grant to another party) are examples of essential rights over rights (Chong 2005). We collectively call these rights administrative to distinguish them from the above core rights. This general functional categorisation of rights fits any digital rights domain.

- **Render rights** refer to the representation of content on some output medium (playing, viewing, printing, etc.).

- **Transport rights** are the rights to move or copy content from one place to another. The differences between copying, moving, and loaning have to do with which users have access to the content at any given time. In copying, user 1 gives a copy of content to user 2, and both have access to the content simultaneously. In moving, user 1 gives up access once content is given to user 2. In loaning, user 1 gives up access to the content temporarily until user 2 gives the rights back.

- **Derivative rights** refer to the manipulation of content to create derivative work: using pieces of content out of their context (extracting), changing the content (editing), or inserting the content into another product (embedding).

- **Administrative rights** are rights over rights (e.g., the right to delegate a certain set of rights to others).

The above categories can be applied to the geospatial domain. For example, an extract feature right is a derivative right, and a regrant right (allowing licensees to grant licenses to sublicensees) is an administrative right. The categories act as units of functional aggregation of rights.

Rights constraints. Constraints on rights are attributes that are attached to each of the above fundamental rights. Attributes include considerations, extents, and types of users (Rosenblatt et al. 2002). Geospatial assets also require spatial constraints.

- **Considerations** are anything a user has to give in exchange for a certain right. An obvious consideration could be money, or it might be a certain form of agreement between the publisher and the user. For example, a publisher might allow someone to use content in return for a copyright notice in the produced work.

- **Extents** specify how long, how many times, or in what places the rights apply.

- **Types of users** can have different rights or right attributes. For example, educational institutions constitute a particular category of users.

Unique identification. Unique identification of licensed content is an integral part of any DRM license. The digital object identifier (DOI) is a standard for online content identification and linking. It is based on URI and URN (uniform resource name) and governed by the International DOI Foundation. A DOI assigned to a digital asset, be it data or service, uniquely identifies that asset. Using regular URLs as a means of identifying digital content is not convenient, because they point to a specific location of an asset or files. If the location of the asset changes, the URLs become invalid. A DOI points to a master table called the DOI directory where each DOI record is assigned a URL that leads to the asset. URL changes are always synchronised with the DOI directory.

Geospatial-data providers can use DOIs to uniquely identify each asset covered by a GeoDRM license.

Rights metadata. In an open distributed network like the Web, users need to be able to search and discover assets based on different types of generalised information.

For example, users can search music based on genre, date of creation, and so on. Although this specific element is not mentioned in the DRM literature, we believe it is an important aspect that is specific to geospatial data. In the geospatial realm, users can search for data through metadata catalogues based on date of creation, specific keywords, and so forth. The role of metadata is to provide information about collections of assets to enable search and discovery. OGC has defined the Catalogue Service for the Web (CSW) to enable clients to search and discover data based on well-defined metadata elements. These elements are defined in the ISO 19115 specification. We recommend that the ISO 19115 metadata standard and its implementation in the CSW be expanded to support searches based on geospatial-data rights metadata. A GeoDRM CSW profile is an important subject that has yet to be considered by the OGC GeoDRM working group. We recommend the following categories for rights metadata:

- **Licensee metadata** specifies the types of users who can have licenses and access rights to an asset (e.g., academic institutions, commercial users).
- **Licensor metadata** identifies the authority that controls the data (as addressed in the ISO 19115 standard).
- **Rights metadata** specifies types of rights and constraints.

Trust information. Trust has both a technical meaning and a social meaning. From a technical perspective, a trusted system is a system that can be trusted to honour the rights, conditions, and fees specified for a digital work. Systems that play, read, or provide access to digital works on a network can be subject to trust. Different implementations of trusted systems have different requirements for security. In the most secure approaches, all of the hardware and software on the platform is certified to honour digital rights. Other approaches focus on the use of so-called secure envelopes or containers, emphasising transmission and storage of information.

From a social perspective, GeoDRM has significant impact on trust between parties participating in geospatial information sharing (Harvey 2003; Harvey et al. 2004)—between government agencies, between the public and private sectors, or between commercial entities. Trust can affect a GeoDRM-enabled SDI on many fronts, including the following:

- Accessing a geospatial service after signing a digital license specifying a level-of-service agreement and terms of use (e.g., "do I trust this service enough to build an emergency application?")
- A government entity licensing a valuable dataset to a user for noncommercial purposes (trust is increased since all participants and rules are well-defined)

A minimum level of information is required in a GeoDRM license to facilitate trust between parties involved in an agreement. Trust would be context sensitive: for example, in certain situations the quality of a dataset could be the major basis of trust, while in other cases the party from which the data originates is the basis of trust.

Asset protection. Digital content can be encrypted and then decrypted by private keys embedded in the license. The encrypted content cannot be used unless a valid license with the corresponding keys is made available (Rosenblatt et al. 2002).

Digital watermarking is an adaptation of the commonly used and well-known paper watermarks to the digital world. Watermarking makes it possible to embed some essential metadata fragments within the content instead of having metadata alongside the content (Rosenblatt et al. 2002). Digital watermarking provides methods and technologies for hiding information (for example, a number or text) in digital media, such as images, video, or audio. The embedding takes place by manipulating the actual digital content. That means that the information is not embedded in the frame around the data. The modifications of the media have to be imperceptible.

Research on geospatial-content watermarking is still in its infancy. However, in digital imagery it is a more mature area than in vector formats, even with multiple commercial vendors offering watermarking protection of imagery. Two-dimensional vector and point datasets have received less attention from the research community (Lopez 2002).

TECHNOLOGIES FOR REPRESENTATION OF RIGHTS

Many technologies for rights representation exist, from metadata standards (for example, the Dublin Core [DC 2005]) to Right Expression Languages (RELs). Also, publishing requirements for industry standard metadata (PRISM) (PR 2005) provide a good base for representing rights information on the metadata level. This representation, however, is meant to be machine interpretable. Even with XML encodings the purpose remains documenting metadata, which does not provide essential facilities for representing digital licenses like security models and digital license structures.

RELs provide a means of managing the expression of contractual-agreement rights in a machine-interpretable format for all sorts of digital assets. They aim to provide a vocabulary and semantics for the expression of terms and conditions and to enable machine-based processing of digital contracts (Guth et al. 2003). RELs have also been defined as a type of policy specification language where the focus of the language is on expressing and transferring rights from one party to another in an interoperable format (LaMacchia 2002).

The concept of RELs originated when the Digital Property Rights Language (DPRL), a LISP 1-based language, was developed by Mark Stefik of Xerox's Palo Alto Research Center (a patent application was filed in 1994, and the patent was granted in 1998). Stefik created DPRL as a machine-readable language that could be used to define access rules and procedures, for use with the trusted PC. Stefik based DPRL 2.0 on XML, because XML is extensible and was interoperable with other emerging standards. From DPRL emerged ODRL and XrML. ISO-REL and MPEG-REL were then developed on the basis of XrML (figure 7).

To be used in the geospatial domain, RELs would naturally need to be extended to accommodate requirements for specifying digital licenses on geospatial assets (for example, without extension, a REL doesn't have the ability to specify a license with spatial extents having x,y dimensions). In our research we have adopted ODRL. Being open source and readily available makes ODRL ideal for research purposes. (The process of extending the ODRL 1.1 standard to accommodate spatial requirements and spatial data types is beyond the scope of this article.)

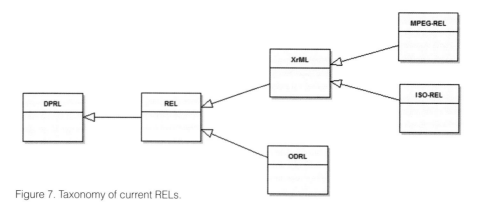

Figure 7. Taxonomy of current RELs.

We offer a concise agenda for future research on GeoDRM which leverages the findings of our research. The items below constitute a high-level research agenda for a full-fledged SDI capable of managing IPR of producers and users of geospatial datasets. More sophisticated business models can be developed for new and innovative products and services across the SDI.

- GeoDRM has legal and technological aspects. A GeoDRM policy is country specific and must use legal and technical means to achieve its goals. Further research into both aspects must define the country-specific guidelines for developing integrated legal and technical measures of implementing GeoDRM.

- Studies are necessary to evaluate how digital licensing and other GeoDRM technologies will affect business models of geoinformation (GI) organisations.

- Harvey (2003) and Harvey et al. (2004) stress the importance of trust as an element in defining geospatial data sharing patterns. Understanding trust and its effects is important in building GeoDRM systems. For example, does higher trust entail lower security measures on GeoDRM systems? And how can the concept of trust leverage GeoDRM's ability to unambiguously define liabilities and duties of parties involved in GI data sharing?

- Each of the elements within the GeoDRM information model needs further study to identify the implementation details in a myriad of situations in the geospatial domain.

- Since digital licenses contain information about geospatial assets (e.g., boundaries), could these licenses be used as service access tickets? For example, do we need GetFeature permissions for WFS, or is sending the license (which constitutes an implicit GetFeature request) sufficient? Users could send licenses to services, which would then release datasets accordingly. Licenses would also be used to enforce the rights on the client machine.

- Current metadata standards and the OGC CSW service need to be extended to accommodate the rights metadata needed for GeoDRM (as discussed above). Also, a GeoDRM product catalogue needs to be established to enable expression of generalised licensing frameworks for negotiating licenses.

- The semantics of digital licensing need to be studied. We believe that ontologies and rules (Web Ontology Language [OWL], Semantic Web Rule Language [SWRL]) will provide reasoning capabilities for resolving conflicts in combining various geospatial assets with heterogeneous licenses.
- Research and development of the GeoDRM architecture is an essential step in establishing a GeoDRM-enabled SDI, facilitating the testing of the technology and of the theoretical concepts.

CONCLUSIONS

An obstacle to the adoption of Web-based geospatial technologies is the intrinsic loss of control over IPR by vendors and providers. Many IPR challenges are apparent in Web-based environments due to the high accessibility afforded by these technologies. The need for adequate IPR management is essential for Web technologies to receive wider acceptance in the geospatial community. Moreover, providers are motivated by more sophisticated geospatial business models that could be developed over the Web and that are hindered by loss of control over IPR downstream in the value chain.

GeoDRM provides an essential legal framework for SDIs, allowing stakeholders to maintain IPR without hampering accessibility to services. In this article, we have proposed a GeoDRM architecture that can integrate with Web services in general and OGC Web services in particular, along with a technology-neutral GeoDRM information model identifying the essential information elements for GeoDRM infrastructures. Further GeoDRM research must leverage current DRM standards. Both the proposed GeoDRM architecture and the information model capitalise on current research in digital licensing infrastructures. The above agenda provides direction for future research on GeoDRM.

REFERENCES

Bishr, Mohamed A. 2006. Geospatial Digital Rights Management with focus on digital licensing of GML datasets. ITC, Enschede, the Netherlands. http://www.itc.nl/library/papers_2006/msc/gim/bishr.pdf.

Chong, C. N. 2005. Experiments in rights control, expression and enforcement. PhD, CTIT, University of Twente, Enschede, the Netherlands. http://www.ctit.utwente.nl/library/phd/2005/chong.pdf.

ContentGuard. 2001. Extensible rights markup language (xrml) v2.0 specification. www.xrml.org.

DC. 2005. http://dublincore.org (accessed December 13, 2005).

Guth, S., G. Neumann, and M. Strembeck. 2003. Experiences with the enforcement of access rights extracted from ODRL-based digital contracts. Workshop-Proceedings of the ACM Workshop on Digital Rights Management, Washington, DC.

Harvey, F. 2003. Developing geographic information infrastructures for local government: The role of trust. The Canadian Geographer/Le Géographe canadien 47: 28–36. http://www.blackwell-synergy.com/doi/abs/10.1111/1541-0064.02e10.

Harvey, F., and D. Tulloch. 2004. How do local governments share and coordinate geographic information? Issues in the United States. Paper presented at the 10th EC GI & GIS Workshop, ESDI State of the Art, Warsaw, Poland, 23–25.

Iannella, R. 2002. Open digital rights language (ODRL) v1.1. ODRL initiative. http://www.odrl.net.

Lopez, C. 2002. Watermarking of digital geospatial datasets: A review of technical, legal and copyright issues. *International Journal of Geographic Information Science* 16 (6): 589–706.

LaMacchia, B. A. 2002. Key challenges in DRM: An industry perspective. 2002 ACM Workshop on Digital Rights Management. Redmond, WA: Microsoft Corporation.

Merriam-Webster's online dictionary. http://www.m-w.com/dictionary/asset (accessed September 12, 2005).

Open Geospatial Consortium. 2002. Catalogue service for the Web implementation specification. http://www.opengeospatial.org/specs/?page=specs (accessed March 22, 2006).

Open Geospatial Consortium. GeoDRM-WG. http://www.opengeospatial.org/groups/?iid=129 (accessed March 22, 2006).

Onsrud, Harlan J., Prudence S. Adler, Hugn N. Archer, Stanley M. Besen, John W. Frazier, Kathleen Green, William S. Holland, Jeff Labonte, Xavier R. Lopez, Stephen M. Maurer, Susan R. Poulter, and Tsering W. Shawa. 2004. *Licensing geographic data and services.* 1st Edition. Washington, DC: The National Academies Press.

Onsrud, H. 1998. Tragedy of the information commons. *Elsevier Science* 141–58.

Park, J., and R. Sandhu. 2004. The uconabc usage control model. *ACM Transactions on Information and System Security* 7 (1): 128–74.

PR. 2005. http://www.prismstandard.org (accessed December 13, 2005).

Rosenblatt, B., B. Trippe, and S. Mooney. 2002. *Digital rights management business and technology.* 2nd Edition. New York: M & T Books.

Rosenblatt, B., and G. Dykstra. 2003. Integrating content management with digital rights management. New York: www.GiantSteps.com.

Thompson, C. W., and R. Jena. 2005. Digital licensing. *IEEE Internet Computing* 1089–7801: 85–88.

Vowles, G., and L. McKee. 2005. Geospatial digital rights management (geoDRM) support for geodata markets and free geodata libraries. *Geoinformatics Magazine* 8 (May 2005).

Spatial Data Needs for Poverty Management

FELICIA O. AKINYEMI

INSTITUTE OF CARTOGRAPHY AND GEOINFORMATICS, UNIVERSITY OF HANNOVER, HANNOVER, GERMANY

ABSTRACT

As spatial determinants are increasingly considered essential in understanding poverty, the use of consistent spatial datasets in developing poverty reduction strategies is a growing requirement. This is because the success of poverty reduction programs (PRPs) depends largely on the use of quality data to help determine the nature and extent of poverty and to properly design and implement strategies for alleviating poverty in a particular context. The growing importance of spatial data in addressing complex social, environmental, and economic issues facing communities around the globe necessitates the establishment of spatial data infrastructures (SDIs) to support the sharing and use of this data locally, nationally, and internationally. SDIs facilitate access to data, data consistency, data sharing, and multinational decision making. They also facilitate the use of spatial data for poverty assessment and mapping and the development of poverty reduction applications that are integrative in nature and substantially generic. This article examines the growing need for spatial data in poverty mapping, particularly for understanding the importance of spatial factors in poverty and food security outcomes. By surveying different poverty mapping studies, the types of spatial data in use, the modes of usage, and data sources were identified. The objective was to identify spatial datasets essential for poverty mapping. The relevance of SDIs to poverty reduction is discussed. The survey reveals that the types of spatial data used for addressing poverty are diverse and the spatial datasets needed vary between programs, depending on the type of poverty measure adopted. Spatial data use is fast becoming a best practice for poverty assessment. The diversity of spatial datasets in use, the huge costs associated with their use, the need for consistency and accuracy in data, and access to spatial data at disaggregated scales are all issues pertinent to poverty reduction that SDIs can help resolve.

Poverty maps are useful in devising policy strategies for tackling poverty (Sachs 2005). They show the spatial distribution of poverty, which can be used to quantify suspected regional disparities in living standards and identify areas falling behind in economic development (Gauci and Steinmayer 2005). With the recognition of the importance of spatial variables as determinants of poverty, spatial data are increasingly needed for use in poverty reduction programs (PRPs), for direct input into assessment, for analysis, and for the making of poverty maps as policy tools.

Pertinent questions are as follows: What are the types of spatial data in use for poverty mapping, and how are they used? What determines the choice of spatial data for poverty analysis? What spatial datasets should be considered essential for decision making and poverty problem solving? What are the practical implications of selecting a particular spatial indicator for use in assessing poverty? Are general-purpose spatial datasets adequate, or are special spatial datasets needed, and what specifications are desirable for poverty mapping applications? Although these pressing questions have no ready answers, they need to be addressed to enhance the usefulness of spatial data infrastructures (SDIs) in poverty management.

"A Spatial Data Infrastructure (SDI) is an initiative intended to create an environment that will ensure that a wide variety of users, who require coverage of a certain area, will be able to access and retrieve complete and consistent datasets in an easy and secure way. Also, it can be viewed as a tool to provide a proper environment in which all stakeholders, both users and producers of spatial data, can cooperate with each other and interact with technology in a cost-effective way to better achieve the objectives at the corresponding political/administrative level" (Rajabifard 2002, p. 1). Within the SDI context, identifying the spatial data requirements for poverty mapping would further facilitate the development of SDI data contents and standards. To achieve this, it is important to identify users and potential applications of SDIs, as this will enhance the development of common solutions for discovery, access, and use of spatial data in response to the needs of diverse communities (DOC 1997).

An SDI cannot exist as an end in itself, since it is essentially an infrastructure that supports the development of spatial data products and services and the needs of diverse decision-making environments (Feeney 2002). Supporting diverse decision making is of utmost concern, since decision making for poverty reduction in particular and sustainable development in general is characterized by multidisciplinary and multiparticipant environments (see Agenda 21 1993 and GSDI 2001, both cited in Feeney et al. 2002). Achieving such support would enhance the benefits accruable from having SDIs established, especially in developing countries.

For poverty reduction, SDIs are of prime importance in facilitating access to data, achieving data consistency, and data sharing. Issues related to the suitability of spatial data such as data quality and the risk of using spatial data from an unreliable source will be lessened with SDIs in place. Furthermore, the costs of data production will be reduced and duplication of effort will be eliminated. Also, SDIs can facilitate decision making across national boundaries. For instance, decision making in dealing with poverty, food insecurity, biodiversity conservation, public health hazards, forced migration of displaced populations resulting from wars, and natural disasters often transcends administrative or national boundaries. With increasing awareness of the importance of SDIs at the regional level, countries are finding it necessary to cooperate with one another in developing multinational SDIs (Rajabifard 2002).

This article seeks to address the growing need for spatial data in poverty reduction efforts. It aims to identify spatial datasets essential for poverty mapping. It discusses the relative importance of SDIs in facilitating access to these datasets and distinguishing between them. Sensitizing spatial data providers to the increasing need for spatial datasets in poverty reduction strategizing and making geospatial scientists aware of the emerging research area of poverty mapping are the secondary purposes of this article. To this end, poverty mapping measures and initiatives were surveyed to identify the spatial datasets in use. Recent poverty mapping studies (conducted within the last eight to ten years) reflecting current methodology and using spatial data in a defined way were selected for the survey.

METHODOLOGY

Four major classes of poverty mapping measures were identified on the basis of data sources, assumptions, and the statistical routines utilized: econometric, social, demographic, and vulnerability-based measures (Davis 2003). Econometric poverty indicators are current consumption expenditures, income, and wealth. Examples of social indicators are nutrition, water, health, and education. Demographic measures may use gender, health, and household age structure indicators. Vulnerability measures are concerned with the level of household exposure to shocks, using indicators such as environmental endowment and hazard, physical insecurity, political change related to empowerment, governance, participation, transparency of the legal system, structural inequities, and skewed processes that become impediments to human well-being (Henninger 1998). Poverty-biodiversity mapping (PBM) is the most recent poverty mapping application area, and it is of growing concern because of the link to conservation. Indicators such as major tropical wilderness and biodiversity hot spots are used (Snel 2004). Although PBM is treated separately in this article, it cannot be said to constitute a separate class of poverty mapping measures, as the methodologies in use vary.

The survey first examines poverty mapping measures and studies in specific contexts in order to identify the spatial datasets in use. The examined studies cover the four major classes of poverty mapping measures and PBM. On the basis of the importance of SDIs in addressing spatial data requirements for poverty mapping, the datasets identified are grouped into four categories: fundamental, geographically derived, project-specific, and additional datasets. The datasets are also categorized as very important, important, or less important (figure 1).

The survey departs from the norm in that, instead of the administration of a questionnaire to solicit answers from respondents, poverty mapping studies and projects that had already been executed were examined in order to identify the types of spatial datasets in use for poverty mapping. This goes beyond identifying what datasets are assumed to be needed, to actually identifying the datasets in use. This is advantageous in that the datasets identified are what is used in actual practice and the results would better reflect what happens in reality, rather than the ideal situation which questionnaire responses would capture. Knowing what

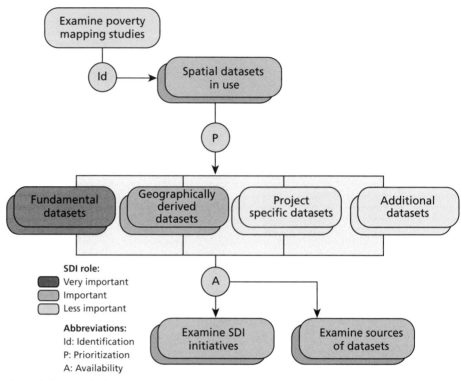

Figure 1. Data analysis procedure used in this study.

datasets are used and how they are used is also helpful, since spatial data providers may not be entirely aware of all the uses to which these datasets are put.

After examining the methodological base of each poverty mapping measure, we take a detailed look at each measure as utilized in country-based studies in order to identify the spatial datasets in use.

Econometrics. The small-area estimation (SAE) method is given prominence in this survey, as it has been implemented in many developing countries for poverty and food security assessment. Snel and Henninger (2002) examined some studies implementing the SAE method, and Hyman et al. (2005) recently reviewed the applications of SAE in poverty mapping research for agriculture and rural areas. Such widespread usage of one methodology in different countries and sociocultural environments looks promising, in that it could bring about some level of uniformity in dealing with poverty across countries, which may serve as a vital input for evolving SDIs aimed at poverty reduction and development.

SAE is an extension of the literature on small-area statistics (Rao 1999) that combines survey and census data to estimate welfare or other indicators for disaggregated geographical units such as municipalities or rural communities. With the use of detailed household survey data, a model of consumption is estimated, and the resulting parameter estimates are used to weight the census-based characteristics of a target population in order to determine its expected poverty level. For example, using a set of explanatory variables that are common to both the household survey and the census (e.g., household size, education, housing and infrastructure characteristics, and demographic variables), parameters of regressions that model consumption are estimated and applied to the census data; the assumption is that the relationship defined by the model holds for the larger population as well as the original sample (Elbers et al. 2002; Davis 2003; World Bank 2004).

SAE can be used with household unit data or community level data. The first method uses census data and survey data of the same period, while the second method uses average values from communities and small towns. The principles utilized in the two methods for calculating total per-capita consumption or any other poverty measure used are the same. In estimating the consumption model, the parameters required can be categorized into household-level characteristics and geographic-level characteristics.

We are more interested in the geographic-level characteristics, as the parameters used are spatial or location based. In estimating these characteristics, spatial indicators such as rainfall amount and timing, evapotranspiration, land cover and land use, and geographically derived indicators such as distance and physical accessibility measures are utilized. With GIS databases, for example, a multitude of these environmental and community characteristics can be geographically defined both comprehensively and with great precision (Elbers et al. 2002). The main virtue of SAE is that it is the only method where the statistical properties have been and continue to be thoroughly investigated (Davis 2003). Because of their methodological rigor, highly complex analytical methods like SAE may be impractical and too expensive in some countries. Moreover, not all intended uses of poverty maps require detailed household survey data and sophisticated econometric analyses (Snel and Henninger 2002).

Social measures. Social measures use nonmonetary variables and aim to capture household well-being as measured by the quality of and access to social services: education, health, nutrition, water, housing, and neighborhood quality. According to the unsatisfied basic needs (UBN) approach, poverty is linked to a state of necessity, a deficiency of income, and a deprivation of the goods and services necessary to sustain life at a minimum standard. Based on the UBN approach, various indices are in use for disaggregated poverty mapping. The indices vary much in the types of variables used and in their weighting methods. Basically, certain needs considered essential are selected, a minimum criterion for satisfying each need is identified or determined (e.g., the corresponding national average value is sometimes used), and the data are then combined into poverty indices. The indices calculated at the household level are aggregated by geographic or administrative units by calculating the proportion of the population in a particular basic-needs stratum (Davis 2003).

Other UBN methods which are based on multivariate statistical techniques such as principal components analysis (PCA), factor analysis, and ordinary least squares are also used in many countries. The most elaborate indices based on the UBN approach are the United Nations Development Program's (UNDP) human development indices—comprising the human development index (HDI), human poverty indices for developing countries (HPI-1) and selected Organization for Economic Cooperation and Development (OECD) countries (HPI-2), gender-related development index, and gender empowerment measure—which are generally available for every country (UNDP 2004).

Demographic measures. Demographic studies of poverty use indicators such as the gender and age structure of households, child nutritional status, and household size. They also focus on various outcome indicators such as calorie intake, low height for age, low weight for age, low weight for height, body mass index, and low birth weight.

Indicators of child nutritional status are often used because they very well indicate the degree of development in a region and can be interpreted as measures of poverty, especially when based on anthropometric measurements such as low height for age. Anthropometric measures are recommended for use as the best general indicators of constraints to the welfare of the poorest, including dietary inadequacies, infectious diseases, and other environmental health risks (UN 1992).

Vulnerability measures. Major vulnerability risk factors are environmental risk (droughts, floods, and pests), market risk (price fluctuations, wage variability, and unemployment), political risk (changes in subsidies or prices, income transfers, and civil strife), social risk (reduction in community support and entitlements), health risk (exposure to diseases that prevent work), and crime risk.

For poverty assessment, indicators focus on the level of household exposure to shocks that can affect poverty status, such as environmental endowment and hazards, physical insecurity, political change, and the diversification and risk of alternative livelihood strategies. Food insecurity encompasses access, availability, risk, and uncertainty, and its indicators include household access to assets, household size and composition, asset liquidity, crop and income diversification, and food production at the household level (Davis 2003). Conceptual frameworks for vulnerability assessments generally include measures of chronic (baseline) vulnerability and acute vulnerability (e.g., reflecting the success or failure of the most recent agricultural season). Most indicators measuring baseline vulnerability could provide appropriate input for poverty assessments. Approaches are adapted to reflect the differences in local risk factors that are caused, for example, by physical geography, agroclimatic conditions, colonial history, agricultural production systems, infrastructure, economic policy environment, and data availability (Henninger 1998).

Hybrid measures. As measures used in poverty-biodiversity mapping and other areas are varied, they are classified as hybrid measures. As noted above, poverty-biodiversity mapping uses all of the above measures rather than one particular class of measures. Snel (2004, p. 2) noted that "While poverty-environment mapping offers a suite of tools for improving the analysis between biodiversity and development issues, it must not be seen as a panacea for understanding or solving poverty-conservation problems. Mapping applications need to be used together, not in lieu of, other approaches including multilevel socio-economic assessments, traditional and community-based knowledge, community mapping, and statistical analyses." Results from SAE, HDIs, or demographic indicators of poverty, such as percentage of underweight children, are examined in conjunction with the occurrence of, for example, major tropical wilderness and biodiversity hot spots. The value of poverty-environment mapping lies in highlighting spatial correlations and disparities for identifying underlying causes and drivers of biodiversity loss. The uniqueness of poverty-biodiversity mapping necessitates separate treatment, and this approach cannot be subsumed under any of the other four major classes of poverty mapping measures.

**SURVEY OF POVERTY-
MAPPING INITIATIVES**

The examination of poverty-mapping initiatives is meant to be broad and to include not only poverty-mapping studies (where research is the main focus) but also programs where operational policy based on results of poverty mapping is important.

SAE method as utilized in country-based studies. In Cambodia, Fujii (2004) utilized a set of geographically derived indicators for assessing poverty at the commune level (municipality): distances from villages to roads, other towns, health facilities, and major rivers; land use and land cover within the commune (agriculture, urban, forested, etc); normalized differential vegetation index (NDVI) (to measure agricultural productivity); climatological variables (elevation, soil quality, and flooding); and the degree of nighttime lighting as a measure of urbanization.

For Mozambique, NDVI and flood risk were also used, alongside data on droughts, water sources, and the number of times the eye of a cyclone passed through the district (Mistiaen et al. 2002). For using SAE, these variables were fitted to the 1993 population census and household survey data (May 1993 and April 1994) collected from 4,508 households.

With recent advances in poverty and food security mapping, SAE is being adapted to calculate welfare indicators relevant to rural areas and food security concerns. An example is its use for estimating the prevalence of child malnutrition in Cambodia (Fujii et al. 2004).

Legg et al. (2005) adapted SAE for regression of poverty and nutrition indicators against biophysical and socioeconomic variables. Spatial attributes such as land cover (percentages of tree cover, grass cover, bare soil, soil fertility [see appendix]), climate (annual rainfall), socioeconomic characteristics (population density, number of households, distances to towns, travel times to markets [calculated for cities of more than 200,000 people, towns of 50,000 to 200,000 people, and towns of 20,000 to 50,000 people]), and amenities were incorporated in the statistical analyses of the datasets. A series of variables indicative of poverty levels, development, and nutrition were also mapped. The authors identified and parameterized key factors and socioeconomic or policy drivers that affect poverty, malnutrition, and undernourishment levels in Nigeria. Parameters that were measured included the core indicators for monitoring food security outcomes recommended by the Food and Agriculture Organization's committee on world food security, which are linked to poverty characteristics. In all, a total of eight variables were selected as independent variables mainly on the basis of being available in digital form.

In these studies, the estimated parameters from the regression are used to compute the probability of each household in the census living in poverty. Household level results can then be aggregated for the geographic or administrative region of interest by taking the mean of the probabilities for the chosen region. This permits the construction of maps illustrating different levels of poverty disaggregated across geographic or administrative units (Davis and Siano 2001). Different kinds of geographically derived indicators are directly utilized as regressors in the consumption model, and cross-terms for household level variables and geographic variables are also included in the model, as in the method described by Fujii (2004).

The reason that SAE studies utilize different types of spatial data is unknown, as no justification for the choice of data is given in the studies, with the exception of Legg et al. (2005), where availability in digital format was reported. The practical implications of selecting a particular geographic variable for SAE analysis should be considered in poverty-mapping studies. It will be instructive to explicitly describe the kinds of spatial data used in SAE.

UBN case studies. In looking at the spatial data components of some UBN-based studies (for detailed descriptions of the UBN procedure, see the cited references), we first examine a sample of studies utilizing the UBN approach in which spatial data is used to determine eligibility for inclusion in PRPs.

For community selection based on factors that influence well-being in Honduras (Leclerc et al. 2000, cited in Davis 2003), three geographic variables—altitudes of villages, population density, and accessibility to urban centers—were among the six factors selected (the others were basic services [like education and water], ethnicity, and gender composition). By combining these factors for every village from the census data and a GIS database, a sample of villages was obtained based on the premise that the indicators of well-being obtained are valid for all villages that have the same combination of factors as the ones in the sample.

In Mexico, a "marginality index" created from PCA was utilized for selection of communities. Communities having a high or very high degree of marginality according to the index were considered priorities for inclusion in the PRP Progresa. To further refine the selection procedure, criteria such as geographic location, distance between communities (connoting the extent of isolation), number of inhabitants, and the existence of health and school infrastructures were imposed. Communities with access to school and health services were considered candidates for selection, taking into account the availability and quality of roads when the services were not located in the same community. The communities selected for inclusion in the program were grouped into "marginality zones" (Skoufias et al. 2001).

Although HDI analyses do not incorporate spatial data as variables, estimates made from HDIs are used in making poverty maps. A very good example is the Atlas of Human Development for Brazil, in which 38 georeferenced variables, including two composite indices based on UNDP methodology (HDI and life conditions index), from the 1970, 1980, and 1991 population censuses were mapped. The atlas has been a tremendous success as a basis for decision making in public investment and targeting of social programs worth billions of dollars (Snel and Henninger 2002).

Demographics-based poverty mapping. Georeferenced indicators of child nutritional status from the Demographic and Health Survey (DHS) for 10 countries in West Africa were aggregated into aridity zones using data on stunting (low height for age), wasting (low weight for age), underweight (low weight for height), and wasting and stunting combined (Henninger 1998). The mean and standard error of the incidence of each indicator per aridity zone (hyperarid, arid, semiarid, dry subhumid, moist subhumid, and humid) were expressed as a percentage of children sampled between 3 and 35 months of age and were disaggregated by urban and rural clusters.

Rural areas had a significantly higher incidence of malnutrition than urban areas, with the highest proportion of stunted children in the semiarid zone for rural clusters and in the dry subhumid zone for urban clusters. Stunting varies greatly across the semiarid zone, ranging from lower values in Senegal and the coastal areas of Ghana and Côte d'Ivoire to the highest values in eastern Niger and northern Nigeria, especially around Kano. Although clear spatial trends were apparent at the aggregated and disaggregated levels, spatial variables such as market access and crop distribution among others were suggested for use in combination with aridity zones to further understand some of the underlying factors contributing to this spatial pattern.

Vulnerability-based poverty mapping. The U.S. Agency for International Development-financed Famine Early Warning System uses food security indicators linked to or reported by geographic area from which the food security situation at the household level is inferred. Data inputs include remotely sensed data, measures of vegetative vigor and land use, as well as agricultural and socioeconomic data at the administrative level.

In rural Malawi, Benson et al. (2005) identified key spatially explicit determinants of differing poverty levels. Using 17 spatially derived variables, they created a subset of potential independent variables for their analysis. These variables had to allow meaningful aggregation and to display variation across the country at the desired aggregated-area scale.

In Honduras, a series of disaster vulnerability indices were developed and mapped: flood and landslide risk areas, total population at risk of flooding and landslides, percentage of very poor people at risk, and roads and electricity lines at risk. These indices were weighted and aggregated into an overall vulnerability index that allowed identification of municipalities for priority intervention (Segnestam et al. 2000, cited in Davis 2003).

Hybrid poverty-mapping studies. Poverty-biodiversity studies are being promoted mainly by the World Conservation Union (IUCN). The IUCN program for 2005 to 2008 highlights the need to evaluate the direct and underlying causes of biodiversity loss and unsustainable practices. It recognizes that wealth/poverty/inequity, human population dynamics, consumption patterns, market failures, and policy distortions are major underlying threats to biodiversity and sustainability (Snel 2004).

Poverty-environment linkages have been examined in Madagascar, Cambodia (food insecurity), and Ecuador (land cover change and human migration) and for identifying the impact of global climate change on poor livestock keepers' livelihoods (Henninger and Snel 2002, Thornton et al. 2002, WWF 2002, all cited in Snel 2004). Maps of indicators of well-being ranging from nutritional status to educational level can be very useful in combination with maps of environmental conditions. They are most commonly used by planning and development agencies to target specific areas with specific needs (Snel 2004).

This survey reveals a broad range of spatial data in use for poverty assessment and policy formulation. The types of spatial data in use differ within and between the four major classes of poverty mapping measures and PBM studies examined. In other words, spatial data types in use are as diverse as the poverty dimensions being examined and the measures utilized. The spatial datasets identified are categorized as fundamental, geographically derived, project specific, and additional. The appendix shows the variety of spatial datasets utilized in poverty mapping studies discussed above.

Fundamental spatial datasets are commonly described as the foundation on which other spatial datasets are built for the production of added-value information, development of applications, and acquisition of other data; they are required by the great majority of users, having widespread usability and constituting the building blocks of any SDI. These sets of data should be widely accessible and distributed under accepted standards (Knippers et al. 2006). This view needs to be broadened to accommodate the differentiation of various levels of fundamental datasets. Geographically derived datasets are derived by spatial analysis from other spatial datasets such as distances to nearest roads or distances to larger urban centers. Project-specific datasets are those used only in specific studies. Additional datasets are nonspatial data for which a geographic reference is needed for use in poverty mapping.

The most common spatial datasets for poverty mapping found by the survey were land cover, NDVI, rainfall data, and soil fertility and quality. This finding is confirmed by Hyman et al. (2005), who noted that soil characteristics, topography, rainfall, evapotranspiration, and vegetative vigor proved to be important explanatory factors in describing poverty in several poverty-mapping studies.

Datasets on travel times to markets and distances to towns and facilities featured in all the studies except demographic and poverty-biodiversity studies. These datasets were found to be important explanatory factors in poverty and food security outcomes for small areas and across countries or provinces in several studies. Utilizing such measures of distance and physical accessibility is increasingly important in poverty mapping studies, since income generation for small-scale farmers, for example, often depends on distances to markets and associated transport costs (Van De Walle 2002; Jacoby 2000; both cited in Hyman et al. 2005).

Spatial data usage in UBN-based studies, for example, is slightly different from that in econometric studies, as exemplified by the SAE method. In econometric studies, spatial data are often incorporated right into the regression model for estimating poverty, whereas in UBN-based studies, census data and/or household survey data are often crossed with spatial data as spatially derived criteria for determining community eligibility. Also, the fact that the two approaches focus on different dimensions of poverty may partly explain the kinds of spatial data chosen and the manner of use. For example, econometric studies capture monetary definitions of poverty, while UBN-based studies focus on nonmonetary poverty dimensions.

Factors determining the choice of spatial data may depend on data availability, access, and format (digital is preferred) and peculiarity of the place in question; the choice could also be a matter of preference. The objective may be the targeting of specific interventions, building a map to convey a political message, or constructing inputs for an analysis, and these may be important in the choice of spatial variables used. The kinds of spatial data utilized in UBN, demographic, and vulnerability-based studies seem to be more determined by program objectives than in econometric studies.

Most SDI initiatives do not include in their list of fundamental datasets natural and environmental datasets such as land cover and land use, and NDVI data that are derived from image classification and analysis. An exception is the Australia–New Zealand Land Information Council (ANZLIC), which specifies a set of natural resource and environmental data as fundamental datasets (ANZLIC 1996). A recent survey of fundamental geospatial datasets in Africa also included land cover but as a level III dataset (fundamental datasets were subdivided into four levels: 0, I, II, and III) (Gyamfi-Aidoo et al. 2006). It is highly recommended that natural and environmental datasets be designated as fundamental in SDIs because of their importance in poverty mapping.

An important benefit of SDIs to poverty mapping would be to facilitate access to cultural and demographic data (CDD). According to the U.S. Federal Geographic Data Committee subcommittee on CDD (SCDD), these data include the characteristics of the people; the nature of the dwellings in which they live; the economic activities they pursue; the facilities they use to support their health and recreational needs; the environmental consequences of their presence; and the boundaries, names, and numeric codes of geographic entities used to report the information collected (see SCDD Web site). Such data also appear on the list of fundamental datasets for Australia and New Zealand (ANZLIC 1996). It is recommended that CDD be included as fundamental datasets in SDIs, although it may be useful to make a distinction between generic and thematic fundamental datasets (see appendix). Generic data is the first priority when establishing fundamental spatial datasets, and thematic data is the second (but still important) priority (Knippers et al. 2006).

The survey showed that some fundamental datasets in an SDI, although important to poverty mapping, may have limited usefulness because of the absence of needed attributes. For example, telecommunication and electricity networks are designated as fundamental datasets in SDIs, but they are less useful for poverty mapping if people's levels of access to telephones and electricity are not indicated. Physical nearness to locations of trunk electricity or telecommunication networks and assets (which is what is described in most SDIs) may not be as important in capturing poverty as the lack of connection to the networks, which is mainly due to the inability to pay for these services.

SDIs would also facilitate access to reference data to which needed indicators can be added. Spatial data such as boundaries of administrative units, road networks, and rivers are needed as reference datasets. Apart from reference datasets, the linking of socioeconomic data to a specific location makes the data available for spatial analysis, which in a sense makes such data spatial. These could be socioeconomic or demographic variables derived, for example, from a census, which are then linked to geographic units appropriate to the levels of data publication.

The capturing, processing, and supplying of spatial data are quite costly and require not only a maximization of benefits but also cost reduction efforts to generate sufficiently positive economic returns on public investments, particularly in developing countries (Byamugisha and Zakout 2000). The huge costs directly associated with producing spatial-poverty maps (because of spatial data creation and analysis) can be substantially reduced with SDIs.

Also, SDIs can link data scattered between agencies and ease the coordination of data capture and maintenance in order to avoid duplication of effort, which is a problem at present. Duplication of effort results when an existing dataset is either unavailable or unknown to project personnel. SDIs can reduce such wastage of resources by providing information on what is available and where (Rajabifard 2002). The development of common content standards for SDIs would further facilitate the use of consistent and accurate spatial datasets, which is a growing requirement for poverty reduction strategies.

Within the framework of SDIs, interoperability and unrestricted access to consistent spatial datasets are essential for poverty reduction and development-related applications. Sometimes data for a particular indicator may be unavailable, only a portion of an area is covered, or portions of an area differ in sampling methodologies, scales, or accuracy levels. Combining inconsistent datasets can produce errors (such as mismatched boundaries) (Snel 2004).

To maximize SDI benefits, obstacles to data sharing need to be overcome. SDIs hold great promise for combating poverty globally by facilitating the use of spatial data for poverty assessment and mapping.

CONCLUSIONS

The main findings of this study are that a great variety of spatial datasets are in use and that the types and uses of spatial data differ between poverty mapping studies. However, land cover and land use, rainfall, NDVI, and soil quality and fertility variables were used in almost all the studies. This is true also of travel times to markets and distances to towns and facilities, which are the most common geographically derived datasets. All these spatial variables were also found to be important explanatory factors in poverty and food security outcomes in many studies. The mode in which spatial data are used is a significant difference between the four major poverty mapping methods. Econometric methods, as exemplified by small-area estimation (SAE), use spatial variables as explanatory factors for direct input in regression analysis, whereas in UBN-based studies, census data and/or household survey data are sometimes crossed with spatial data for community eligibility determinations. The types of spatial data utilized in UBN, demographic, and vulnerability-based studies seem more consistent and largely determined by program objectives.

Spatial data can assist in providing wider policy options for reducing poverty. The link between geography and poverty requires further investigation, and the techniques described above need to be applied more widely in poverty mapping studies, especially with respect to agriculture, rural issues, and biodiversity concerns (Hyman et al. 2005; Snel 2004).

SDIs can help with the huge costs of creating spatial data, the diversity of datasets used for poverty mapping, the need for consistent and accurate data, the availability of disaggregated spatial data for examining poverty at the local level, and the regional nature of poverty reduction decision making. By providing uniform datasets, SDIs can facilitate comparisons over time and across countries. At present, the choice of spatial data is determined largely by availability and format (Benson et al. 2005; Legg et al. 2005).

ACKNOWLEDGMENTS

Useful comments and thoughtful reviews were received from Professor Monika Sester and anonymous reviewers. The Research Fellowship of the Alexander von Humboldt Foundation and the supportive research environment provided by the Institute of Cartography and Geoinformatics, University of Hannover, are gratefully acknowledged.

Appendix. Spatial data in poverty assessment[a]				
Category	Datasets	Sources	Coverage	Comment(s)
Fundamental				
Generic				
Hydrography	Lakes, waterfalls, water sources, rivers	NDW, UNEP, WMO, GIWA, NOAA	National, provinces	
Height	Steep slopes, elevation, 90-m DEM	GTOPO30	Global	A global DEM at 1-km resolution
Transport	Road track and network, quality of road	NMA	National	
Administration	Administrative boundaries, centers	NMA	National	
Geographic names	Village names and locations, geographic coordinates	National statistical offices	National	
Thematic				
Land cover	Land cover/land use, % grass cover/forage land, forests, % tree cover, NDVI deviation from average	National departments of agriculture (NDA), defence, forestry, and environment; FAO; UNEP; ESRI's Digital Chart of the World; NOAA, WRI, AVHRR	National	Land cover change could signify an increase or decrease in agricultural activity and forest encroachment
Soils and geology	Soil fertility/quality, flood risk, bare soil, landslide risk, dominant soils subject to flood or having high agricultural potential, geomorphology	Remotely sensed data, FAO soil classification, NDA, FAO, UNEP 1996, NDW	Villages, provinces	Useful for locating areas of food insecurity or low agricultural potential
Cultural and demographic	Population size and demographic structure; population density/flows; organized social groups; market, mill, brick factory; sex, dependency ratio, gender; low height for age, low weight for age, low weight for height, stunted growth; calorie intake; age, HH size and structure, HHs headed by women	Census, markets database, LandScan[b], first-graders' height (FAO 2004)	Villages, national, administrative units	All poverty mapping methods utilise some kind of socioeconomic data
Basic services	Health care center locations and status, prevalence of guinea worm	NDW, WHO	Villages, national	
Utilities	Access to telephones and electricity	Census		
Climate	Rainfall amount and timing, 30-year averages of climatological variables, temperature	Global Climate Dataset of University of East Anglia, meteorological offices		
Hydrological	Description and coordinates of water points, pumps, dams, irrigation systems; water basins, water table, water points management, pollution problems	National river authorities, State of the Rivers Report, pollution control authorities		Surface and groundwater maps are useful for assessing the effect of water quality on biodiversity
Biodiversity	Endemic birds, amphibian species per 0.25° grid cell, major tropical wilderness and biodiversity hot spots, aridity zones	Stattersfield et al.1998, Conservation International[c], national departments of the environment, UNEP, IUCN, WRI		Data are usually based on forest cover or protected areas, although specific species or habitat atlases also exist
Geographically derived				
Distance and physical accessibility	Distance to larger urban centers, distance to nearest roads, average weighted road density (m/km2), distance to major rivers, travel time to and accessibility of markets, distance to nearest school and distribution of schools, distance to nearest drinking water point, average travel time to nearest major forest reserve or national park, distance/average travel time to nearest health center, percentage of students living a certain distance from school, average distance from school to place of residence	NDW; maps of settlements, facilities, services, and transportation networks	Villages, national	For measuring the extent of isolation or access to common property resources
Agroclimatology	Average rainfall coefficient of variation, highest and lowest quintiles of rainfall deviation from long-term mean, evapo-transpiration, climate variability, pests, droughts	Meteorological offices, DMA, Mistiaen et al. 2002	National sample	Climate indicators

Continued on next page

Appendix (continued)				
Category	Datasets	Sources	Coverage	Comment(s)
Project specific				
Urban	Nighttime lighting	Defense Meteorological Satellite Program, National Geophysical Data Center		City lights satellite data at 1-km resolution for 1994–95
Natural hazard zones	Number of times the eye of a cyclone passed through a district[d]			
Environmental laws and management	Remotely sensed data, e.g., on land cover change and land degradation, related to environmental treaties	CIESIN global mapping project		
Additional				
Education	Literacy rate differences between men and women, mean educational attainment for HHs, number of classrooms and benches, sex and age ratios of schoolchildren for each class, age and diplomas for school teachers	Ministries of education, enrollment data		
Agriculture and livelihoods	Annual yield and crop cultivation area, mean maize yield and coefficient of variation, cultivable area not used for a staple crop, cost and sale prices of crops, dips in livestock numbers, workers not in agriculture	Markets database	Provinces, national	
Other	Income per capita, children under 15 years of age with at least one parent dead, Gini coefficient, cattle numbers, native language, political party	World Bank, CIESIN, FAO Food Insecurity and Vulnerability Information and Map, WRI		

a NDW, national departments of water; WMO, World Meteorological Organization; GIWA, Global International Waters Assessment; AVHRR, NASA's Advanced Very High Resolution Radiometer; WRI, World Resources Institute, CIESIN, Center for International Earth Science Information Network; NMA, national mapping agencies; DEM, digital elevation model; HH, household.
b LandScan 2002, cited in Snel 2004.
c Christ et al. 2003.
d Mistiaen et al. 2002.

Sources: Bigman 1997, Henninger 1998, Benson et al. 2005, Snel 2004, NDW, WMO, GIWA, AVHRR, WRI, and CIESIN.

ANZLIC. 1996. Spatial Data Infrastructure for Australia and New Zealand, May 2006, http://anzlic.org.au/get/2374268456.

Benson, T., J. Chamberlin, and I. Rhinehart. 2005. An Investigation of the Spatial Determinants of the Local Prevalence of Poverty in Rural Malawi. Food Policy, 30(5–6), 532–50.

Bigman, David. 1997. Community Targeting of Public Projects in Burkina Faso, The World Bank, Washington, DC (mimeo).

Byamugisha, F., and W. Zakout. 2000. World Bank support for land-related projects in developing countries: experiences and implications for international cooperation. Paper prepared for the UN Regional Cartographic Conference for Asia and the Pacific, 11–14 April 2000, Kuala Lumpur, Malaysia.

Christ, C., O. Hillel, S. Matus, and J. Sweeting. 2003. Tourism and biodiversity: Mapping tourism's footprint. Washington, DC: UNEP and Conservation International.

Davis, Benjamin, and R. Siano. 2001. Issues and Concepts for the Norway-funded Project-Improving Methods for Poverty and Food Insecurity Mapping and Its Use at Country Level, Prepared for the Expert Consultation, FAO, February 2005, http://www.povertymap.net/publications/doc/Poverty_mapping_issues_paper.pdf.

Davis, Benjamin. 2003. Choosing a method for poverty mapping, March 2006. http://www.povertymap.net/publications/doc/CMPMDAVIS13apr03sec.pdf.

DOC (Department of Commerce). 1997. Department of Commerce accomplishment report: National Spatial Data Infrastructure (NSDI) Federal Geographic Data Committee (FGDC) 1997 activities. March 2006. http://www.esdim.noaa.gov/doc97rpt.html.

Elbers, C., J. Lanjouw, and P. Lanjouw. 2002. Micro-Level Estimation of Welfare. Policy Research Working Paper 2911. Washington, DC: World Bank.

FAO. 2004. Chronic Undernutrition among Children. Rome, Italy: UN Food and Agriculture.

Organization Nutrition Planning, Assessment and Evaluation Service.

Feeney, M., I. Williamson, and I. Bishop. 2002. SDI development to Support Spatial Decision-Making. Proceedings of the GSDI 6 Conference on From Global to local. Budapest, Hungary.

Fujii, T. 2004. Commune Level Estimation of Poverty Measures and its Application in Cambodia. Research Paper No. 2004/48. Helsinki, Finland: UNU-WIDER.

Fujii, T., P. Lanjouw, S. Alayon, and L. Montana. 2004. Micro-Level Estimation of Prevailence of Child Malnutrition in Cambodia. http://siteresources.worldbank.org/INTPGI/Resources/3426741092157888460/Fujii.MicrolevelCambodia.pdf.

Gauci, A., and V. Steinmayer. 2005. Poverty maps: A useful tool for policy design to reduce poverty. Poverty and Social Policy Team, ESPD, UN Economic Commission for Africa working paper.

http://www.uneca.org/espd/publications/poverty_maps_a_useful_tool_for_policy_design_to_reduce.pdf.

Gyamfi-Aidoo, Jacob, Craig Schwabe, and Sives Govender, eds. 2006. Determination of the fundamental geo-spatial datasets for Africa through a user needs analysis: A synthesis report. Human Sciences Research Council and EIS AFRICA. http://www.gsdi.org/SDIA/docs2006/may06links/africa-fundamentaldata.zip.

Henninger, N. 1998. *Mapping and Geographic Analysis of Human Welfare and Poverty: Review and Assessment.* Washington, DC: World Resources Institute. http://www.povertymap.net/publications/doc/henninger.

Hyman, G., C. Larrea, and A. Farrow. 2005. Methods, Results and Policy Implications of Poverty and Food Security Mapping Assessments. *Food Policy* 30 (5–6): 453–60.

Knippers, R. A., J. E. Stoter, and M. J. Kraak. 2006. Fundamental geospatial datasets for Africa: A global perspective. In Determination of the Fundamental geo-spatial datasets for Africa through a user needs analysis, ed. C. Schwabe, J. Gyamfi-Aidoo, and S. Govender. Human Sciences Research Council and EIS AFRICA. http://www.gsdi.org/SDIA/docs2006/may06links/africa-fundamentaldata.zip.

Legg, C., P. Kormawa, B. Maziya-Dixon, R. Okechukwu, S. Ofodile, and T. Alabi. 2005. Report on mapping livelihoods and nutrition in Nigeria using data from the National Rural Livelihoods Survey and the National Food Consumption and Nutrition Survey. http://gisweb.ciat.cgiar.org/povertymapping/Nigeria.pdf.

Mistiaen, J. A., B. Özler, T. Razafimanantena, and J. Razafindravonona. 2002. Putting welfare on the map in Madagascar. *Africa region working paper series* 34. World Bank publication.

Rajabifard, A. 2002. Diffusion of regional spatial data infrastructures: With particular reference to Asia and the Pacific. PhD thesis. Geomatics, University of Melbourne. http://www.geom.unimelb.edu.au/research/publications/Rajabifard_thesis.pdf.

Rao, J. N. K. 1999. Some recent advances in model-based small area estimation. *Survey Methodology* 25:175–86.

Sachs, J. 2005. *The end of poverty: Economic possibilities for our time.* New York: The Penguin Press.

Skoufias, E., B. Davis, and S. de la Vega. 2001. Targeting the poor in Mexico: An evaluation of the selection of households into PROGRESA. *World Development* 29 (10): 1769–84.

Snel, M., and N. Henninger. 2002. *Where are the poor? Experiences with the development and use of poverty maps.* Washington, DC: World Resources Institute.

Snel, M. 2004. Poverty-conservation mapping applications. IUCN World Conservation Congress, Bangkok, Thailand. http://www.povertymap.net/publications/doc/iucn_2004/poverty-biodiversity.pdf.

Stattersfield, A. J., M. J. Crosby, A. J. Long, and D. C. Wege. 1998. Endemic bird areas of the world: Priorities for biodiversity conservation. *BirdLife Conservation Series* 7. Cambridge, UL: BirdLife International.

UN. 1992. Global and regional results. Second report on the world nutrition situation. Committee on Coordination: Subcommittee on Nutrition Vol.1. Geneva: United Nations.

UNDP. 2004. Human development report 2004: Cultural liberty in today's diverse world. http://hdr.undp.org/reports/global/2004.

World Bank. 2004. Mapping Poverty. http://siteresources.worldbank.org/-INTPGI/Resources/ 3426741092157888460/poverty_mapping.pdf.

SCDD website: http://www.census.gov/geo/www/standards/scdd.

Spatial Data Infrastructures in Management of Natural Disasters

KWABENA O. ASANTE,[1,2] JAMES P. VERDIN,[2] MICHAEL P. CRANE,[2] SEZIN A. TOKAR,[3] AND JAMES ROWLAND[1,2]

SCIENCE APPLICATIONS INTERNATIONAL CORPORATION (SAIC),[1] AND U.S. GEOLOGICAL SURVEY (USGS) CENTER FOR EARTH RESOURCES OBSERVATION AND SCIENCE (EROS),[2] SIOUX FALLS, SOUTH DAKOTA, AND U.S. AGENCY FOR INTERNATIONAL DEVELOPMENT, WASHINGTON, DC,[3] UNITED STATES

ABSTRACT

High-profile natural disasters like the Asian tsunami of 2004 and Hurricane Katrina of 2005 emphasize the need for a systematic approach to the integration of geospatial technology in support of disaster management. In this article, the data and information needs of various users involved in the disaster management cycle are analyzed for the preparedness, early warning, response, recovery, and reconstruction phases. A risk characterization process that is evolving in southern Africa is presented. The process involves using historical geospatial data to develop baseline hazard and vulnerability profiles during preparedness planning. The profiles are integrated into a response plan for reducing the risk associated with various hazard scenarios. During actual hazard events, an appropriate risk profile is selected based on real-time hazard and vulnerability information, and the associated activities included in the response plan are implemented. The article emphasizes the diversity of data sources that must be integrated both in developing baseline risk profiles and in real-time event analysis. It also highlights the diversity of data producers and users who must be linked through the risk characterization and communication process. Integrated products (such as the Atlas for Disaster Preparedness and Response in the Limpopo Basin) and national contingency plans are presented as useful for communicating risk and for initiating personal and communal preparedness activities. However, these tools must be kept current, as vulnerability and hazard profiles frequently change. Data production, integration, and dissemination systems that are dynamically linked using service architecture are proposed as the most practical means of ensuring that multisectoral disaster managers have access to the most current hazard and vulnerability information to minimize loss of life and property damage.

In December 2004, the Asian tsunami produced a widespread disaster (Inoue 2005; Aitchison 2005) with unprecedented death and destruction spanning the continents of Asia, Australia, and Africa. Less than a year later, in August 2005, Hurricane Katrina, the strongest storm to hit the U.S. mainland in the last hundred years, caused massive destruction and flooding along the Gulf Coast, particularly in the city of New Orleans (Graumann et al. 2005). When the total cost of cleanup is assessed, Hurricane Katrina is expected to become the most expensive storm in U.S. history, with a total cost of over $100 billion. Concurrently, multiyear droughts have devastated food and water supplies on which agropastoralists in the Greater Horn of Africa rely for their livelihoods, resulting in human and animal deaths. High-profile natural disasters like these emphasize the need for a systematic and integrated approach to the development of disaster preparedness and response systems in order to minimize loss of life and damage to property (Wilson and Oyola-Yemaiel 2001).

The need for geospatial data in natural hazard characterization and response planning has never been greater. For the Hurricane Katrina event, weather scientists relied on remotely sensed data from satellites such as the Tropical Rainfall Measuring Mission (TRMM) to monitor the evolution of the storm. Meteorological forecasts of the storm's path and intensity based on this technology are widely acknowledged to be more accurate and reliable than was possible prior to the TRMM satellite. Taken together, the Asian tsunami and Hurricane Katrina resulted in the transfer of a record volume of satellite imagery from the U.S. Geological Survey (USGS) Center for Earth Resources Observation and Science (EROS) to end users around the world in 2005. These satellite datasets were used in a wide variety of applications, one of the most significant of which was mapping the spatial extent of the disasters. Yet few would attempt to argue that the Katrina disaster was well managed. A report on the federal response to the disaster found that information available to the various response agencies and field units was inconsistent, poorly integrated, and often inadequate to facilitate an effective response (The White House 2006). Data integration problems were documented at the federal, state, and local levels. Although there were other shortcomings in response coordination, it is clear that a major problem was the ineffective conversion of meteorological hazard information into response actions based on assessments of physical, humanitarian, and economic impacts. As a result, several individuals and institutions charged with managing different facets of the response were forced to rely on incomplete or unfamiliar data in determining the actions that needed to be taken.

Remote sensing technology has provided us with a variety of geospatial data, such as optical and radar imagery, LIDAR elevation grids, and global positioning systems (GPS), that aid the identification of areas and features impacted by natural hazards (Tralli et al. 2005). In addition, large databases of digital orthophotography and infrastructure data such as transportation, storm drains, canals and levees, population and socioeconomic variables, soils, and land use and land cover have all been created. These datasets have been linked together into a national spatial data infrastructure (SDI) through the Geospatial One-Stop initiative (http://gos2.geodata.gov/wps/portal/gos). Clearinghouse technology, data and metadata standards, procedures, and policies are used to promote coordination among data providers and dissemination of geospatial data among users around the nation. Yet no integrated products describing the impacts of the hazard and actions that needed to be taken were available to disaster managers. It could consequently be argued that poor vertical integration of geospatial data hampered the response to Hurricane Katrina.

The impact of Hurricane Katrina on the U.S. Gulf Coast was not unprecedented in scope. During the first three months of 2000, the Indian Ocean coast of Mozambique was struck by a series of tropical storms, four of which achieved hurricane force winds (Christie and Hanlon 2001). These cyclones (as Indian Ocean hurricanes are called) created a series of flood inundation waves along many large rivers in the region. The most severe of these storms was Cyclone Eline, which resulted in record flooding in the lower reaches of the Limpopo River. Record expanses of inundation extended over 30 kilometers in width, and water depths greater than 10 meters were recorded in places that are normally dry. In the aftermath of these storms, Mozambican authorities worked with staff from a number of international agencies, including the USGS, to improve their disaster preparedness (INGC et al. 2003). The USGS is also involved in disaster preparedness in southern Africa through its involvement with the regional flood monitoring network of the Southern African Development Community (SADC) and the Famine Early Warning Systems Network (FEWSNET), as well as the Space and Major Disasters International Charter and the Global Earth Observing System of Systems (GEOSS).

This article examines information needs at various points in the disaster management cycle and the role that spatial data infrastructures can play in meeting these needs. It draws on the southern Africa experience, where robust approaches for living with a wide range of natural disasters are evolving. Relevant datasets generated before and during a major hazard event are described, as are ways of effectively integrating these datasets in order to disseminate pertinent information to the various entities responsible for disaster management.

DISASTER MANAGEMENT CYCLE

Appreciation of the data integration challenges associated with disaster management requires a basic understanding of the phases of the disaster management cycle as it applies to various types of hazard events. Although alternate classification schemes exist, this study uses a disaster management cycle composed of five phases: preparedness, early warning, response, recovery, and reconstruction (figure 1). The preparedness phase should begin long before any specific hazard event is detected. It begins with the perception of an elevated potential for a natural hazard to occur at a particular location. This perception of hazard potential could be attributed to historical precedent, proximity to a source of hazards, or the condition of leading indicators.

In the case of historical precedent, the local historical time series of events is the primary dataset used to determine hazard potential. For example, tornado preparedness activities are undertaken regularly in the central United States (commonly referred to as Tornado Alley) because of a well-documented history of severe storms in the region. In contrast, there have been renewed efforts to implement tsunami preparedness activities in many coastal areas of the world that have no history of tsunami events due to the perceived tsunami hazard potential associated with proximity to oceans (Collins 2005). This perception

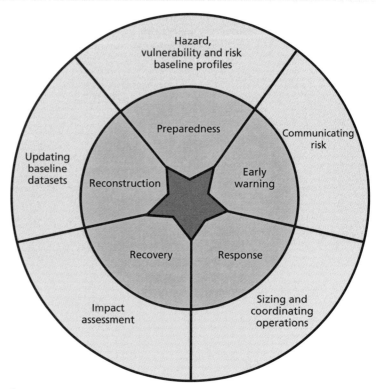

Figure 1. Spatial data infrastructure roles in relation to the disaster management cycle.

may have been fuelled by data from a tsunami event occurring elsewhere, like the Indian Ocean tsunami of 2004. When leading indicators are the source of elevated hazard potential, a determination is made based on prior understanding of the natural system that a hazard is likely to develop. An example of a leading indicator is the use of sea surface temperatures to forecast seasonal precipitation totals and the level of cyclonic activity. In this setting, the sea surface temperatures are not hazards in themselves. Rather, they point to the potential for a hazard to develop sometime in the future. Whether initiated by historical or current data, preparedness activities are triggered when it is determined that there is a need to reduce vulnerability to a future hazard.

The early warning phase of the disaster management cycle is initiated after a specific hazard has been observed. The emphasis during this phase is on generating and transferring actionable information to disaster managers and vulnerable individuals. A variety of geospatial information is also required to track the position of the hazard relative to vulnerable population centers, physical infrastructure, and zones of economic activity. The integration of hazard and vulnerability information forms the basis for determining which individuals and infrastructure components are at risk, and warning messages are generated with recommendations for urgent action to minimize loss of life and damage to property.

After the hazard has occurred, a response phase is initiated to minimize damage to property and loss of life by removing any secondary hazards such as fallen power lines and to provide humanitarian assistance to those in urgent need. Baseline geospatial information is invaluable for identifying secondary hazards and for planning response activities such as rescue efforts and the delivery of humanitarian supplies. However, such information must be accurate and available in a form that is easily integrated with other information such as damage reports and the location of temporary accommodation centers. Poor-quality baseline information becomes a hindrance to response efforts and is quickly abandoned by field crews as they become more familiar with the landscape and develop their own geographic baseline based on field observations. Opportunities to further assist the disaster management effort grow progressively smaller as the disaster enters the recovery and reconstruction phases. During these phases, focus shifts to logistics associated with the movement of people and supplies, financial resources, and construction materials. Information management during these phases centers on tracking transactions and stocks of materials rather than on spatial information. The challenge for the geospatial community is therefore to develop and integrate spatial databases well before any hazard events or face the risk of becoming irrelevant to disaster management efforts.

DISASTER RISK ANALYSIS FRAMEWORK

Given the importance of spatial data in the preparedness and early warning phases of disasters, it seems prudent to identify the main players and their data requirements. There are numerous types of natural disasters, including floods, cyclones, tornadoes, droughts, hurricanes, fires, snowstorms, avalanches, cold spells, heat waves, landslides, earthquakes, and volcanic eruptions. While we cannot hope to characterize each of these natural disasters in this article, we will develop a common framework within which some general spatial data integration concepts can be explored. Disasters such as droughts evolve slowly over a period of months or even years, while others such as floods develop in days or even minutes. Some disasters such as earthquakes may even occur with no apparent warning at all.

As a starting point, we adopt the definition of risk as a measure of vulnerability to a hazard. This definition is well suited to natural disaster management because it implies an analysis of both the hazard and the vulnerabilities. Alternate loss-based definitions of risk that are frequently applied in the banking and insurance industries are less suitable for natural disaster management because they tend to emphasize financial recovery after a hazard as the only basis for action. Risk management based on economic loss control can consequently obscure the benefits of early warning systems and risk reduction actions such as evacuation from the path of the hazard. The hazard-vulnerability approach also allows noneconomic losses and humanitarian considerations, such as number of displaced people, to be taken into account in quantifying risk.

Determination of the level of risk posed by any specific hazard requires an understanding of complex spatial and temporal interactions among multiple datasets. These interactions may even involve analysis of unprecedented events. It is unreasonable to expect individuals involved in disaster response management to analyze these complex interactions under emergency conditions without major omissions or errors. The preparedness phase therefore represents the best opportunity for effective risk analysis. In this phase, there is no specific event to prepare for; consequently, a range of possible events must be analyzed to establish baseline profiles for both hazards and vulnerabilities, to which future events can be compared (INGC et al. 2003).

SDIs IN HAZARD CHARACTERIZATIONS

A baseline hazard profile is a series of realizations of hazard events of varying severity that is generated for preparedness planning. These event realizations are usually classified on the basis of one particular attribute of the hazard. For example, cyclones are classified on the basis of intensity attributes such as maximum sustained wind speed (Saffir-Simpson scale), while earthquakes are usually classified using the Richter scale, which measures the energy emitted by the vibrations. Floods are classified by frequency of occurrence, while droughts are typically classified in terms of duration and magnitude.

These severity classifications are adopted more for the convenience of the agencies responsible for monitoring the hazards than for those estimating their impacts. Primary hazard data producers are interested mainly in improving their understanding of the dynamics of the hazard in terms of its genesis, propagation, and disintegration. They are considered to have characterized the hazard successfully if they are able to accurately forecast its spatial location and intensity. There are a few exceptional cases in which hazards are classified on the basis of impact. An example is the Fujita tornado classification scale, in which a tornado is classified on the basis of field surveys of damage to man-made structures. The limitation of such classification systems is that hazard severity information is not available during the early warning phase, as severity can be estimated only after the event. For this reason, impact-based classification systems are used only for monitoring hazards for which preimpact severity estimation technologies have not yet been implemented. For other hazards, a single physical attribute of the hazard is the basis for severity classification. An example of a single-attribute classification system used for cyclones is shown in figure 2. Inferences on impacts of the hazard from severity classifications alone are incomplete, as they refer solely to potential wind damage and not to coastal damage from a storm surge or inland flooding or to the spatial and temporal distribution of damage.

A secondary layer of data producers is required to assess the impacts of the hazard on the natural and built environments. The secondary hazard data producers consist of agencies that are responsible for the monitoring of natural resources such as forests, rivers, and lakes, as well as physical infrastructure such as dams, levees, and canals. Information from these impact assessments should be presented in a form that allows disaster managers to make quantitative determinations of actions needed to minimize vulnerability to the hazard. For example, given forecasting information on an approaching cyclone from a meteorological agency, the hydrological agency should be able to supply disaster managers with forecasts of inundation extent. Without such information, the disaster manager is forced to relate cyclone wind speeds to the number of individuals who would be flooded out of their houses when no scientific, discernible relationship exists between the two variables.

In Mozambique, a dual-classification approach has been adopted in which severity is characterized using a traditional numbering system (categories 1 to 5) based on maximum sustained wind speed, while imminence is communicated through

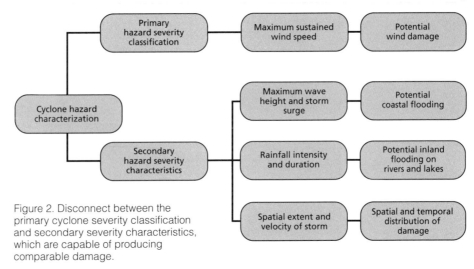

Figure 2. Disconnect between the primary cyclone severity classification and secondary severity characteristics, which are capable of producing comparable damage.

a color-coding system with blue, yellow, and red indicating less than 48, 24, or 6 hours to landfall, respectively. Warnings based on this system are generated by the national meteorological agency. Separate flood warnings are generated by the water management agencies, which similarly characterize flood severity on the basis of expected depth of water and imminence based on expected time of arrival of peak flows. The importance of such multistage warnings is demonstrated by the TRMM-based satellite rainfall fields from Cyclone Eline over Mozambique. The first image in figure 3 shows the cyclone making landfall on February 22, while the second image shows the cyclone being degraded to a tropical storm with the decline in its maximum sustained winds on February 23. However, the last image shows that the most severe rainfall actually occurred on February 24, after the downgrading of the storm. The rainfall pattern emphasizes the need to incorporate both primary and secondary severity characteristics into response planning, impact analysis, and decision making.

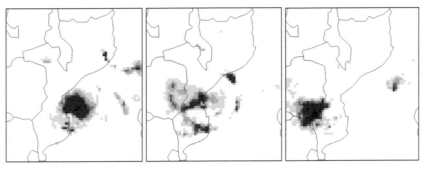

Figure 3. Daily rainfall from Cyclone Eline between February 22 and 24, 2000. The cyclone made landfall on the 22nd (first image) and was degraded to tropical storm status after the decline of maximum sustained winds on the 23rd (second image), but heavy rainfall returned on the 24th (last image), contributing to widespread flooding.
From ArcView 3.2a 1999, courtesy of ESRI

Disaster managers are the primary users of meteorological and water hazard characterizations. They must design and implement hazard response plans to translate hazard analyses into response actions to be taken by secondary data users, who consist of at-risk communities of individuals, businesses, and other organizations such as government agencies. The translation of hazard characterizations into recommended actions requires an understanding of the vulnerability of at-risk communities to the hazard. A detailed description of the vulnerability assessment process and the role of spatial data infrastructures is the subject of the next section.

SDIs IN VULNERABILITY ASSESSMENTS

Two equally severe hazards that pass over areas of similar geomorphology will not necessarily cause equal levels of damage or elicit similar responses. Differences in physical infrastructure and in socioeconomic factors determine vulnerability to a hazard (Wisner et al. 2004), and these must be taken into account in response planning. Vulnerability is a dynamic factor that can change significantly in a very short time. In the Mozambique floods of 2000, the residents of the lower Limpopo Valley were rendered more vulnerable to Cyclone Eline than they would otherwise have been because of damage to their flood monitoring infrastructure caused by another cyclone less than a month earlier (Christie and Hanlon 2001). The value of a well-developed and up-to-date spatial data infrastructure quickly becomes apparent in vulnerability assessment, even during the preparedness phase.

Let us begin by examining the data needed to address the physical infrastructure variable in the vulnerability equation. At the most basic level, a determination has to be made of the names, locations, and sizes of population centers. National census data are often a good source of this information, but the data have to be compiled in geospatial format to allow for integration into the vulnerability analysis. The next step is to identify the location and status of infrastructure used to monitor potential hazards to population centers. This information typically comes from the various agencies responsible for monitoring meteorological, hydrologic, oceanic, and subsurface events such as earthquakes and volcanoes. In the United States, meteorological and oceanic infrastructure is maintained by the National Oceanic and Atmospheric Administration (NOAA), while infrastructure for monitoring hydrologic and subsurface events is maintained by the USGS. If the location and condition of these monitoring systems are part of an SDI, then vulnerabilities arising out of malfunctions of the monitoring infrastructure can be assessed in near real time.

Next, the condition of physical infrastructure such as roads, residential housing, and commercial buildings within these urban areas must be assessed to evaluate their vulnerability to the hazard. Information on the condition of individual structures is often maintained by federal, state, and local agencies responsible for issuing building permits and maintaining structures such as roads and bridges. Again, the value of linking local, state, and national SDIs becomes apparent. If databases are dynamically linked and updated, it is possible to make an assessment of how many, or even which, specific structures could be vulnerable to a particular hazard. Newly constructed or demolished structures and roads under repair could be taken into account in such an assessment. It also becomes possible to determine which structures can serve as temporary shelters during an emergency. Similarly, assessment of industrial facilities that could pose a secondary hazard to population centers or the natural environment would be performed to identify vulnerabilities associated with secondary hazards such gas leaks, oil spills, or other chemical releases.

The determination of socioeconomic vulnerability likewise requires a number of data inputs. A national land-use data layer is the most basic input in this analysis. It allows hazard zones to be subdivided by economic activities such as agriculture, forestry, mining, recreational, transportation, industrial, commercial, and residential (INGC et al. 2003). Economic data identifying the human, material, and energy resource inputs for these activities and the resulting production figures allow for the quantification of economic vulnerability. These vulnerabilities may also be cyclical in nature, with seasonal or interannual patterns of variation (Patwardhan and Sharma 2005). For example, in agricultural land uses, a hailstorm occurring before the initiation of planting or after the harvest will have a much smaller economic impact than one that occurs in the middle of the growing season. These variations in vulnerability can be captured by seasonal crop calendars. Other land uses such as tourism may likewise have seasonal patterns which can be easily represented in a database as part of the vulnerability information to allow for the assessment of economic disruptions.

Other economic indices such as per capita income, unemployment rate, per capita motor vehicle ownership, and proportion of residents that have personal property and commercial insurance are all important indicators of the ability of the population to recover after the event (Kumar and Newport 2005). Some of these data are captured in census tabulations or annual tax returns and could also be incorporated into the SDI. The ability to recover, also termed resilience, is an important parameter for response planning. It is often closely linked to the willingness of populations to leave their possessions behind should an evacuation become necessary. Members of resilient communities are more willing to abandon their primary residences because of their ability to replace damaged property using personal income reserves, personal or business credit, or risk management tools such as insurance. They are also more likely to have access to personal transportation during the evacuation. By contrast, members of communities with low resilience are unlikely to want to evacuate their residences and may require additional personnel to encourage, assist, and if necessary, enforce such evacuation. They are also more likely to require public transportation, food, and accommodation, and provision should be made to meet these needs in the disaster response plan. Figure 4 shows a summary of the main data inputs for hazard and vulnerability assessments.

SDIs IN PREPAREDNESS AND RESPONSE PLANNING

Given the assessments of hazards and vulnerabilities, disaster managers must recommend a set of actions to end users in at-risk communities that will minimize risk to the community as a whole. The disaster managers may recommend that residents stay at home, congregate at more secure locations, or evacuate the risk zone completely. A combination of measures may be adopted on the basis of stated individual vulnerability criteria. In making these decisions, the disaster manager is often faced with a large number of geospatial variables and may

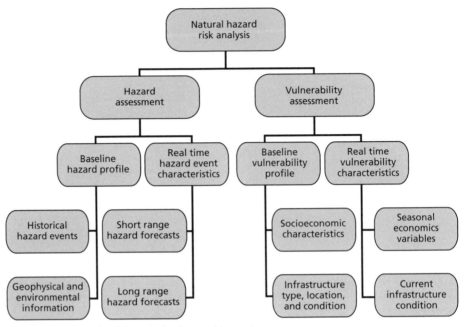

Figure 4. Data input for risk analysis of natural hazards.

even require the use of complex decision analysis models (Levy et al. 2005) to integrate the data and arrive at the appropriate decision. Individual end users are motivated to action (or inaction) by both their personal perception of the magnitude of risk and their need to preserve and protect their personal welfare. Perception can be influenced by a variety of factors including age, gender, health, and socioeconomic condition (Brilly and Polic 2005). Differences between objective risk analysis and subjective risk perception mean that not all actions recommended by disaster managers will be adhered to by end users. Disaster managers must therefore estimate the likelihood of compliance and make provision for it in response planning through measures to improve compliance or provide additional assistance to noncompliant groups. Measures to improve compliance could include increased emphasis on preparedness training, insurance plans, and improved warning dissemination. For those hazard events in which evacuations are necessary, provision of security, coordination of transportation, location of accommodation centers, and assurance of humanitarian supplies (food, water, health care, etc.) can be important determinants of compliance (Gall 2004).

Disaster managers must also plan and coordinate their own response efforts. One study of the Mozambique floods of 2000, for example, identified over 65 nongovernmental organizations and 10 government agencies involved in the response (Moore et al. 2003). Figure 5 shows the extensive process that has been established for planning and coordinating response activities among these key institutions in Mozambique and the southern Africa region. At the beginning of each rainy season, the agrometeorological and water sectors of the respective countries initiate the process with a forum to generate regional consensus of both prevailing climatic conditions and forecasts for the upcoming season. After negotiation to reconcile differences, a consensus outlook is generated and distributed to the various agrometeorological and hydrological sectors in each country. Within Mozambique, each of the application sectors is required to use the consensus forecasts to update disaster preparedness plans, which are then integrated into a national contingency plan. This process has contributed to the development of an integrated national spatial data infrastructure, as discrepancies in datasets used for contingency planning quickly become apparent (Asante and Verdin 2005). Postseason assessment forums involving both regional and national hydrological, meteorological, agricultural, and disaster management agencies are held each year. These forums revisit the seasonal climate outlooks and assess their validity. They also assess water resource, food security, and vulnerability conditions as a basis for recommending additional response activities to address lingering climate-related problems from the previous season. Policy recommendations are made and submitted to the respective national and regional policy-making organs in southern Africa for implementation. Past policy recommendations have included requests for review of import and export regulations, requests for unified field assessments, and regional disaster relief appeals. Other actions have included updating of agency contact lists and communication channels, the redesign of specific hazard monitoring products, and requests for specific flow measuring gauges to be repaired or installed.

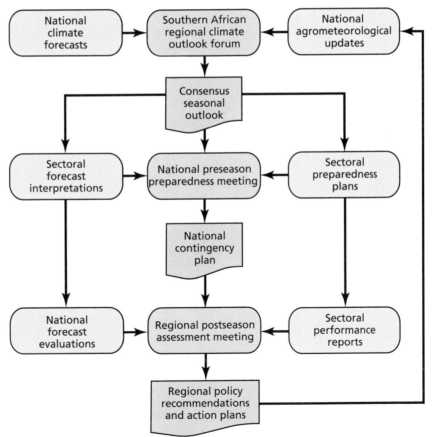

Figure 5. Integration of Mozambican disaster preparedness and response planning into the southern Africa regional planning processes.

Disaster response plans also help mitigate the physical and economic impacts of specific hazard events on communities in the hazard zone. For slow-onset hazards such as drought, long-range forecasts can be combined with real-time model runs to provide early warning. Spatial data integration efforts can contribute to adoption of long-range actions to manage risk. Financial risk management tools such as insurance, economic diversification, and modified marketing may be adopted as coping strategies. At the local scale, modified water use policies, culling of livestock, and modification of planting cycles and crop types are examples of long-range actions that may be taken as a result of such policy analysis. With rapid onset events such as floods, response actions are usually defined on the basis of scenario analyses of a priori model runs. The focus of early warning systems shifts to communicating risk information to at-risk communities, while response focuses on providing impact assessments to multisectoral disaster managers. Real-time hazard information is the basis for determining which of the scenarios described in the disaster response plan should be put into effect. In Mozambique, the disaster preparedness and response atlas has proven to be a useful tool for communicating policy analysis associated with various disaster scenarios (INGC et al. 2003). End users are able to quickly make policy recommendations by relating real-time hazard information to one of the hazard scenarios described in the atlas. However, the information in the atlas can quickly become outdated with changes in hazard profiles, population centers, and other socioeconomic conditions.

The physical infrastructure components of SDIs could play an important role in ensuring that end users have access to the most up-to-date versions of geospatial datasets, through the integration of Web mapping technologies and service architecture. Existing online geospatial data search services such as the USGS Earth Explorer (http://earthexplorer.usgs.gov), Internet map servers such as Google Earth, and data distribution services such as the USGS National Map (http://nationalmap.gov) are invaluable for transmitting real-time imagery and baseline data to users (Mansourian et al. 2006). Map servers have enabled the rapid distribution of real-time hazard information (Asante and Verdin 2005). Satellite-based communication systems have also been used for transmitting information from computer models, socioeconomic databases, and field reports from central repositories to end users operating in the field during a disaster (Brandon 2002). In the commercial sector, automated teller machines and point-of-sale units rely on service architecture for remote financial data access and transaction processing. However, there are few examples of integrated policy assessments that are generated by automatic processing of data available from geospatial data servers. The integration of service architecture and map server technology with satellite-based communication systems could bring about the integration of available geospatial data and field reports into a coherent picture of disasters in near real time and would significantly increase opportunities for SDIs to contribute directly to disaster response and to policy analysis in general.

SDIs IN RECOVERY AND RECONSTRUCTION

Following the initial response to a natural disaster event, attention shifts to restoring livelihoods and economic activities to a semblance of pre-event levels. An important part of this recovery and reconstruction effort is the estimation of the extent of damage and the overall cost of the disaster. A detailed handbook on how to perform such assessments has been developed by the United Nations Economic Commission for Latin America and the Caribbean (ECLAC 2003), on the basis of experiences in dealing with natural disasters in that region. The publication outlines methods of estimating the cost of disruptions to livelihoods, housing, and social services such as education, health, energy, water, sanitation, and transportation. Economic impacts in agriculture, trade, industry, and tourism are estimated as well as costs associated with cleanup within both the built-up environment and the natural environment. Damage assessments must also be performed to enable private individuals, businesses, and policy makers to make determinations of whether to repair, replace, or completely abandon damaged infrastructure. Many of the datasets used in the preparedness phase of disasters are applicable during the damage assessment portion of the recovery phase.

The cost of long-term impacts on the environment, population distribution, and socioeconomic dynamics of the region are also taken into account. These long-term costs underscore the disruptive effects of major disasters on both the physical and the cultural landscapes. Recovery from environmental impacts such as beach erosion, salt water intrusion, and sediment deposition can take several decades. These impacts, together with the psychological trauma from the event, may cause many individuals to move away permanently from an impacted region. Because of such changes, pre-event datasets may no longer be representative of postevent settlements and population distribution, socioeconomic baselines, hazard baselines, and environmental conditions. During the Asian tsunami of 2004, for example, mangrove forests served as a bioshield, blunting the force of the tsunami as it propagated inland (Danielsen et al. 2005; Kathiresan and Rajendran 2005). Many communities living behind the mangroves were spared the catastrophic impacts experienced by other coastal communities. However, many of the mangrove forests were also damaged by the tsunami, and the communities behind them are more vulnerable now than they were before the tsunami. New hazard and vulnerability profiles that reflect postevent conditions must therefore be developed after each major disaster. If a distributed, integrated SDI exists (or is developed as part of reconstruction efforts), then updates to population distribution, socioeconomic, and environmental datasets that are produced by the different agencies involved in reconstruction work become available for preparedness planning as they are generated. Development of this kind of feedback mechanism provides an invaluable opportunity for knowledge gained from the management of a disaster to systematically contribute to planning for future disasters.

CONCLUSIONS   Management of natural disasters requires the analysis and integration of both hazard and vulnerability data produced by many different agencies at the federal, state, and local levels. The southern African experience suggests that such data integration is best performed during the preparedness phase by teams representing both primary and secondary hazard data producers as well as end users who could be impacted by the natural disasters. Integrated products such as the Atlas for Disaster Preparedness and Response in the Limpopo Basin (INGC et al. 2003) and national contingency plans are useful for communicating risk and initiating personal and communal preparedness activities. However, such integrated products can also become outdated as vulnerability and hazard profiles change. Only when data producers, disaster managers, and end users are linked through an integrated national spatial data infrastructure can changing risk profiles be updated in near real time to support effective decision making during natural disasters. An integrated national spatial data infrastructure must incorporate online databases, data processing, real-time mapping capabilities, and information collection and dissemination technologies. Recent natural disasters have served to illustrate that, while the geospatial databases are well developed, the data integration and dissemination components are still lagging behind in many countries.

REFERENCES

Aitchison, Jonathan C. 2005. The great Indian Ocean tsunami disaster. *Gondwana Research* 8 (2): 107–8.

Asante, Kwabena O., and James P. Verdin. 2005. Regional flood and drought management arrangements: Lessons learned and challenges. In Implementing the SADC Regional Strategic Action Plan for Integrated Water Resource Management (1999-2004): Lessons and Best Practice, eds. Brian E. Hollingworth, and Thomas Chiramba. Bonn, Germany: Inwent Press.

Brandon, William T. 2002. An introduction to disaster communications and information systems. *Space Communication* 18 (3-4): 133–38.

Brilly, M., and M. Polic. 2005. Public perception of flood risks, flood forecasting and mitigation. *Natural Hazards and Earth System Sciences* 5 (April): 345–55.

Christie, Frances, and Joseph Hanlon. 2001. *Mozambique and the Great Flood of 2000.* Bloomington, Indiana: Indiana University Press.

Collins, Larry. 2005. Tsunamis: A wakeup call for the US, Part 1. *Fire Engineering* (September): 63–74.

Danielsen, Finn, Mikael K. Sørensen, Mette F. Olwig, Vaithilingam Selvam, Faizal Parish, Neil D. Burgess, Tetsuya Hiraishi, Vagarappa M. Karunagaran, Michael S. Rasmussen, Lars B. Hansen, Alfredo Quarto, and Nyoman Suryadiputra. 2005. The Asian Tsunami: A protective role for coastal vegetation. Science 310 (October): 643.

Economic Commission for Latin America and the Caribbean (ECLAC). 2003. *Handbook for estimating the socio-economic and environmental effects of disasters.* Mexico City: ECLAC.

Gall, Melanie. 2004. Where to go? Strategic modelling of access to emergency shelters in Mozambique. *Disasters* 28 (March): 82–97.

Graumann, Axel, Tamara Houston, Jay Lawrimore, David Levinson, Neal Lott, Sam McCown, Scott Stephens, and David Wuertz. 2005. Hurricane Katrina: A climatological perspective. Preliminary report. Ashville, NC: National Climatic Data Center.

INGC, UEM–Dept. of Geography and FEWSNET MIND. 2003. Atlas for Disaster Preparedness and Response in the Limpopo Basin. Cape Town, South Africa: Creda Communications Ltd.

Inoue, Kazuo. 2005. Massive tsunami in Indian Ocean coasts. *Disaster Management and Response* 3 (April): 33.

Kathiresan, Kandasamy, and Narayanasamy Rajendran. 2005. Coastal mangrove forests mitigated tsunami. *Estuarine Coastal and Shelf Science* 65 (November): 601–6.

Kumar, Anand T. S., and Jeyanth K. Newport. 2005. Role of microfinance in disaster mitigation. *Disaster Prevention and Management* 14:176–82.

Levy, Jason K., Chennat Gopalakrishnan, and Zhaohui Lin. 2005. Advances in decision support systems for flood disaster management: Challenges and opportunities. *International Journal of Water Resources Development* 21 (December): 593–612.

Mansourian, A., A. Rajabifard, M. Valadan Zoej, and I. Williamson. 2006. Using SDI and web-based system to facilitate disaster management. *Computers & Geosciences* 32 (April): 303–15.

Moore, Spencer, Eugenia Eng, and Mark Daniel. 2003. International NGOs and the role of network centrality in humanitarian aid operations: A case study of coordination during the 2000 Mozambique floods. *Disasters* 27 (December): 305–18.

Patwardhan, Anand, and Upasna Sharma. 2005. Improving the methodology for assessing natural hazard impacts. *Global & Planetary Change* 47 (July): 253–65.

The White House. 2006. The federal response to Hurricane Katrina: Lessons learned. Washington, DC: The White House.

Tralli, David M., Ronald G. Blom, Victor Zlotnicki, Andrea Donnellan, and Dianne L. Evans. 2005. Satellite remote sensing of earthquake, volcano, flood, landslide and coastal inundation hazards. *ISPRS Journal of Photogrammetry and Remote Sensing* 59 (June): 185–98.

Wilson, Jennifer, and Arthur Oyola-Yemaiel. 2001. The evolution of emergency management and the advancement towards a profession in the United States and Florida. *Safety Science* 39 (October): 117–31.

Wisner, Ben, Piers Blaikie, Terry Cannon, and Ian Davis. 2004. *At risk: Natural hazards, people's vulnerability and disasters.* 2nd ed. London: Routledge.

Related titles from ESRI Press

GIS Worlds: Creating Spatial Data Infrastructures
ISBN 978-1-58948-122-0

Spatial Portals: Gateways to Geographic Information
ISBN 978-1-58948-131-2

Building European Spatial Data Infrastructures
ISBN 978-1-58948-165-7

ESRI Press publishes books about the science, application, and technology of GIS. Ask for these titles at your local bookstore or order by calling 1-800-447-9778. You can also read book descriptions, read reviews, and shop online at www.esri.com/esripress. Outside the United States, contact your local ESRI distributor.